# Modeling, Control and Coordination of Helicopter Systems

Beibei Ren • Shuzhi Sam Ge • Chang Chen
Cheng-Heng Fua • Tong Heng Lee

# Modeling, Control and Coordination of Helicopter Systems

Springer

Beibei Ren
Department of Electrical
and Computer Engineering
National University of Singapore
Singapore

Department of Mechanical
and Aerospace Engineering
University of California, San Diego
CA, USA
helenren.ac@gmail.com

Shuzhi Sam Ge
Department of Electrical
and Computer Engineering
National University of Singapore
Singapore

The Robotics institue
University of Electronic Science and
Technology of China, Chengdu, China
samge@nus.edu.sg

Chang Chen
DSO National Laboratories
Science Park Drive 20
118230 Singapore
cchang@dso.org.sg

Cheng-Heng Fua
Institute of High Performance Computing
Agency for Technology, Science
and Technology
1 Fusionopolis Way
Connexis
Singapore
fuach@ihpc.a-star.edu.sg

Tong Heng Lee
Department of Electrical
and Computer Engineering
National University of Singapore
4 Engineering Drive 3
117576 Singapore
eleleeth@nus.edu.sg

ISBN 978-1-4614-1562-6 e-ISBN 978-1-4614-1563-3
DOI 10.1007/978-1-4614-1563-3
Springer New York Dordrecht Heidelberg London

Library of Congress Control Number: 2011941211



Printed on acid-free paper

Springer is part of Springer Science+Business Media (www.springer.com)

For the subject under study:

by modeling, we have a deeper and better understanding;

by control, we make our course of action steady and solid;

by coordination, we appreciate our friends for their supports in our mission.

# Preface

Unmanned aerial vehicles (UAVs) are finding wider applications including surveillance, rescue, navigation, formation, coordination, among others. Helicopter-type UAVs offer the best option in many open and/or built-up areas for their maneuverability through narrow alleys and sharp beds, and their ability to hover in place if, for example, there is a need to have a close look at a place of interest.

As with any vehicular design, control design for helicopters is non-trivial. For example, ensuring stability in helicopter flight is a challenging problem for nonlinear control design and development. Unlike many classes of mechanical systems that naturally possess desirable structural properties such as passivity or dissipativity, helicopter systems are inherently unstable without closed-loop control, especially when hovering. In addition, helicopter dynamics are highly nonlinear and strongly coupled such that disturbances along a single degree of freedom can easily propagate to the other degrees of freedom and lead to loss of performance, even destabilization.

The fundamental requirement for control system design is to guarantee the stability for helicopter systems. Many techniques have been proposed in the literature for the motion control of helicopter systems, ranging from feedback linearization to model reference adaptive control and dynamic inversion, and these techniques typically require reasonably precise knowledge of the dynamic models in order to achieve satisfactory performance. An important concern when designing controllers for helicopters is the manner on how to deal with unknown perturbations to the nominal model, in the form of parametric and functional uncertainties, unmodelled dynamics, and disturbances from the environment. Helicopter control applications are characterized by time-varying aerodynamical disturbances, which are generally difficult to model accurately. The presence of uncertainties and disturbances could disrupt the function of the feedback controller and lead to degradation of performance.

While autonomous vehicles performing solo missions can yield significant benefits, greater efficiency and operational capability can be realized from teams of autonomous vehicles operating in a coordinated fashion. Potential applications for multi-vehicle systems include space-based interferometers, future combat systems, surveillance and reconnaissance, hazardous material handling, and distributed reconfigurable sensor networks. To realize the potential of a multi-helicopter team, coordination techniques between helicopters are needed. One fundamental problem in multi-helicopter cooperation is formation control, in which the helicopters are required to keep a desired formation configuration to complete the assigned tasks.

The proposed book is meant to provide a comprehensive treatment of helicopter systems, ranging from related nonlinear flight dynamic modeling and stability analysis, to advanced control design for a single UAV, to the coordination and formation control of multiple helicopters to achieve high performance tasks. This book can be a good reference and introduction to the modeling, control and coordination of helicopter systems.

The book starts with a brief introduction to the development of helicopter systems modeling, control and coordination in Chap. 1.

Chapter 2 focuses on nonlinear fight dynamic modeling for a general class of rotary-wing aircraft. Centred around the general rigid body equations of motion, the modeling process is based on modules that follow closely with physical subsystems on the rotary-wing aircraft. These modules may include main and tail rotors, propeller, horizontal and vertical tails, wings, fuselage, and the flight control system. Aerodynamic interference among different modules can also be established. Besides the general modeling, performance calculation is also illustrated as it is a critical index in understanding the aircraft.

Following the flight dynamic modeling, trim study and stability analysis are conducted in Chap. 3. Trim is a state of the aircraft in which forces and moments reach equilibrium. Trim is the basis of many fundamental analysis, including stability analysis, control system design, handling qualities assessment, and simulation. A numerical procedure on how to obtain trim is described. With the trim condition established, a linearization process can be carried out to obtain a linearized model for the rotary-wing aircraft. Elements in the linearized model are called stability and control derivatives. A detailed description on the derivatives is provided, with the emphasis of their physical meanings. To illustrate, three examples are also presented in the chapter, which include a simplified single helicopter system and two hobby helicopters at hover.

Linearized models are not only useful for stability analysis, but also for control design, by blending them over different operating points with gain scheduling. However, this requires extensive modeling, which is expensive and time-consuming. There is a need for control systems that can operate with minimal model information and handle nonlinearities over the entire flight regime. Chapter 4 aims to address this need, by presenting a robust adaptive neural network (NN) control for helicopters. In particular, we focus on vertical flight, which can be represented in the single-input–single-output (SISO) nonlinear nonaffine form, because the coupling between longitudinal and lateral directional equations in this flight regime is weak. Although a nonaffine system can be rendered affine by adding an integrator to the control input, allowing many control methods for affine nonlinear system to be used, the disadvantage of this approach is that the dimension of the system is increased, and control efforts are not direct and immediate either. Subsequently, effective control for the system may not be achieved. In this chapter, we focus on control design for the nonaffine system directly, without adding any integrators to the input. Differing from the approaches in the literature, which were based on approximate dynamic inversion with augmented NNs, we utilize the Mean Value Theorem and the Implicit Function Theorem as mathematical tools to handle the nonaffine nonlinearities in the helicopter dynamics. In cases where reasonably accurate knowledge of the dynamic inversion model is available, the method using approximate dynamic inversion has been shown to provide an effective solution to the problem. However, the construction of the dynamic inversion for a nonaffine system may not be an easy task in general. For such cases, our approach offers a feasible means of tackling the problem, since a priori knowledge of the inversion is not required.

In Chap. 5, neural network control is redesigned to track both altitude and yaw angle at the same time. We consider a scale model helicopter, mounted on an experimental platform, in the presence of model uncertainties, which may be caused by unmodelled dynamics, or aerodynamical disturbances from the environment. Two different types of NN, namely a multilayer neural network (MNN) and a radial basis function neural network (RBFNN) are adopted in control design and stability analysis. Based on Lyapunov synthesis, the proposed adaptive NN control ensures that both the altitude and the yaw angle track the given bounded reference signals to a small neighborhood of zero, and guarantees the Semi-Globally Uniformly Ultimate Boundedness (SGUUB) of all the closed-loop signals at the same time. The effectiveness of the proposed control is illustrated through extensive simulations. Compared with the model-based control, approximation-based control yields better tracking performance in the presence of model uncertainties.

Unlike previous chapters, which consider actuators with instantaneous response, Chap. 6 deals with a more realistic scenario where actuator dynamics are present. Backstepping technique, combined with NNs, is employed to design the robust attitude control for uncertain multi-input multi-output (MIMO) nonlinear helicopter dynamics. To the best of our knowledge, there are few works in the literature that take the actuator dynamics into account in the helicopter control, which is practically relevant but more challenging as well. In this chapter, helicopter models are considered as MIMO nonlinear dynamic systems, where the actuator dynamics in the first-order low-pass filter form are considered. Secondly, the possible singularity problem of the control coefficient matrix for the model-based attitude control case is tackled effectively by introducing a design matrix. Thirdly, approximation-based attitude control is developed to handle the model uncertainties (e.g. unknown moment coefficients and mass) and external disturbances. Rigorous stability analysis and extensive simulations results show the effectiveness and robustness of the proposed attitude control.

Beyond single-helicopter control, this book also touches on formation control of multiple helicopters. In Chap. 7, the concept of the Q-structure is introduced as a novel and flexible methodology to define and support a large variety of formations. The Q-structure allows automatic scaling of formations according to changes in the overall size of the helicopter team. The chapter begins by exploring the use of the Q-structure for formation control where perfect communication is present between all members of the team. The second part of the chapter focuses on how the Q-structure can be adapted and used for teams where communication is imperfect. In this chapter, we examine the properties of Q-structures in relation to other formation representation schemes, and look at the ways Q-structures can be used with artificial potential trenches to improve the scalability of the formations and support a large number of different formations. In particular, the Q-structure does not require explicit representation of every single node of the formation and is able to ensure the formation maintenance of a large number of helicopters. The formation is also robust against possible communication breakdown and/or limited wireless communication ranges.

The Q-structure formation control is useful for motion planning at the kinematics level. For formation control that takes into account the dynamics of helicopters, a different approach is proposed. Chapter 8 presents synchronized altitude tracking control of helicopters with unknown dynamics by graph theory, while the desired trajectory is available to a portion of the team. Since only the neighbors' information is available to the helicopter, we use the weighted average of neighbors' states as the reference state of the helicopter in the control design. We prove that if the extended communication graph contains a spanning tree with the virtual vehicle as its root, then its Laplacian will be positive definite. This property is essential for the stability proven and it also makes the proof of the stability for the results easy and direct. The mathematical stability proof, which makes use of the positive definite property of the graph Laplacian, is provided for both full-state and output feedback cases.

This book is primarily intended for academics, researchers and engineers who are interested in modeling, control and coordination of helicopter systems, and autonomous systems at large because many of the techniques and concepts could be extended further and applied directly to many other systems of interest. It can also serve as complementary reading for nonlinear systems, robotics, and adaptive control at the postgraduate level.

The book summarizes the research works carried out by the authors and a number of their close collaborators including Keng Peng Tee, Mou Chen and Rongxin Cui. For the final completion of the book, we gratefully acknowledge the unreserved support, constructive comments, and fruitful discussions from Miroslav krstic, University of California, San Diego; Frank L. Lewis, University of Texas at Arlington; Masayoshi Tomizuka, University of California at Berkeley; lan Postlethwaite, University of Northumbria; J.V.R. Prasad, Daniel Schrage, Lakshmi N. Sankar and George Vachtsevanos, Georgia Institute of Technology; Ben M. Chen, National Univesity of Singapore; Xianbin Cao and Jun Zhang, Beijing University of Aeronautics and Astronautics; Lei Guo and Yiguang Hong, Chinese Academy of Sciences; Trung T. Han and Rui Li, University of Electronic Science and Technology of China and Hailong Pei and Cong Wang, South China University of Technology.

Last but not the least, we would like to thank Gang Wang, Qun Zhang, Zhen Zhong, Zhenzhen Zhang, Peng Zhou and Jeffrey Sampson for their time and efforts in proofreading, and providing numerous useful comments and suggestions to improve the quality of this book.

Special thanks go to Steven Elliot for bringing the work to the attention of Springer-Verlag. We are also grateful to Alex Greene, Editorial Director of Springer-New York, Merry Stuber, Editorial Assistant of Springer-New York and Andrew Leigh, Assistant Editor of Springer-New York, for their reliable help and patience in the process of publishing this book.

## Financial Support

The first two authors would like to acknowledge the partial financial supports from the Basic Research Program of China (973 Program) under Grant 2011CB707005, Project 263-000-404-112/646/731 Development of a Fully Autonomous Mobile Urban Robot for TechX Challenge of the National University of Singapore, and the Grant of the Thousand Talents Scheme (TTS) Professorship at the University of Electronic Science and Technology of China.

Beibei Ren
*The National University of Singapore*
*Singapore*
*University of California, San Diego, CA, USA*

Shuzhi Sam Ge
*The National University of Singapore*
*Singapore*
*University of Electronic Science and Technology of China*
*Chengdu, China*

Chang Chen
*DSO National Laboratories*
*Singapore*

Tong Heng Lee
*The National University of Singapore*
*Singapore*

Cheng-Heng Fua
*Agency for Science, Technology and Research, Singapore*
*Singapore*

# Contents


**1 Introduction** ........ 1
- 1.1 Background ........ 1
- 1.2 Outline of the Book ........ 3

**2 Building a Nonlinear Rotary-Wing Aircraft Model** ........ 7
- 2.1 Introduction ........ 7
- 2.2 General Equations of Motion ........ 7
  - 2.2.1 Force Equation ........ 7
  - 2.2.2 Moment Equation ........ 8
  - 2.2.3 Kinematic Equation ........ 9
  - 2.2.4 Navigation Equation ........ 9
- 2.3 Modular-Based Modeling ........ 10
  - 2.3.1 Main Rotor ........ 10
  - 2.3.2 Transformation from Body to Hub ........ 12
  - 2.3.3 Blade Flapping Dynamics ........ 12
  - 2.3.4 Inflow Dynamics ........ 17
  - 2.3.5 Main Rotor Forces and Moments ........ 18
  - 2.3.6 Transformation from Hub to Body ........ 22
  - 2.3.7 Tail Rotor ........ 23
  - 2.3.8 Propeller ........ 24
  - 2.3.9 Horizontal Tail ........ 26
  - 2.3.10 Wing ........ 29
  - 2.3.11 Vertical Tail ........ 29
  - 2.3.12 Fuselage ........ 30
  - 2.3.13 Aerodynamic Interference ........ 31
  - 2.3.14 Rotor Rotational Degree of Freedom ........ 31
  - 2.3.15 Flight Control System ........ 32
- 2.4 Helicopter Performance Prediction ........ 34
- 2.5 Conclusion ........ 37

**3 Stability Analysis for Rotary-Wing Aircraft** .............................. 41
3.1 Introduction .............................................................. 41
3.2 Trim ....................................................................... 41
3.3 Linearization ............................................................. 43
3.4 Description of Stability and Control Derivatives ..................... 49
3.5 Yamaha R50 Helicopter at Hover ....................................... 51
3.6 Copterworks AF25B Helicopter in Forward Flight ................... 55
3.7 Conclusion ............................................................... 58

**4 Altitude Control of Helicopters with Unknown Dynamics** .............. 59
4.1 Introduction .............................................................. 59
4.2 Problem Formulation and Preliminaries ............................... 61
4.3 Function Approximation with Neural Networks ..................... 64
4.3.1 Function Approximation with RBFNN ..................... 64
4.3.2 Function Approximation with MNN ........................ 65
4.4 Adaptive NN Control Design ........................................... 66
4.4.1 Full State Feedback Control ................................ 66
4.4.2 Output Feedback Control .................................... 73
4.5 Simulation Study ........................................................ 82
4.5.1 Linear Models .................................................. 83
4.5.2 Nonlinear Model ............................................... 86
4.6 Conclusion ............................................................... 92

**5 Altitude and Yaw Control of Helicopters with Uncertain Dynamics** ... 93
5.1 Introduction .............................................................. 93
5.2 Problem Formulation and Preliminaries ............................... 94
5.3 Control Design ........................................................... 96
5.3.1 RBFNN-Based Control ....................................... 97
5.3.2 MNN-Based Control .......................................... 106
5.4 Simulation Study ........................................................ 114
5.4.1 Internal Dynamics Stability Analysis ...................... 115
5.4.2 Performance Comparison Results Between Approximation-Based Control and Model-Based Control .. 116
5.5 Conclusion ............................................................... 119

**6 Attitude Control of Uncertain Helicopters with Actuator Dynamics** .. 121
6.1 Introduction .............................................................. 121
6.2 Problem Formulation .................................................... 122
6.3 Model-Based Attitude Control for Nominal Plant ..................... 126
6.4 Robust Attitude Control of Helicopters with Uncertainties and Disturbances ........................................................... 129
6.5 Approximation-Based Attitude Control of Helicopters .............. 133
6.6 Simulation Results ....................................................... 139
6.6.1 Model-Based Attitude Control ............................... 140
6.6.2 Robust Attitude Control ....................................... 142
6.6.3 Approximation-Based Attitude Control ..................... 143
6.7 Conclusion ............................................................... 146

**7 Kinematic Formation Control Using Q-structures** ..... 147
7.1 Introduction ..... 147
7.2 Q-Structures and Formations ..... 148
7.2.1 Assumptions ..... 148
7.2.2 Division of Information Flow ..... 150
7.2.3 Elements of the Q-structure ..... 150
7.2.4 Properties of the Q-structure ..... 153
7.3 Q-Structure with Perfect Communication ..... 156
7.3.1 Changing Queues ..... 156
7.3.2 Potential Trench Functions ..... 158
7.3.3 Helicopter Behaviors ..... 162
7.3.4 Simulation Experiments ..... 163
7.4 Q-Structure with Imperfect Communication ..... 175
7.4.1 Determination of Target on Queue ..... 175
7.4.2 Navigation of Helicopters to Positions in Formations ..... 178
7.4.3 Simulation Studies ..... 186
7.5 Conclusion ..... 194

**8 Dynamic Altitude Synchronization Using Graph Theory** ..... 195
8.1 Introduction ..... 195
8.2 Problem Formulation ..... 196
8.2.1 Helicopter Dynamics ..... 196
8.2.2 Formation Control of Helicopters ..... 197
8.3 Control with Full Information ..... 200
8.4 Control with Partial Information ..... 207
8.5 Simulation Study ..... 211
8.6 Conclusion ..... 216

**References** ..... 217

**Index** ..... 223

# Chapter 1
# Introduction

## 1.1 Background

Unmanned helicopters pose a challenge to nonlinear control design and development. Unmanned helicopters are inherently unstable without closed loop control, setting it apart from many other classes of mechanical systems that possess desirable structural properties such as passivity or dissipativity. In addition, helicopter dynamics are highly nonlinear and strongly coupled, and therefore disturbances along a single degree of freedom can easily propagate to the other degrees of freedom, leading to a loss of performance or even destabilization. Furthermore, unrestrained helicopter motion is governed by underactuated configuration, that is, the number of control inputs is less than the number of degrees of freedom to be stabilized, which makes it difficult to apply the conventional approach (usually in the field of robotics) for controlling Euler–Lagrange systems. Due to these reasons, substantial research efforts have focused on control design that guarantee stability and robustness for unmanned helicopters.

Most helicopters systems are intrinsically nonlinear and their dynamics are described by nonlinear differential equations. However, for the purpose of analysis, researchers commonly derive linearized models of helicopter systems. More specifically, a linear model that approximates a particular nonlinear helicopter system can be obtained if this system operates around an operating point and the signals involved are small. Many techniques for the design of controllers and analysis techniques for linear systems have been proposed in literature, for example, linear $H_\infty$ control theory has been applied to a linear helicopter model in [10,82,95,105]. However, control laws based on the linearized helicopter dynamics do not have global applicability as they only exhibit desirable behavior around an operating point. Because of this, many researchers have applied nonlinear control techniques to dynamic models of helicopter systems. Based on a reduced order subsystem of a Lagrangian helicopter model, feedback linearization control was applied for altitude tracking [104]. Dynamic sliding mode control was proposed for altitude regulation in [93]. By modifying the internal dynamics to remove the nonhyperbolicity

B. Ren et al., *Modeling, Control and Coordination of Helicopter Systems*,
DOI 10.1007/978-1-4614-1563-3_1, © Springer Science+Business Media, LLC 2012 

followed by stable inversion, the problem of near nonhyperbolic internal dynamics in helicopter control was addressed in [19]. In [55], approximate input–output linearization was employed to obtain a dynamically linearizable helicopter system without zero dynamics, and which possessed the desirable property of differential flatness. Internal model-based control was suitably applied to the problem of helicopter landing on an oscillating ship deck in [40]. In [45, 47], a novel method is presented for the design of a fuzzy flight controller for the unmanned APID-MK3 helicopter.

The control techniques proposed in the research work mentioned above require reasonably precise knowledge of dynamic models involved in order to achieve satisfactory performance. In the control of helicopters, an important concern is how one can deal with unknown perturbations to the nominal model. Such unknown perturbations include parametric and functional uncertainties, unmodelled dynamics, and disturbances from the environment. The presence of uncertainties and disturbances could disrupt the function of the feedback controller and lead to degradation of performance. To deal with the presence of model uncertainties, approximation-based techniques using neural networks (NNs) have been proposed. Approximate dynamic inversion with augmented NNs was proposed to handle unmodelled dynamics in [38, 54], while neural dynamic programming was shown to be effective for the tracking and trimming control of helicopters in [21].

A significant portion of this book is dedicated to a comprehensive treatment of the control of a single helicopter. However, multi-agent systems, consisting of huge numbers of interacting agents and autonomous unmanned vehicles, linked together in complex networks, are becoming increasingly prevalent. The advancement of technology in robotics, vehicular control, computer science and communications, has made possible the deployment of large teams of mobile UAVs in real life scenarios. These systems have been applied to a wide variety of areas, from manufacturing and warehouse automation, to construction and shipping industries, to autonomous UAV humanitarian demining, and to surveillance and urban search-and-rescue with UAVs/helicopters. A comprehensive overview of issues in multi-vehicle cooperation may be found in [5].

Formation control of UAV teams in dynamic and uncertain environments has been intensively studied in recent years. The amount of interest in this topic can be attributed to at least two reasons: (1) it is a good platform (and task) for studying multi-agent cooperation techniques and (2) there are practical applications for formation control techniques (e.g., in autonomous demining where a team of demining-vehicles need to systematically "sweep" through a plot of land). The need to use UAVs in real applications means considering real-world constraints, such as imperfect communications between agents, in the design of coordination techniques.

A host of different techniques for formation control in multi-UAV teams is available in the literature, for example, [3, 17, 18, 22, 59]. In particular, many of these techniques deal, in some way or other, with the robust operation of UAVs. For example, Balch and Hybinette [4] used the concept of social potentials that allow UAVs to "snap" into the correct relative positions, based on a predetermined set of "attachment sites" around each UAV. This method however offers no control over the geometry of the formation, and more dispersed formations, like the wedge and diamond formations, cannot be achieved. Carpin and Parker [11] described a distributed approach, based on explicit communication between UAVs, for coordinated motion in a linear pattern. The framework is also able to handle the emergence of unexpected obstacles within the formation. Kostelnik et al. [56] solved the problem by using a communication network between the UAVs, so that each UAV may be assigned social roles by a dynamically chosen leader. Kang et al. [49, 50] proposed general methods for the controller design for the formation maintenance of multiple vehicles tracking a desired path. It was assumed that the desired trajectories are known, and each UAV is associated with a specific node in the formation. The formation controllers can be easily altered by the designer when vehicles are added or removed from the system.

Another pressing concern in formation control is that of ensuring convergence and stability within the proposed schemes. Many researchers have focused on this aspect, and have derived provably stable convergent algorithms for formation control. In the work by Song and Kumar [96], specific artificial potential fields for each UAV relative to the other UAVs are calculated to produce stable formations in equilibrium. The concept of Leader-to-Formation Stability is introduced in [100], and the paper examined how errors propagate from the leader and influence the stability of formations (those based on leader following). Navigation functions, first proposed in [86], have also been used for centralized control of formations [98]. The problem of ensuring formation stability in a UAV team while it moves in formation through an obstacle field between two points was investigated in [75].

The later chapters in this book dealing with the coordination within UAV teams present control techniques that are specifically designed to take into account heterogeneity within teams and the variability of the environments these teams are likely to operate in. Specifically, they will focus on improving the scalability and robustness of agent teams. These chapters will complement earlier chapters in the book by providing an overarching coordination framework within which agents (UAVs) will operate, where each agent is subject to control laws described in earlier chapters of this book.

The main objectives of the book is to present and discuss various control strategies for helicopters and algorithms for coordinating groups of helicopters.

## 1.2 Outline of the Book

The book starts with a brief introduction of development of helicopter systems modeling, control and coordination in Chap. 1.

Chapter 2 focuses on nonlinear fight dynamic modeling for a general class of rotary-wing aircraft. Centered around the general rigid body equations of motion, the modeling process is based on modules that follow closely with physical subsystems on the rotary-wing aircraft. These modules may include main and tail rotors, propeller, horizontal and vertical tails, wings, fuselage, and the flight control system. Aerodynamic interference among different modules can also be established. Besides the general modeling, performance calculation is also illustrated as it is a critical index in understanding the aircraft.

Following the flight dynamic modeling, trim study and stability analysis are conducted in Chap. 3. Trim is a state of the aircraft in which forces and moments reach equilibrium. Trim is the basis of many fundamental analyses, including stability analysis, control system design, handling qualities assessment, and simulation. A numerical procedure on how to obtain trim is described. With trim condition established, linearization process can be carried out to obtain a linearized model for the rotary-wing aircraft. Elements in the linearized model are called stability and control derivatives. A detailed description on the derivatives is provided, with the emphasis of their physical meanings. To illustrate, three examples are also presented in the chapter, which include a simplified single helicopter system and two hobby helicopters at hover.

Linearized models are not only useful for stability analysis, but also for control design, by blending them over different operating points with gain scheduling. However, this requires extensive modeling, which is expensive and time-consuming. There is a need for controllers that can operate with minimal model information and handle nonlinearities over the entire flight regime. Chapter 4 aims to address this need, by presenting a robust adaptive NN control for helicopters. In particular, we focus on vertical flight, which can be represented in the single-input–single-output (SISO) nonlinear nonaffine form, because the coupling between longitudinal and lateral directional equations in this flight regime is weak. Although a nonaffine system can be rendered affine by adding an integrator to the control input, thus allowing many control methods for affine nonlinear system to be used, the disadvantage of this approach is that the dimension of the system is increased, and control efforts are not direct and immediate either. Subsequently, effective control for the system may not be achieved. In this chapter, we focus on control design for the nonaffine system directly, without adding any integrators to the input. Differing from the approaches in the literature, which were based on approximate dynamic inversion with augmented NNs, we utilize the Mean Value Theorem and the Implicit Function Theorem as mathematical tools to handle the nonaffine nonlinearities in the helicopter dynamics. In cases where reasonably accurate knowledge of the dynamic inversion model is available, the method using approximate dynamic inversion has been shown to provide an effective solution to the problem. However, the construction of the dynamic inversion for a nonaffine system may not be an easy task in general. For such cases, our approach offers a feasible means of tackling the problem, since a priori knowledge of the inversion is not required.

In Chap. 5, NN control is redesigned to track both altitude and yaw angle at the same time. We consider a scale model helicopter, mounted on an experimental platform, in the presence of model uncertainties, which may be caused by unmodelled dynamics, or aerodynamical disturbances from the environment. Two different types of NN, namely a multilayer neural network (MNN) and a radial basis function neural network (RBFNN) are adopted in control design and stability analysis. Based on Lyapunov synthesis, the proposed adaptive NN control ensures that both the altitude and the yaw angle track the given bounded reference signals to a small neighborhood of zero, and guarantees the Semi-Globally Uniformly Ultimate boundedness (SGUUB)of all the closed-loop signals at the same time. The effectiveness of the proposed control is illustrated through extensive simulations. Compared with the model-based control, approximation-based control yields better tracking performance in the presence of model uncertainties.

Unlike previous chapters which consider actuators with instantaneous response, Chap. 6 deals with a more realistic scenario where actuator dynamics are present. Backstepping technique, combined with NNs, is employed to design the robust attitude control for uncertain multi-input–multi-output (MIMO) nonlinear helicopter dynamics. To the best of our knowledge, there are few works in the literature that take the actuator dynamics into account in the helicopter control, which is practically relevant but more challenging as well. In this chapter, helicopter models are considered as MIMO nonlinear dynamic systems, where the actuator dynamics in the first-order low-pass filter form are considered. Secondly, the possible singularity problem of the control coefficient matrix for the model-based attitude control case has been tackled effectively by introducing a design matrix. Thirdly, approximation-based attitude control is developed to handle the model uncertainties (e.g., unknown moment coefficients and mass) and external disturbances. Rigorous stability analysis and extensive simulations results show the effectiveness and robustness of the proposed attitude control.

Beyond single-helicopter control, this monograph also touches on formation control of multiple helicopters. In Chap. 7, the concept of the Q-structure is introduced as a novel and flexible methodology to define and support a large variety of formations. The Q-structure allows automatic scaling of formations according to changes in the overall size of the helicopter team. The chapter begins by exploring the use of the Q-structure for formation control where perfect communication is present between all members of the team. The second part of the chapter focuses on how the Q-structure can be adapted and used for teams where communication is imperfect. In this chapter, we examine the properties of Q-structures in relation to other formation representation schemes, and look at the ways Q-structures can be used with artificial potential trenches to improve the scalability of the formations and support a large number of different formations. In particular, the Q-structure does not require explicit representation of every single node of the formation and is able to ensure the formation maintenance of a large number of helicopters. The formation is also robust against possible communication breakdown and/or limited wireless communication ranges.

The Q-structure formation control is useful for motion planning at the kinematics level. For formation control that takes into account the dynamics of helicopters, a different approach is proposed. Chapter 8 presents synchronized altitude tracking control of helicopters with unknown dynamics by graph theory, while the desired trajectory is available to a portion of the team. Since only the neighbors' information is available to the helicopter, we use the weighted average of neighbors' states as the reference state of the helicopter in the control design. We prove that if the extended communication graph contains a spanning tree with the virtual vehicle as its root, then its Laplacian will be positive definite. This property is essential to prove the stability and it also makes the proof of the stability for the results easy and direct. The mathematical stability proof, which makes use of the positive definite property of the graph Laplacian, is provided for both full-state and output feedback cases.

# Chapter 2
# Building a Nonlinear Rotary-Wing Aircraft Model

## 2.1 Introduction

This chapter focuses on nonlinear flight dynamic modeling of rotary-wing aircraft. The intention is to establish a general modeling framework, that applies to not only traditional single main-rotor helicopters, but also unconventional rotary-wing aircraft. Apart from general equations of motion, the modeling framework relies heavily on and is extended from the results in [97]. Differing from a traditional modeling description, the inflow dynamics of a rotor module include latest result from [13] to handle a rotor in descent motion. In addition, a propeller thrust and torque calculation procedure is illustrated using the typical nomenclature in helicopter theory books. For the purposes of easy implementation and simple writing format, vector description of equations is utilized whenever possible.

## 2.2 General Equations of Motion

Assumptions associated with the classical six degree-of-freedom (DOF) equations of motion include rigid body dynamics and insignificant effects from Earth's curvature and rotation. These two assumptions are usually satisfied for applications in performance study, stability analysis, dynamic simulation, control system design, and handling qualities assessment.

Typically, the general equations of motion include four sets of equations: force equation, moment equation, kinematic equation, and navigation equation.

### *2.2.1 Force Equation*

Figure 2.1 shows typical coordinate systems for both aircraft body and the main rotor hub. Let $u_B$, $v_B$, and $w_B$ be three velocity components and $p_B$, $q_B$, and $r_B$

B. Ren et al., *Modeling, Control and Coordination of Helicopter Systems*,
DOI 10.1007/978-1-4614-1563-3_2, © Springer Science+Business Media, LLC 2012

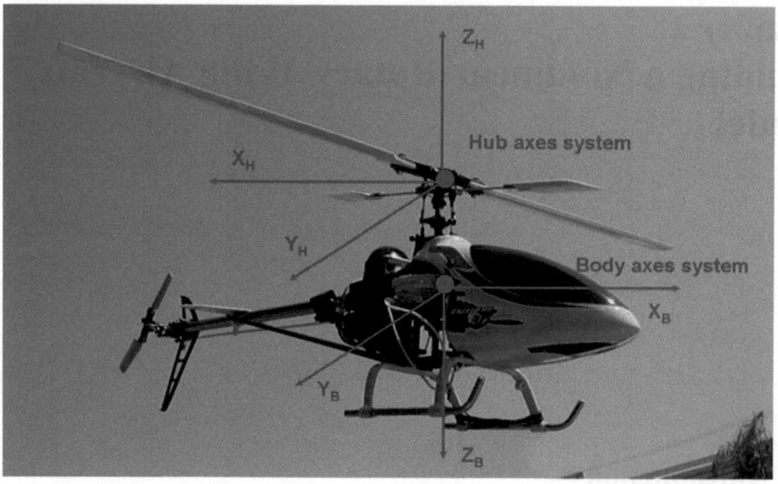


**Fig. 2.1** Definitions of body axes and hub axes system (Image with courtesy of http://cheaprchelicopters.net)

be three angular rate components along the $x$-, $y$-, $z$-axis in the body axes system, respectively. Furthermore, define the Euler angle representation as roll ($\phi$), pitch ($\theta$), and yaw ($\psi$). The force equation is listed as follows:

$$\begin{bmatrix} \dot{u}_{\mathrm{B}} \\ \dot{v}_{\mathrm{B}} \\ \dot{w}_{\mathrm{B}} \end{bmatrix} = -\begin{bmatrix} p_{\mathrm{B}} \\ q_{\mathrm{B}} \\ r_{\mathrm{B}} \end{bmatrix} \times \begin{bmatrix} u_{\mathrm{B}} \\ v_{\mathrm{B}} \\ w_{\mathrm{B}} \end{bmatrix} + \left( \begin{bmatrix} -g\sin\theta \\ g\sin\phi\cos\theta \\ g\cos\phi\cos\theta \end{bmatrix} + \frac{1}{m} \begin{bmatrix} F_x \\ F_y \\ F_z \end{bmatrix} \right) \tag{2.1}$$

The cross-product term is the Coriolis acceleration due to body rotation. There are two contributions to the force term of the equation: one is the gravitational force resolved in the body axes system, and another is all other forces in the body axes system, represented as $F_x$, $F_y$, and $F_z$.

### 2.2.2 *Moment Equation*

Define $L$, $M$, and $N$ as the resultant rolling, pitching, and yawing moments along the $x$-, $y$-, $z$-axis in the body axes system, respectively. The moment equation is:

$$\begin{bmatrix} \dot{p}_B \\ \dot{q}_B \\ \dot{r}_B \end{bmatrix} = -I^{-1} \left( \begin{bmatrix} p_{\mathrm{B}} \\ q_{\mathrm{B}} \\ r_{\mathrm{B}} \end{bmatrix} \times \left( I \begin{bmatrix} p_{\mathrm{B}} \\ q_{\mathrm{B}} \\ r_{\mathrm{B}} \end{bmatrix} \right) \right) + I^{-1} \begin{bmatrix} L \\ M \\ N \end{bmatrix} \tag{2.2}$$

The inertial matrix, $I$, can be represented as

$$I = \begin{bmatrix} I_{xx} & 0 & -I_{xz} \\ 0 & I_{yy} & 0 \\ -I_{xz} & 0 & I_{zz} \end{bmatrix}, \qquad I^{-1} = \frac{1}{\Gamma} \begin{bmatrix} I_{zz} & 0 & I_{xz} \\ 0 & \frac{\Gamma}{I_{yy}} & 0 \\ I_{xz} & 0 & I_{xx} \end{bmatrix} \tag{2.3}$$

where $\Gamma = I_{xx}I_{zz} - I_{xz}^2$. In the moment equation, the first term on the right-hand side includes both inertial coupling and gyroscopic effect. Notice that in the $I$ matrix, there are no coupling terms $I_{yz}$ and $I_{xy}$. This is due to the fact that for most flying vehicles, the $x - z$ plane is a plane of symmetry. Although, strictly speaking, a classical single main-rotor helicopter has no plane of symmetry due to its tail rotor, it is often used to simplify the resultant equations. The inertial coupling term associated with $I_{xz}$ will be significant in the case of highly maneuverable motions. In some special cases with three symmetric planes like quad-rotor system, there is no inertial coupling presented in their motions. While gyroscopic effect is evident for a single-engine fixed-wing aircraft in maneuvering flight, it is even more significant in a rotary-wing vehicle due to its main rotor. In fact, the moment due to gyroscopic effect is the main source of roll and pitch dampings from the rotor in an unaugmented rotorcraft.

### *2.2.3 Kinematic Equation*

It should be pointed out that the three Euler angle derivatives, $\dot{\phi}$, $\dot{\theta}$, and $\dot{\psi}$, are not orthogonal to each other. The relationship between Euler angle derivatives and body angular rates is provided as follows:

$$\begin{bmatrix} \dot{\phi} \\ \dot{\theta} \\ \dot{\psi} \end{bmatrix} = \begin{bmatrix} 1 & \tan\theta\sin\phi & \tan\theta\cos\phi \\ 0 & \cos\phi & -\sin\phi \\ 0 & \frac{\sin\phi}{\cos\theta} & \frac{\cos\phi}{\cos\theta} \end{bmatrix} \begin{bmatrix} p_{\mathrm{B}} \\ q_{\mathrm{B}} \\ r_{\mathrm{B}} \end{bmatrix} \tag{2.4}$$

It is noted that the kinematic equation has a singularity at $\theta = 90^0$. For all-attitude flight, a quaternion representation is recommended.

### *2.2.4 Navigation Equation*

The navigation equation is represented in the local NED (North–East–Down) frame. The transformation from the vehicle's body frame to the local NED frame follows the predefined rotational sequence: roll, pitch, and yaw. Therefore, the navigation equation is

$$\begin{bmatrix} \dot{x}_\mathrm{N} \\ \dot{x}_\mathrm{E} \\ \dot{x}_\mathrm{D} \end{bmatrix} = \begin{bmatrix} \cos\psi & -\sin\psi & 0 \\ \sin\psi & \cos\psi & 0 \\ 0 & 0 & 1 \end{bmatrix} \begin{bmatrix} \cos\theta & 0 & \sin\theta \\ 0 & 1 & 0 \\ -\sin\theta & 0 & \cos\theta \end{bmatrix} \begin{bmatrix} 1 & 0 & 0 \\ 0 & \cos\phi & -\sin\phi \\ 0 & \sin\phi & \cos\phi \end{bmatrix} \begin{bmatrix} u_\mathrm{B} \\ v_\mathrm{B} \\ w_\mathrm{B} \end{bmatrix} \tag{2.5}$$

In practical applications, position information is often obtained in terms of GPS data in the Geodetic form (longitude, latitude, and height). In such cases, transformations can be carried out first from Geodetic frame to ECEF (Earth centered Earth fixed) frame, and then to local NED frame.

In both force and moment equations, the resultant forces ($F_x$, $F_y$, and $F_z$) and moments ($L$, $M$, and $N$) are contributions from various modules of a rotary-wing aircraft, including main rotor, tail rotor (for a classical helicopter), propeller (if mounted, typically with a compound helicopter or an autogyro), fuselage, horizontal and vertical tails, wing (if mounted), landing gear, and slung loads (for underslung operation). In the case of aerodynamic forces and moments, they are typically computed in the local wind axes system. This process requires that the relevant motion variables be transformed to the wind axes system and resultant forces and moments be transformed back to the body axes system.

A block diagram for a general rotorcraft model is provided in Fig. 2.2, showcasing the relationship between each individual module and general equations of motion. Typically, the general equations of motion take forces and moments transformed from each module and update the motion variables at each time step. The motion variables are fed back into each module through appropriate transformations to generate the updated forces and moments.

In the next section, the main characteristics of each module in Fig. 2.2 will be analyzed. It shall be emphasized that although the study uses a helicopter as a baseline platform, it can be extended to unconventional rotorcraft platforms. Readers are encouraged to explore the use of modeling and simulation to evaluate advanced or novel rotorcraft concepts and applications.

## 2.3 Modular-Based Modeling

### *2.3.1 Main Rotor*

General assumptions associated with flight dynamic modeling for the main rotor are [97]

- Small angle approximation. This is applied to the blade flapping angle and inflow angle used during the derivation.

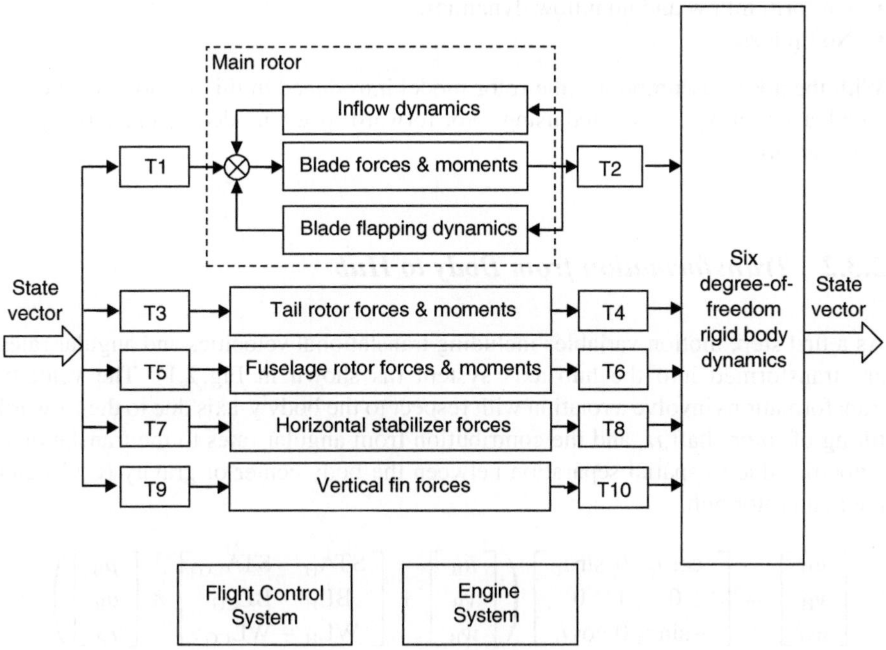


**Fig. 2.2** Block diagram for a general rotorcraft model [97]

- Rigid blade. This assumption is valid for flight dynamic applications including performance estimation, trim study, stability and control analysis, handling quality assessment, and dynamic response simulation. Elastic blade modeling or even finite element blade modeling are only justified when flexibility of the blade is essential.
- Reverse flow region, stall, and compressibility effects are ignored. These effects are only appreciable during high forward speed. A criteria for judging the boundary speed is that the maximum tip speed for the advancing blade should not exceed $60--70\%$ of the speed of sound.
- First harmonic approximation to the blade flapping. In this case, the blade flapping angle is determined by coning angle $\beta_0$, longitudinal first harmonic term $\beta_{1c}$ (positive backward), and lateral first harmonic term $\beta_{1s}$ (positive toward the advancing side). Mathematically, the blade flapping angle is represented by

$$\beta = \beta_0 - \beta_{1c} \cos \psi - \beta_{1s} \sin \psi \tag{2.6}$$

where $\psi$ is the blade azimuthal angle.
- Quasi-steady flapping dynamics. The rotor would respond to continuously changing motions as if they were a sequence of steady conditions. Mathematically, it assumes that the terms $\frac{dp}{d\psi}$ and $\frac{dq}{d\psi}$ are insignificant in determining the rotor first-harmonic terms.

- Uniform inflow and no inflow dynamics.
- No tip losses.

With the above assumptions, the rotor model introduced in this section is valid for an advance ratio $\mu$, computed using rotor forward speed divided by blade tip speed at hover, up to 0.3.

### 2.3.2 Transformation from Body to Hub

As a first step, motion variables including translational velocities and angular rates are transformed into the hub axis system (as shown in Fig. 2.1). The velocity transformations involve a rotation with respect to the body y-axis due to the forward tilting of rotor shaft $i_s$, and the contribution from angular rates to the translational velocities due to spatial separation between the body center of gravity (C.G.) and the main rotor hub.

$$\begin{bmatrix} u_H \\ v_H \\ w_H \end{bmatrix} = \begin{bmatrix} \cos i_s & 0 & \sin i_s \\ 0 & 1 & 0 \\ -\sin i_s & 0 & \cos i_s \end{bmatrix} \left( \begin{bmatrix} u_B \\ v_B \\ w_B \end{bmatrix} + \begin{bmatrix} \mathrm{STA_H} - \mathrm{STA_{CG}} \\ \mathrm{BL_H} - \mathrm{BL_{CG}} \\ \mathrm{WL_H} - \mathrm{WL_{CG}} \end{bmatrix} \times \begin{bmatrix} p_B \\ q_B \\ r_B \end{bmatrix} \right) \tag{2.7}$$

where STA, BL, and WL are the stationline, buttline, and waterline for a sub-system (*e.g.*, the main rotor hub), respectively. The subscript $H$ stands for the main rotor hub. Similarly, the angular rate transformation involves the rotation with respect to body $y$-axis due to forward tilting of the rotor shaft,

$$\begin{bmatrix} p_H \\ q_H \\ r_H \end{bmatrix} = \begin{bmatrix} \cos i_s & 0 & \sin i_s \\ 0 & 1 & 0 \\ -\sin i_s & 0 & \cos i_s \end{bmatrix} \begin{bmatrix} p_B \\ q_B \\ r_B \end{bmatrix} \tag{2.8}$$

### 2.3.3 Blade Flapping Dynamics

The blade flapping dynamics are described by a second-order differential equation [97]

$$\begin{bmatrix} \ddot{\beta}_0 \\ \ddot{\beta}_{1c} \\ \ddot{\beta}_{1s} \end{bmatrix} + D \begin{bmatrix} \dot{\beta}_0 \\ \dot{\beta}_{1c} \\ \dot{\beta}_{1s} \end{bmatrix} + K \begin{bmatrix} \beta_0 \\ \beta_{1c} \\ \beta_{1s} \end{bmatrix} = F \tag{2.9}$$

where

$$D = \Omega \begin{bmatrix} \frac{\gamma}{2}(\frac{1}{4} - \frac{2}{3}\epsilon + \frac{\epsilon^2}{2}) & 0 & -\frac{\gamma\mu}{4}(\frac{1}{3} - \epsilon + \epsilon^2) \\ 0 & \frac{\gamma}{2}(\frac{1}{4} - \frac{2}{3}\epsilon + \frac{\epsilon^2}{2}) & 2 \\ -\frac{\gamma\mu}{2}(\frac{1}{3} - \epsilon + \epsilon^2) & -2 & \frac{\gamma}{2}(\frac{1}{4} - \frac{2}{3}\epsilon + \frac{\epsilon^2}{2}) \end{bmatrix} \tag{2.10}$$

$$K = \Omega^2 \begin{bmatrix} \begin{matrix} P^2 + \\ \frac{\gamma K_1 \mu^2}{4}(\frac{1}{2} - \epsilon + \frac{\epsilon^2}{2}) \end{matrix} & -\frac{\gamma\mu}{4}(\frac{\epsilon}{2} - \epsilon^2) & -\frac{\gamma K_1 \mu}{4}(\frac{2}{3} - \epsilon) \\ -\frac{\gamma\mu}{2}(\frac{1}{3} - \frac{\epsilon}{2}) & \begin{matrix} P^2 - 1 + \\ \frac{\gamma K_1 \mu^2}{8}(\frac{1}{2} - \epsilon + \frac{\epsilon^2}{2}) \end{matrix} & \begin{matrix} \frac{\gamma}{2}(\frac{1}{4} - \frac{2}{3}\epsilon + \frac{\epsilon^2}{2}) + \\ \frac{\gamma\mu^2}{8}(\frac{1}{2} - \epsilon + \frac{\epsilon^2}{2}) \end{matrix} \\ -\frac{\gamma K_1 \mu}{2}(\frac{2}{3} - \epsilon) & \begin{matrix} -\frac{\gamma}{2}(\frac{1}{4} - \frac{2}{3}\epsilon + \frac{\epsilon^2}{2}) + \\ \frac{\gamma\mu^2}{8}(\frac{1}{2} - \epsilon + \frac{\epsilon^2}{2}) \end{matrix} & \begin{matrix} P^2 - 1 + \\ \frac{3}{8}\gamma K_1 \mu^2(\frac{1}{2} - \epsilon + \frac{\epsilon^2}{2}) \end{matrix} \end{bmatrix} \tag{2.11}$$

$$F = \Omega^2 \left\{ \begin{bmatrix} -\frac{M_\beta}{I_\beta \Omega^2} \\ 0 \\ 0 \end{bmatrix} + \begin{bmatrix} \frac{\gamma}{2}[(\frac{1}{5} - \frac{\epsilon}{4}) + \frac{\mu^2}{2}(\frac{1}{3} - \frac{\epsilon}{2})] \\ 0 \\ -\frac{\gamma\mu}{2}(\frac{1}{2} - \frac{2}{3}\epsilon) \end{bmatrix} \theta_t + \begin{bmatrix} \frac{\gamma}{2}(\frac{1}{3} - \frac{\epsilon}{2}) \\ 0 \\ -\frac{\gamma\mu}{2}(\frac{1}{2} - \epsilon + \frac{\epsilon^2}{2}) \end{bmatrix} \lambda \right.$$

$$+ \begin{bmatrix} \begin{matrix} \frac{\gamma}{2}[(\frac{1}{4} - \frac{\epsilon}{3}) + \\ \frac{\mu^2}{2}(\frac{1}{2} - \epsilon + \frac{\epsilon^2}{2})] \end{matrix} & 0 & -\frac{\gamma}{2}[\mu(\frac{1}{3} - \frac{\epsilon}{2})] \\ 0 & \begin{matrix} \frac{\gamma}{2}[(\frac{1}{4} - \frac{\epsilon}{3}) + \\ \frac{\mu^2}{4}(\frac{1}{2} - \epsilon + \frac{\epsilon^2}{2})] \end{matrix} & 0 \\ -\frac{\gamma}{2}\mu(\frac{2}{3} - \epsilon) & 0 & \begin{matrix} \frac{\gamma}{2}[(\frac{1}{4} - \frac{\epsilon}{3}) + \\ \frac{3\mu^2}{4}(\frac{1}{2} - \epsilon + \frac{\epsilon^2}{2})] \end{matrix} \end{bmatrix} \begin{bmatrix} \theta_0 \\ \theta_{1c} \\ \theta_{1s} \end{bmatrix}$$

$$\left. + \begin{bmatrix} \frac{\gamma}{8}\mu(\frac{2}{3} - \epsilon) & 0 \\ -2(1 + \frac{eM_\beta}{gI_\beta}) & \frac{\gamma}{2}(\frac{1}{4} - \frac{\epsilon}{3}) \\ -\frac{\gamma}{2}(\frac{1}{4} - \frac{\epsilon}{3}) & -2(1 + \frac{eM_\beta}{gI_\beta}) \end{bmatrix} \begin{bmatrix} \frac{p_H}{\Omega}\cos\beta_w + \frac{q_H}{\Omega}\sin\beta_w \\ \frac{p_H}{\Omega}\sin\beta_w - \frac{q_H}{\Omega}\cos\beta_w \end{bmatrix} \right\} \tag{2.12}$$

with the parameters defined as

$\theta_0$ = collective pitch angle
$\theta_{1s}$ = longitudinal cyclic pitch angle
$\theta_{1c}$ = lateral cyclic pitch angle
$\Omega$ = rotor rotational speed
$\gamma$ = blade lock number
$\theta_t$ = blade twist
$\lambda$ = total inflow ratio
$\beta_w$ = rotor sideslip angle, $a\sin(\frac{v_H}{\sqrt{u_H^2+v_H^2}})$, defined as 0 if $u_H = v_H = 0$.

In (2.11), $P$ represents the ratio of flapping frequency to rotor speed

$$P^2 = 1 + \frac{eM_\beta}{gI_\beta} + \frac{K_\beta}{I_\beta \Omega^2} + \frac{\gamma K_1}{8}\left(1 - \frac{4}{3}\epsilon\right) \tag{2.13}$$

where $e$, $K_\beta$, and $K_1$ represent flapping hinge offset, flap spring, and pitch-flap coupling, respectively. $\epsilon = \frac{e}{R}$ is the non-dimensionalized hinge offset ratio. The term $M_\beta$ in the above equation represents blade weight moment. For a blade with uniform mass distribution, the simple relationship between $M_\beta$ and blade moment of inertia $I_\beta$ can be established [44]:

$$M_\beta = \frac{3g}{2R} I_\beta \tag{2.14}$$

Without hinge offset, flap spring, and pitch-flap coupling, the flap frequency is the same as the rotor frequency. With either hinge offset (typically with articulated rotor, or hingeless rotor), or flap spring (normally coupled with a teetering rotor), or the pitch-flap coupling (normally with a tail rotor), the flap frequency is raised above the rotor frequency, with typical value $P = 1.0 - 1.2$ [44].

While (2.9) is derived based on a general rotor configuration, simplifications can be made on a number of cases, which in turn provides important physical insights on the rotor flapping motion. In the first simplification, it is assumed that the rotor has zero hinge offset, no flap spring, and no pitch-flap coupling. Equation (2.9) can thus be re-written:

$$\begin{bmatrix} \ddot{\beta}_0 \\ \ddot{\beta}_{1c} \\ \ddot{\beta}_{1s} \end{bmatrix} + \Omega \begin{bmatrix} \frac{\gamma}{8} & 0 & -\frac{\gamma\mu}{12} \\ 0 & \frac{\gamma}{8} & 2 \\ -\frac{\gamma\mu}{6} & -2 & \frac{\gamma}{8} \end{bmatrix} \begin{bmatrix} \dot{\beta}_0 \\ \dot{\beta}_{1c} \\ \dot{\beta}_{1s} \end{bmatrix} + \Omega^2 \begin{bmatrix} 1 & 0 & 0 \\ -\frac{\gamma\mu}{6} & 0 & \frac{\gamma}{8} + \frac{\gamma\mu^2}{16} \\ 0 & -\frac{\gamma}{8} + \frac{\gamma\mu^2}{16} & 0 \end{bmatrix} \begin{bmatrix} \beta_0 \\ \beta_{1c} \\ \beta_{1s} \end{bmatrix}$$

$$
= \Omega^2 \begin{bmatrix} -\dfrac{M_\beta}{I_\beta \Omega^2} + \dfrac{\gamma}{8}(1+\mu^2)\theta_0 - \dfrac{\gamma\mu}{6}\theta_{1s} + \dfrac{\gamma}{2}(\dfrac{1}{5} + \dfrac{\mu^2}{6})\theta_t \\ +\dfrac{\gamma}{6}\lambda + \dfrac{\gamma\mu}{12}(\dfrac{p_H}{\Omega}\cos\beta_w + \dfrac{q_H}{\Omega}\sin\beta_w) \\ \\ -2(\dfrac{p_H}{\Omega}\cos\beta_w + \dfrac{q_H}{\Omega}\sin\beta_w) + \dfrac{\gamma}{2}(\dfrac{1}{4} + \dfrac{\mu^2}{4})\theta_{1c} \\ +\dfrac{\gamma}{8}(\dfrac{p_H}{\Omega}\sin\beta_w - \dfrac{q_H}{\Omega}\cos\beta_w) \\ \\ -2(\dfrac{p_H}{\Omega}\sin\beta_w - \dfrac{q_H}{\Omega}\cos\beta_w) - \dfrac{\gamma\mu}{3}\theta_0 \\ -\dfrac{\gamma\mu}{4}\theta_t + \dfrac{\gamma}{2}(\dfrac{1}{4} + \dfrac{3\mu^2}{8})\theta_{1s} \\ -\dfrac{\gamma\mu}{4}\lambda - \dfrac{\gamma}{8}(\dfrac{p_H}{\Omega}\cos\beta_w - \dfrac{q_H}{\Omega}\sin\beta_w) \end{bmatrix} \tag{2.15}
$$

A notable observation can be made from the above equation on the flapping response due to body angular rates. When the rotor is at hover and under zero roll and pitch controls, the steady state response of $\beta_{1s}$ and $\beta_{1c}$ are

$$
\beta_{1s} = -\frac{16}{\gamma}\frac{p_H}{\Omega} - \frac{q_H}{\Omega} \tag{2.16}
$$

$$
\beta_{1c} = \frac{p_H}{\Omega} - \frac{16}{\gamma}\frac{q_H}{\Omega} \tag{2.17}
$$

This demonstrates that rotor first harmonic variation of flapping angles is proportional to angular rates of the body during steady pull-up or steady roll flight.

A second simplification involves steady state response of (2.9). In this case, let both first and second derivatives of three blade flapping terms be zero, as follows:

$$
K \begin{bmatrix} \beta_0 \\ \beta_{1c} \\ \beta_{1s} \end{bmatrix} = F \tag{2.18}
$$

where matrices $K$ and $F$ are from (2.11) and (2.12). For studies that focus on low frequency, (2.18) is more commonly used. Equation (2.18) may not be sufficient for studies involving high fidelity dynamic response validation and aeroelastic analysis.

In a third simplification, steady state response of $\beta_0$, $\beta_{1c}$, and $\beta_{1s}$ are derived from (2.15) for the rotor having zero hinge offset, no flap spring and no pitch-flap coupling, and under zero angular rates;

$$
\beta_0 = -\frac{M_\beta}{I_\beta \Omega^2} + \gamma\left[\frac{\theta_0}{8}(1+\mu^2) + \frac{\theta_t}{10}\left(1 + \frac{5}{6}\mu^2\right) + \frac{\lambda}{6} - \frac{\mu}{6}\theta_{1s}\right] \tag{2.19}
$$

$$
\beta_{1s} = \frac{\frac{\mu}{6}\beta_0 + \frac{1}{8}(1+\mu^2)\theta_{1c}}{\frac{1}{8}(1+\frac{\mu^2}{2})} \tag{2.20}
$$

$$
\beta_{1c} = \frac{\frac{\mu}{3}\theta_0 + \frac{\mu}{4}\theta_t + \frac{\mu}{4}\lambda - \frac{1}{8}(1+\frac{3}{2}\mu^2)\theta_{1s}}{\frac{1}{8}(1-\frac{\mu^2}{2})} \tag{2.21}
$$

At hover, the steady state responses of longitudinal and lateral flapping to the cyclic control inputs can be reduced even further:

$$\beta_{1s} = \theta_{1c} \tag{2.22}$$

$$\beta_{1c} = -\theta_{1s} \tag{2.23}$$

This shows the one-to-one correspondence between the first harmonic flapping and their corresponding cyclic controls.

In the last simplification, it should be noted that for many types of helicopters, their rotor structure is teetering configuration. For teetering configuration, the angle $\beta_0$ is treated as a constant pre-cone angle, $\beta_{0p}$. It thus follows that $\ddot{\beta}_0 = \dot{\beta}_0 = 0$. The two first harmonic terms $\beta_{1c}$ and $\beta_{1s}$ can be solved using two formulae in (2.9):

$$\begin{aligned}
&\begin{bmatrix} \ddot{\beta}_{1c} \\ \ddot{\beta}_{1s} \end{bmatrix} + \Omega \begin{bmatrix} \frac{\gamma}{8} & 2 \\ -2 & \frac{\gamma}{8} \end{bmatrix} \begin{bmatrix} \dot{\beta}_{1c} \\ \dot{\beta}_{1s} \end{bmatrix} \\
&+\Omega^2 \begin{bmatrix} P^2 - 1 + \frac{\gamma K_1 \mu^2}{16} & \frac{\gamma}{8}(1 + \frac{\mu^2}{2}) \\ -\frac{\gamma}{8}(1 - \frac{\mu^2}{2}) & P^2 - 1 + \frac{3}{16}\gamma K_1 \mu^2 \end{bmatrix} \begin{bmatrix} \beta_{1c} \\ \beta_{1s} \end{bmatrix} \\
&= \Omega^2 \begin{bmatrix} \frac{\gamma\mu}{6} \\ \frac{\gamma K_1 \mu}{3} \end{bmatrix} \beta_{0p} + \Omega^2 \begin{bmatrix} 0 & 0 \\ -\frac{\gamma\mu}{3} & -\frac{\gamma\mu}{4} \end{bmatrix} \begin{bmatrix} \theta_0 \\ \theta_t \end{bmatrix} \\
&+\Omega^2 \begin{bmatrix} 0 \\ -\frac{\gamma\mu}{4} \end{bmatrix} \lambda + \Omega^2 \begin{bmatrix} \frac{\gamma}{8}(1 + \frac{\mu^2}{2}) & 0 \\ 0 & \frac{\gamma}{8}(1 + \frac{3}{2}\mu^2) \end{bmatrix} \begin{bmatrix} \theta_{1c} \\ \theta_{1s} \end{bmatrix} \\
&+\Omega \begin{bmatrix} -2 & -\frac{\gamma}{8} \\ -\frac{\gamma}{8} & 2 \end{bmatrix} \begin{bmatrix} p_{\mathrm{H}} \cos\beta_w + q_{\mathrm{H}} \sin\beta_w \\ p_{\mathrm{H}} \sin\beta_w - q_{\mathrm{H}} \cos\beta_w \end{bmatrix}
\end{aligned} \tag{2.24}$$

### 2.3.4 Inflow Dynamics

Induced velocity at the rotor can be derived from Momentum Theory. The formulation is represented as rotor inflow ratio $\lambda$:

$$\lambda = \frac{w_{\mathrm{H}}}{\Omega R} - \frac{C_T}{2\sqrt{\mu^2 + \lambda^2}} \tag{2.25}$$

When $\mu = 0$, $\lambda$ can be solved directly from the equation. When $\mu \neq 0$, $\lambda$ needs to be solved implicitly through the Newton–Raphson iterative technique.

Equation (2.25) is valid when the rotor operates under hover, climb, steep descent, and forward flight conditions. It is not valid under moderate descent rate with slow forward speed when vortex ring state (VRS) occurs. In VRS, the rotor encounters its own wake resulting in a doughnut-shaped ring around the rotor disk. Typical phenomena associated with VRS include flow unsteadiness, excessive thrust and torque fluctuations, uncommanded drop in descent rate, an increase in required power, loss of control effectiveness, and a significant increase in vibration.

One drawback from Momentum Theory is its ignorance of flow interaction between the rotor wake and surrounding airflow in descending condition [44]. Effects of the interaction may be less significant at hover or in climb. Nevertheless, as a rotor increases its descent rate, the interaction becomes more and more intense due to larger velocity gradients between the upflow outside the wake and the downflow inside the wake. As such, a new inflow model was proposed to take into account the flow interactions, known as ring vortex model [13, 15]. The ring vortex model supposes that, due to the flow interaction, there exists a series of vortex rings located at the rotor periphery. Vortex rings move downward along the wake when descending at a low rate. As the rate of descent increases, vortex rings tend to accumulate near the rotor tip. When the rate of descent further increases, vortex rings move upward along the wake. A new vortex ring is formed at every blade rotation, i.e., $\frac{2\pi}{\Omega N_b}$ second. The convection speed of the vortex rings, $V_{\mathrm{con}}$, is approximated as follows:

$$V_{\mathrm{con}} = V_i - \frac{5}{3} w_{\mathrm{H}} \tag{2.26}$$

where the induced velocity $V_i = \frac{C_T}{2\sqrt{(\mu^2+\lambda^2)}} \Omega R$. The locations of these discrete vortex rings can thus be determined by the product of convection velocity of the vortex rings and $\frac{2m}{\Omega N_b}$ ($m$: an integer representing the numbering of vortex rings).

Each vortex ring induces additional normal velocity at the rotor disk. The vortex strength of each ring, $\Gamma$, is approximated as follows:

$$\Gamma = k_\Gamma V_i R \tag{2.27}$$

where $N_{\mathrm{ring}} k_\Gamma = 0.2167$. The number of vortex rings, $N_{\mathrm{ring}}$, is typically set to two. The flow field of a vortex ring can thus be computed based on elliptic integrals

[12]. One advantage of utilizing vortex rings is that the effect of vortex rings is non-uniform with respect to relative distance between the rings and the rotor disk. The closer a vortex ring is to the rotor disk, the larger the magnitude of normal velocity at the disk. The resulting non-uniform effect conforms to what has been observed in test data.

In the ring vortex model, downward velocity due to vortex rings is integrated into the induced velocity calculated by Momentum Theory. The concept works well in the VRS and windmill phases. Nevertheless, in axial descent and inclined descent at low forward speed ($\frac{\mu}{\sqrt{C_T/2}}$ up to 0.6204), Momentum Theory fails to predict a transition phase between the helicopter and the windmill branches. To address this problem, (2.25) is modified as

$$\lambda = \frac{w_{\mathrm{H}}}{\Omega R} - \frac{C_T}{2\sqrt{k_{\mathrm{RVM}}^2 + \mu^2 + \lambda^2}} \tag{2.28}$$

where

$$k_{\mathrm{RVM}} = \frac{\lambda}{2.72(1 + \frac{\mu^2}{C_T/2})} \tag{2.29}$$

The additional term $k_{\mathrm{RVM}}$ is analogous to the parachute drag term and it modifies equilibrium curves for inflow dynamics, creating a steady state transition between helicopter and windmill branches in axial and steep descents. Its effect diminishes at other flight conditions.

To demonstate the effectiveness of the ring vortex model, comparison are made between the numerical predictions and wind-tunnel test results from [108] in terms of induced velocity variation (normalized by hover induced velocity, $V_{\mathrm{h}}$). It is shown from Fig. 2.3 that good agreement is reached with respect to various descent angles $\alpha_{\mathrm{D}}$.

### *2.3.5 Main Rotor Forces and Moments*

Main rotor forces and moments are calculated in the hub-wind axes system. The quantities include rotor thrust $T$, H-force $H_w$, Y-force $Y_w$, rolling moment $L_w$, pitching moment $M_w$, and rotor torque $Q$.

A closed-form thrust equation is thus provided in the hub-wind axes system [97]:

$$\begin{aligned} T = \frac{n_b}{2}\rho a c R(\Omega R)^2 \Big\{ &\frac{1}{2}(1-\epsilon^2)\lambda + \theta_0\left[\frac{1}{3} + \frac{\mu^2}{2}(1-\epsilon)\right] + \theta_t\left[\frac{1}{4} + \frac{\mu^2}{4}(1-\epsilon^2)\right] \\ &- \frac{\mu}{2}(1-\epsilon^2)(\theta_{1s} - K_1\beta_{1s}) - \beta_0\left[\frac{1}{3} + \frac{\mu^2}{2}(1-\epsilon)\right]K_1 + \beta_{1c}\left[\frac{\mu}{2}\epsilon(1-\epsilon)\right]\Big\} \end{aligned} \tag{2.30}$$

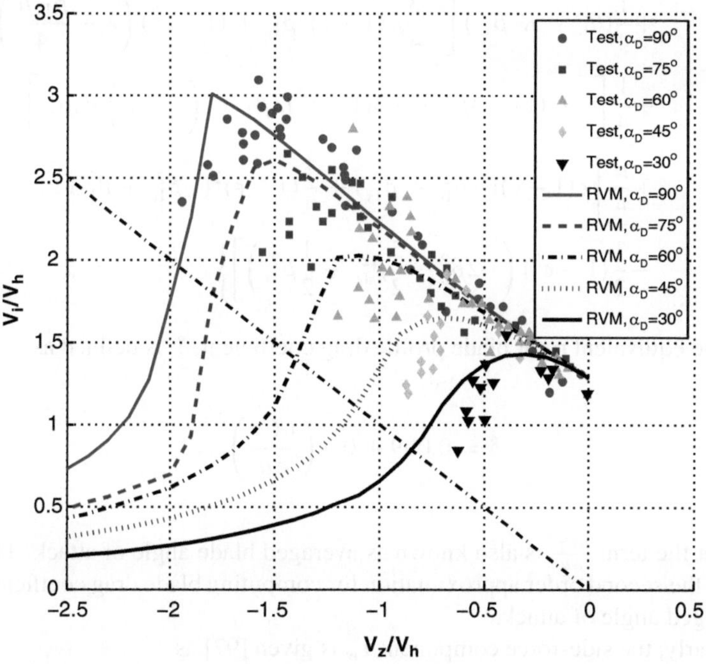


**Fig. 2.3** Induced velocity variations from Yaggy's wind-tunnel test [108] and predictions from the ring vortex model: axial and nonaxial flow

In many hobby helicopter models, their blades are teetering ($\epsilon = 0$), have untwisted blades ($\theta_t = 0$), and are without pitch flap coupling ($K_1 = 0$). The thrust equation can thus be further simplified as:

$$T = \frac{n_b}{2}\rho acR(\Omega R)^2\left\{\frac{\lambda}{2} + \theta_0\left[\frac{1}{3} + \frac{\mu^2}{2}\right] - \frac{\mu}{2}\theta_{1s}\right\} \tag{2.31}$$

Horizontal force component $H_w$ is provided [97] as

$$\begin{aligned} H_w = \frac{n_b}{2}\rho acR(\Omega R)^2 &\left\{\frac{\delta\mu}{2a}(1-\epsilon^2)\right.\\ &-\frac{1}{4}(\theta_0 - K_1\beta_0)\left[2\lambda\mu(1-\epsilon) + \left(\epsilon - \frac{2}{3}\right)\beta_{1c} - \frac{2}{3}\beta_{1c}\right]\\ &-\frac{\theta_t}{4}\left[\mu\lambda(1-\epsilon^2) + 2\left(\frac{\epsilon}{3} - \frac{1}{4}\right)\beta_{1c} - \frac{\beta_{1c}}{2}\right]\\ &+\frac{1}{4}(\theta_{1c} - K_1\beta_{1c})\left[-\frac{\beta_{1s}\mu}{4}(1-\epsilon^2) + \frac{1}{4}\mu(1-\epsilon)^2\beta_{1s} + \frac{2}{3}\beta_0\right] \end{aligned}$$

$$
+\frac{1}{4}(\theta_{1s}-K_1\beta_{1s})\left[-\frac{3}{4}\mu(1-\epsilon)^2\beta_{1c}+(1-\epsilon^2)\left(\lambda-\frac{\beta_{1c}\mu}{4}\right)\right]
$$
$$
+\frac{1}{4}\left[-4\epsilon(1-\epsilon)\beta_{1c}\lambda-3(1-\epsilon^2)\lambda\beta_{1c}-\left(\frac{2}{3}-\epsilon\right)\beta_0\beta_{1s}\right]
$$
$$
+\frac{\mu}{4}\left[\epsilon(1-\epsilon)(-\beta_{1c}^2+\beta_{1s}^2)+\frac{1}{4}(1-\epsilon)^2(-\beta_{1c}^2+\beta_{1s}^2)\right.
$$
$$
\left.\left.-\frac{1}{2}(1-\epsilon^2)\left(-2\beta_0^2-\frac{5}{2}\beta_{1c}^2+\frac{1}{2}\beta_{1s}^2\right)\right]\right\} \tag{2.32}
$$

where the equivalent rotor blade profile drag coefficient, $\delta$, is defined as

$$
\delta = 0.009+0.3\left(\frac{6C_T}{\sigma a}\right)^2 \tag{2.33}
$$

Note that the term $\frac{6C_T}{\sigma a}$ is also known as averaged blade angle of attack. Therefore, (2.33) is the second order approximation for computing blade drag coefficient based on averaged angle of attack.

Similarly, the side-force component $Y_w$ is given [97] as

$$
Y_w = \frac{n_b}{2}\rho acR(\Omega R)^2\left(-\frac{1}{4}(\theta_0-K_1\beta_0)\left[\left(\epsilon-\frac{2}{3}\right)\beta_{1s}\right.\right.
$$
$$
\left.-\frac{2}{3}\beta_{1s}+3\beta_0(1-\epsilon^2)\mu-2\beta_{1s}(1-\epsilon)\mu^2\right]
$$
$$
-\frac{\theta_t}{4}\left[\left(\frac{2}{3}\epsilon-\frac{1}{2}\right)\beta_{1s}-\frac{\beta_{1s}}{2}+2\beta_0\mu-\beta_{1s}(1-\epsilon^2)\mu^2\right]
$$
$$
-\frac{1}{4}(\theta_{1c}-K_1\beta_{1c})\left\{\lambda(1-\epsilon^2)+\mu\left[\frac{5}{4}\beta_{1c}(1-\epsilon^2)-\frac{1}{4}(1-\epsilon)^2\beta_{1c}\right]\right\}
$$
$$
-\frac{1}{4}(\theta_{1s}-K_1\beta_{1s})\left[-\frac{2}{3}\beta_0+\mu\left(\frac{7}{4}\beta_{1s}(1-\epsilon^2)+\frac{1}{4}(1-\epsilon)^2\beta_{1s}\right)\right.
$$
$$
\left.-2\beta_0\mu^2(1-\epsilon)\right]-\frac{1}{4}\left[-2\lambda(1-\epsilon)^2\beta_{1s}-2\beta_0\beta_{1c}\left(\frac{1}{3}-\frac{\epsilon}{2}\right)-\lambda\beta_{1s}(1-\epsilon^2)\right]
$$
$$
-\frac{\mu}{4}\left[6\beta_0\lambda(1-\epsilon)-\frac{1}{2}\beta_{1c}\beta_{1s}(1-\epsilon^2)-\frac{7}{4}(1-\epsilon)^2\beta_{1c}\beta_{1s}\right.
$$
$$
\left.\left.+\frac{5}{4}(1-\epsilon)^2\beta_{1c}\beta_{1s}\right]-\mu^2\beta_0\beta_{1c}(1-\epsilon)\right) \tag{2.34}
$$

The calculated rotor forces are transformed back from hub-wind axes system to hub axes system:

$$\begin{bmatrix} H_{\mathrm{H}} \\ Y_{\mathrm{H}} \\ T \end{bmatrix} = \begin{bmatrix} \cos\beta_w & \sin\beta_w & 0 \\ -\sin\beta_w & \cos\beta_w & 0 \\ 0 & 0 & 1 \end{bmatrix} \begin{bmatrix} H_w \\ Y_w \\ T \end{bmatrix} \tag{2.35}$$

Rolling moment $L_w$ is computed based on the following formula [97]:

$$\begin{aligned} L_w = {} & \frac{n_b}{2}\left(k_\beta \beta_{1s} + \frac{eM_\beta}{g}\beta_{1s}\Omega^2\right) - \frac{n_b}{2} I_\beta \Omega^2 \gamma \epsilon \Bigg\{ \frac{\mu}{2}(1-\epsilon^2)(\theta_0 - K_1\beta_0) \\ & - \left[\frac{1}{6} + \frac{3}{8}\mu^2(1-\epsilon)\right](\theta_{1s} - K_1\beta_{1s}) + \frac{\mu}{3}\theta_t + \frac{\mu}{2}(1-\epsilon)\lambda \\ & + \frac{\mu^2}{8}(1-\epsilon)\beta_{1c} - \left(\frac{1}{6} - \frac{\epsilon}{4}\right)\beta_{1c} \Bigg\} \end{aligned} \tag{2.36}$$

Similarly, pitching moment $M_w$ is calculated [97] as

$$\begin{aligned} M_w = {} & \frac{n_b}{2}\left(k_\beta \beta_{1c} + \frac{eM_\beta}{g}\beta_{1c}\Omega^2\right) \\ & - \frac{n_b}{2} I_\beta \Omega^2 \gamma \epsilon \Bigg\{ -\left[\frac{1}{6} + \frac{\mu^2}{8}(1-\epsilon)\right](\theta_{1c} - K_1\beta_{1c}) \\ & - \frac{\mu}{4}(1-\epsilon^2)\beta_0 + \frac{\mu^2}{8}(1-\epsilon)\beta_{1s} + \left(\frac{1}{6} - \frac{\epsilon}{4}\right)\beta_{1s} \Bigg\} \end{aligned} \tag{2.37}$$

For helicopter rotors with teetering blades and no flapping spring constraint, it follows that $L_w = 0$ and $M_w = 0$. It is also interesting to study rotor blades with hinge offset and/or flapping spring constraint. In this case, let $C_{L_w} = L_w/(\rho A R(\Omega R)^2)$ and $C_{M_w} = M_w/(\rho A R(\Omega R)^2)$ be coefficients of rolling moment and pitching moment, respectively. From both (2.36) and (2.37), it results in:

$$C_{L_w} = \frac{\sigma a}{2}\frac{P^2 - 1}{\gamma}\beta_{1s} \tag{2.38}$$

$$C_{M_w} = \frac{\sigma a}{2}\frac{P^2 - 1}{\gamma}\beta_{1c} \tag{2.39}$$

Recall that $P$ is known as the non-dimensional flap frequency. This demonstrates that for a rotor with hinge offset or flap spring, the direct hub rolling and pitching moments are proportional to rotor first-harmonic tilting. As for the resultant rolling and pitching moments acting on the vehicle's C.G., they are the summation of direct hub moment and the moments caused by the tilting of the rotor thrust about the body's C.G. For an articulated rotor, direct hub moments are about the same as the moments from the thrust tilting. For a hingeless or rigid rotor, the direct hub moments may be two to four times larger than the moments from the thrust tilting.

Main rotor torque can either be estimated from rotor power consumption (as will be illustrated in Sect. 2.4), or from the closed-form expression given as follows [97]:

$$
\begin{aligned}
Q = \frac{n_b}{2}\rho a c R^2(\Omega R)^2 \Bigg\{ & \frac{\delta}{4a}\left[1 + (1-\epsilon^2)\mu^2\right] - (\theta_0 - K_1\beta_0)\left(\frac{\lambda}{3} + \frac{\mu}{4}\epsilon\beta_{1c}\right) \\
& + (\theta_{1c} - K_1\beta_{1c})\left[\left(\frac{1}{8} - \frac{\epsilon}{6}\right)\beta_{1s} - \frac{\mu}{6}\beta_0 - \frac{\beta_{1s}}{16}(1-\epsilon^2)\mu^2\right] \\
& + (\theta_{1s} - K_1\beta_{1s})\left[-\left(\frac{1}{8} - \frac{\epsilon}{6}\right)\beta_{1c} + \frac{1}{2}(1-\epsilon^2)\left(\frac{\mu\lambda}{2} + \frac{\beta_{1c}}{8}\mu^2\right)\right] \\
& - \theta_t\left(\frac{\lambda}{4} + \frac{\epsilon\mu}{6}\beta_{1c}\right) - \frac{1}{2}(1-\epsilon^2)\left[\lambda^2 + \lambda\mu\beta_{1c} + \mu\epsilon\beta_0\beta_{1s} + \mu^2\left(\frac{\beta_0^2}{2} + \frac{3}{8}\beta_{1c}^2\right.\right. \\
& \left.\left. + \frac{1}{8}\beta_{1s}^2\right)\right] + \frac{\mu}{3}\beta_0\beta_{1s} - \left(\frac{1}{4} - \frac{2}{3}\epsilon + \frac{\epsilon^2}{2}\right)\left(\frac{1}{2}\beta_{1s}^2 + \frac{1}{2}\beta_{1c}^2\right)\Bigg\}
\end{aligned}
\tag{2.40}
$$

The calculated rotor moments are transformed from the hub-wind axes system back to the hub axes system:

$$
\begin{bmatrix} L_H \\ M_H \\ Q \end{bmatrix} = \begin{bmatrix} \cos\beta_w & -\sin\beta_w & 0 \\ \sin\beta_w & \cos\beta_w & 0 \\ 0 & 0 & 1 \end{bmatrix} \begin{bmatrix} L_w \\ M_w \\ Q \end{bmatrix}
\tag{2.41}
$$

## 2.3.6 *Transformation from Hub to Body*

The forces at the rotor hub are further transformed into the body axes system involving the forward tilting of the rotor shaft $i_s$:

$$
\begin{bmatrix} X_{MR} \\ Y_{MR} \\ Z_{MR} \end{bmatrix} = \begin{bmatrix} \cos i_s & 0 & -\sin i_s \\ 0 & 1 & 0 \\ \sin i_s & 0 & \cos i_s \end{bmatrix} \begin{bmatrix} -H_H \\ Y_H \\ -T \end{bmatrix}
\tag{2.42}
$$

Similarly, the moments at the rotor hub are transformed back to the body axes system, involving the forward tilting of the rotor shaft as well as additional moments about the C.G. from the rotor forces due to spatial separation between the C.G. and the rotor hub:

$$
\begin{bmatrix} L_{MR} \\ M_{MR} \\ N_{MR} \end{bmatrix} = \begin{bmatrix} \cos i_s & 0 & -\sin i_s \\ 0 & 1 & 0 \\ \sin i_s & 0 & \cos i_s \end{bmatrix} \begin{bmatrix} L_H \\ M_H \\ Q \end{bmatrix} - \begin{bmatrix} \mathrm{STA_H} - \mathrm{STA_{CG}} \\ \mathrm{BL_H} - \mathrm{BL_{CG}} \\ \mathrm{WL_H} - \mathrm{WL_{CG}} \end{bmatrix} \times \begin{bmatrix} X_{MR} \\ Y_{MR} \\ Z_{MR} \end{bmatrix}
\tag{2.43}
$$

### 2.3.7 Tail Rotor

A tail rotor usually serves two purposes for a conventional single main-rotor helicopter: to provide anti-torque for the main rotor and to support directional control.

The analysis of a tail rotor is very similar to that of a main rotor [97]. In fact, it is often treated as a special case of the main rotor in terms of both flapping and force calculations. However, there are some special features associated with the tail rotor which may make the analysis easier. These features include:

- The tail rotor has only collective for the thrust, and hence there is no cyclic pitch control ($\theta_{1c} = \theta_{1s} = 0$);
- The tail rotor rotates at a much faster speed, hence a steady state blade flapping response in the non-rotating frame is sufficient.

The tail rotor may come in various configurations, including teetering, articulated, hingeless, bearingless, Fenestrons (fan-in-fin), and even NOTAR (No Tail Rotor). As the teetering is still the most commonly used configuration, the study in this section will focus on the teetering tail rotor.

As a first step, the velocity components at the body C.G. frame is transformed to the local velocity at the hub of the tail rotor:

$$\begin{bmatrix} u_{\mathrm{TR}} \\ v_{\mathrm{TR}} \\ w_{\mathrm{TR}} \end{bmatrix} = \begin{bmatrix} u_{\mathrm{B}} \\ v_{\mathrm{B}} \\ w_{\mathrm{B}} \end{bmatrix} + \begin{bmatrix} \mathrm{STA_{TR}} - \mathrm{STA_{CG}} \\ \mathrm{BL_{TR}} - \mathrm{BL_{CG}} \\ \mathrm{WL_{TR}} - \mathrm{WL_{CG}} \end{bmatrix} \times \begin{bmatrix} p_{\mathrm{B}} \\ q_{\mathrm{B}} \\ r_{\mathrm{B}} \end{bmatrix} + \begin{bmatrix} 0 \\ 0 \\ w_{i_{\mathrm{TR}}} \end{bmatrix} \tag{2.44}$$

where $w_{i_{\mathrm{HT}}}$ is the velocity that accounts for the interference from the main rotor to the tail rotor. The advance ratio and sideslip angle for the tail rotor can thus be obtained as

$$\mu_{\mathrm{TR}} = \frac{\sqrt{u_{\mathrm{TR}}^2 + w_{\mathrm{TR}}^2}}{\Omega_{\mathrm{TR}} R_{\mathrm{TR}}} \tag{2.45}$$

$$\beta_{\mathrm{TR}} = \tan^{-1}\left(\frac{w_{\mathrm{TR}}}{u_{\mathrm{TR}}}\right) \tag{2.46}$$

Since a steady state solution is sufficient for the teetering tail rotor, it follows that $\ddot{\beta}_{1c} = \ddot{\beta}_{1s} = \dot{\beta}_{1c} = \dot{\beta}_{1s} = 0$. The two first harmonic terms $\beta_{1c}$ and $\beta_{1s}$ can be solved by using (2.24). Notice that $\theta_{1c} = \theta_{1s} = 0$ in the case of a tail rotor. The inflow ratio for the tail rotor is very similar to that for the main rotor:

$$\lambda_{\mathrm{TR}} = -\frac{v_{\mathrm{TR}}}{\Omega_{\mathrm{TR}} R_{\mathrm{TR}}} - \frac{C_{T_{\mathrm{TR}}}}{2\sqrt{\mu_{\mathrm{TR}}^2 + \lambda_{\mathrm{TR}}^2}} \tag{2.47}$$

For the tail rotor thrust, H-force, and Y-force, (2.30), (2.32), and (2.34) can be applied by selecting corresponding tail rotor parameters and setting $\epsilon = 0$. As mentioned earlier, the rolling and pitching moments of the tail rotor are zero due to its teetering configuration. For the tail rotor torque, (2.40) can be applied.

All forces and moments for the tail rotor are transformed back from the hub wind frame to body frame:

$$\begin{bmatrix} X_{\mathrm{TR}} \\ Y_{\mathrm{TR}} \\ Z_{\mathrm{TR}} \end{bmatrix} = \begin{bmatrix} \cos\beta_{\mathrm{TR}} & -\sin\beta_{\mathrm{TR}} & 0 \\ 0 & 0 & 1 \\ \sin\beta_{\mathrm{TR}} & \cos\beta_{\mathrm{TR}} & 0 \end{bmatrix} \begin{bmatrix} -H_{W\mathrm{TR}} \\ Y_{W\mathrm{TR}} \\ T_{\mathrm{TR}} \end{bmatrix} \tag{2.48}$$

$$\begin{bmatrix} L_{\mathrm{TR}} \\ M_{\mathrm{TR}} \\ N_{\mathrm{TR}} \end{bmatrix} = \begin{bmatrix} 0 \\ -Q_{\mathrm{TR}} \\ 0 \end{bmatrix} - \begin{bmatrix} \mathrm{STA}_{\mathrm{TR}} - \mathrm{STA}_{\mathrm{CG}} \\ \mathrm{BL}_{\mathrm{TR}} - \mathrm{BL}_{\mathrm{CG}} \\ \mathrm{WL}_{\mathrm{TR}} - \mathrm{WL}_{\mathrm{CG}} \end{bmatrix} \times \begin{bmatrix} X_{\mathrm{TR}} \\ Y_{\mathrm{TR}} \\ Z_{\mathrm{TR}} \end{bmatrix} \tag{2.49}$$

### *2.3.8 Propeller*

The analysis of a propeller is very similar to a helicopter rotor. In fact, the basic analytical tools are the same: Momentum theory and Blade Element theory. However, caution must be taken when applying closed-form equations in Sect. 2.3.1 to the propeller performance calculation. This is because a few assumptions in reaching the closed-form equations are no longer valid. For example, it is common that a propeller blade can have fairly large pitch, in which case we can not make a small angle assumption. Moreover, the chordwise blade profile can be much more complicated than that of an ordinary rotor blade.

A propeller thrust and torque calculation procedure (based on Vortex theory [68]) will be provided in the following sections. Historically, the terminology used in the helicopter rotor and the propeller are quite different. For example, the advance ratio for the helicopter rotor is defined as $\frac{V}{\Omega R}$, while the same term for the propellers is defined as $\frac{V}{nD} = \frac{\pi V}{\Omega R}$ ($n$ is the rotational speed in revolutions per second and $D$ is the propeller diameter). The procedure will adopt the typical nomenclature from helicopter theory books.

1. Break a single propeller blade into $n$ elements. Let $r$ and $x$ be the collection of dimensional and dimensionless radial stations along the blade.
2. Let $w_t$ and $w_a$ be tangential and axial components of the induced velocity. Assume $\frac{w_t}{V_T} = 0$, where $V_T = \Omega R$ is the tip velocity of the propeller blade.
3. Calculate $w_a$:

$$\frac{w_a}{V_T} = \frac{1}{2}\left[-\mu + \sqrt{\mu^2 + 4\frac{w_t}{V_T}\left(x - \frac{w_t}{V_T}\right)}\right] \tag{2.50}$$

where $\mu = \frac{V}{V_T}$ is the advance ratio for the propeller.

4. Obtain $\phi = \tan^{-1}\frac{\mu}{x}$.
5. Calculate induced angle of attack $\alpha_i$:

$$\alpha_i = \tan^{-1}\left(\frac{V + w_a}{\Omega r - w_t}\right) - \phi \tag{2.51}$$

6. Compute a correction term for finite thickness of the blade:

$$\Delta\alpha = \frac{4\mu\sigma}{15(\mu^2 + x^2)}\frac{t_{\max}}{c} \tag{2.52}$$

where $t_{\max}$ is the maximum thickness of the blade and $c$ is the sectional blade chord.
7. Obtain local angle of attack:

$$\alpha = \theta - \alpha_i - \phi - \Delta\alpha \tag{2.53}$$

8. Compute a second correction term for effective camber due to flow curvature:

$$\Delta C_l = \frac{a\Delta\theta}{4} \tag{2.54}$$

where $\Delta\theta = \tan^{-1}\left(\frac{V+w_a}{\Omega r-2w_t}\right) - \tan^{-1}\left(\frac{V+w_a}{\Omega r}\right)$.
9. Obtain sectional lift coefficient $C_l$:

$$C_l = a(\alpha + \alpha_{0l}) - \Delta C_l \tag{2.55}$$

where $\alpha_{0l}$ is the angle between the zero-lift line and the chord line for the propeller airfoil.
10. Calculate sectional total velocity $V_e$:

$$\frac{V_e}{V_T} = \sqrt{\left(\mu + \frac{w_a}{V_T}\right)^2 + \left(x - \frac{w_t}{V_T}\right)^2} \tag{2.56}$$

11. Obtain bound circulation $\Gamma$:

$$\Gamma = \frac{1}{2}\nu C_l V_e \tag{2.57}$$

12. Update $w_t$:

$$w_t = \frac{n_b\Gamma}{4\pi r F} \tag{2.58}$$

where $F$ is the Prandtl's tip loss factor

$$F = \frac{2}{\pi} \cos^{-1} \exp\left[-\frac{n_b(1-x)}{2\sin\theta_T}\right] \tag{2.59}$$

13. Return to Step 3 until the absolute difference (vector norm) between the calculated $w_t$ and that from the previous calculation reduces to a desired value.
14. Obtain differential lift $\mathrm{d}L$ and differential drag $\mathrm{d}D$:

$$\mathrm{d}L = \frac{1}{2}\rho V_e^2 c C_l \mathrm{d}r \tag{2.60}$$

$$\mathrm{d}D = \frac{1}{2}\rho V_e^2 c C_d \mathrm{d}r \tag{2.61}$$

where sectional drag coefficient$C_d$ can be represented as a function of $C_l$.
15. Compute sectional thrust $\mathrm{d}T$ and torque $\mathrm{d}Q$:

$$\mathrm{d}T = \mathrm{d}L\cos(\phi+\alpha_i) - \mathrm{d}D\sin(\phi+\alpha_i) \tag{2.62}$$

$$\mathrm{d}Q = r[\mathrm{d}L\sin(\phi+\alpha_i) + \mathrm{d}D\cos(\phi+\alpha_i)] \tag{2.63}$$

16. Numerically integrate the above two equations in order to obtain the total thrust $T$ and torque $Q$.
17. Finally, compute the propeller efficiency:

$$\eta = \frac{TV}{Q\Omega} \tag{2.64}$$

The procedure listed above is used to compute thrust and power for a three-bladed propeller with Clark-Y section from a wind-tunnel test [37]. The predicted non-dimensional thrust and torque coefficients from Vortex Theory are compared with the experimental results in Fig. 2.4 for two different blade pitch angles at 0.75R: 15° and 35°. The prediction clearly matches with the test very well. In addition, another set of predictions using combined Momentum-Blade Element Theory is also provided (for computational procedure with this method, refer to [68]). Prediction with Vortex Theory is considerably precise as it takes into account flow rotation, tip loss, finite thickness of a blade, and effective camber of the blade section due to flow curvature.

### 2.3.9 *Horizontal Tail*

In steady forward flight, the horizontal tail can generate a trim load to compensate main rotor longitudinal flapping [97]. More importantly, the horizontal tail is able

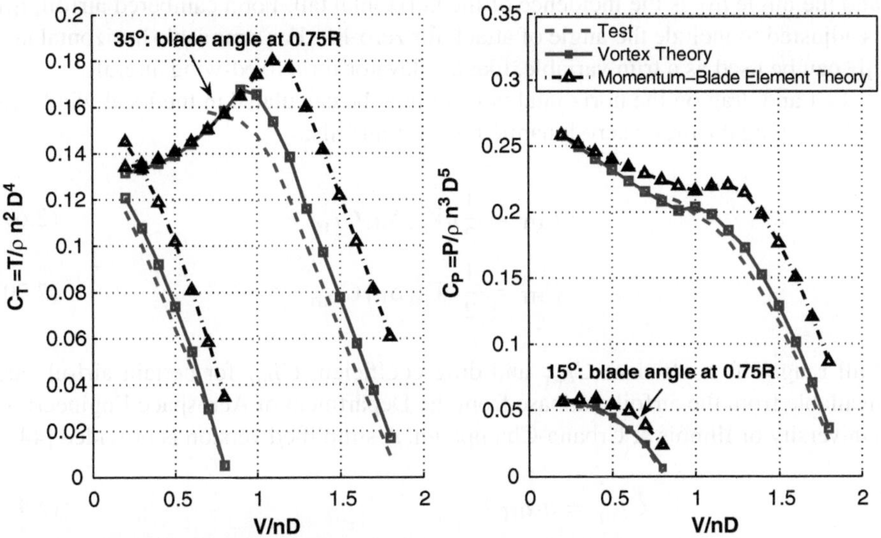


**Fig. 2.4** Comparison of predictions from Vortex Theory and Momentum-Blade Element Theory with experiment from [37]

to provide a stabilizing pitch moment due to angle of attack variation to enhance the pitch stability. Primary forces from the horizontal tail are lift and drag.

The velocity components acting on the horizontal tail in its local body axes system are as follows:

$$\begin{bmatrix} u_{\mathrm{HT}} \\ v_{\mathrm{HT}} \\ w_{\mathrm{HT}} \end{bmatrix} = \begin{bmatrix} u_{\mathrm{B}} \\ v_{\mathrm{B}} \\ w_{\mathrm{B}} \end{bmatrix} + \begin{bmatrix} \mathrm{STA_{HT}} - \mathrm{STA_{CG}} \\ \mathrm{BL_{HT}} - \mathrm{BL_{CG}} \\ \mathrm{WL_{HT}} - \mathrm{WL_{CG}} \end{bmatrix} \times \begin{bmatrix} p_{\mathrm{B}} \\ q_{\mathrm{B}} \\ r_{\mathrm{B}} \end{bmatrix} + \begin{bmatrix} 0 \\ 0 \\ w_{i_{\mathrm{HT}}} \end{bmatrix} \tag{2.65}$$

where $w_{i_{\mathrm{HT}}}$ is the velocity that accounts for the interference angle from the main rotor to the horizontal tail. The angle of attack and sideslip angle of the horizontal tail are then obtained:

$$\alpha_{\mathrm{HT}} = \tan^{-1}\left(\frac{w_{\mathrm{HT}}}{u_{\mathrm{HT}}}\right) + i_{\mathrm{HT}} \tag{2.66}$$

$$\beta_{\mathrm{HT}} = \sin^{-1}\left(\frac{v_{\mathrm{HT}}}{V_{\mathrm{HT}}}\right) \tag{2.67}$$

where

$$V_{\mathrm{HT}} = \sqrt{u_{\mathrm{HT}}^2 + v_{\mathrm{HT}}^2 + w_{\mathrm{HT}}^2} \tag{2.68}$$

and the angle $i_{HT}$ is the incidence of the horizontal tail. For a cambered airfoil, $i_{HT}$ is adjusted to include the angle of attack for zero-lift. For a movable horizontal tail, $i_{HT}$ can be used as a trim variable, like an elevator for a fixed-wing aircraft.

Lift and drag on the horizontal tail can then be calculated in the local wind axis system using quadratic aerodynamic form for airfoils:

$$L_{HT} = \frac{1}{2}\rho V_{HT}^2 S_{HT} C_{L_{HT}} \tag{2.69}$$

$$D_{HT} = \frac{1}{2}\rho V_{HT}^2 S_{HT} C_{D_{HT}} \tag{2.70}$$

Full range lift coefficient $C_{L_{HT}}$ and drag coefficient $C_{D_{HT}}$ for certain airfoils are available from the airfoil database from the Department of Aerospace Engineering, University of Illinois at Urbana-Champaign. A simplified version is provided [44]:

$$C_{L_{HT}} = a\alpha_{HT} \tag{2.71}$$

$$C_{D_{HT}} = 0.0087 - 0.0216\alpha_{HT} + 0.4\alpha_{HT}^2 \tag{2.72}$$

where $a$ is the lift curve slope, and $C_{D_{HT}}$ is obtained for the NACA 23012 airfoil and considered accurate up to its stall angle (12°).

Subsequently, lift and drag can be transformed from local wind axes system to local body axes system:

$$\begin{bmatrix} X_{HT} \\ Y_{HT} \\ Z_{HT} \end{bmatrix} = \begin{bmatrix} \cos\alpha_{HT} & 0 & -\sin\alpha_{HT} \\ 0 & 1 & 0 \\ \sin\alpha_{HT} & 0 & \cos\alpha_{HT} \end{bmatrix} \begin{bmatrix} \cos\beta_{HT} & -\sin\beta_{HT} & 0 \\ \sin\beta_{HT} & \cos\alpha_{HT} & 0 \\ 0 & 0 & 1 \end{bmatrix} \begin{bmatrix} -D_{HT} \\ 0 \\ -L_{HT} \end{bmatrix} \tag{2.73}$$

Moments associated with the horizontal tail are computed based on the location of the aerodynamic center on the tail with respect to the body C.G.:

$$\begin{bmatrix} L_{HT} \\ M_{HT} \\ N_{HT} \end{bmatrix} = -\begin{bmatrix} \mathrm{STA_{HT}} - \mathrm{STA_{CG}} \\ \mathrm{BL_{HT}} - \mathrm{BLCG} \\ \mathrm{WL_{HT}} - \mathrm{WL_{CG}} \end{bmatrix} \times \begin{bmatrix} X_{HT} \\ Y_{HT} \\ Z_{HT} \end{bmatrix} \tag{2.74}$$

In general, lift on the horizontal tail is the dominant force while pitch moment is the dominant moment. It, therefore, may be sufficient just to compute the lift and pitch moment from the horizontal tail in certain applications.

### 2.3.10 Wing

Wing is sometimes added to a rotor-wing aircraft to produce lift during forward flight. The wing is treated in the same manner as the horizontal tail. All the formulae listed in Sect. 2.3.9 are portable in the calculation of forces and moments for the wing module.

### 2.3.11 Vertical Tail

The vertical tail can be treated in a similar fashion as the horizontal tail with minor differences. In a standard single main rotor configuration, an additional interference effect needs to be taken into account from the tail rotor to the vertical tail [97]. As such, the velocity components acting on the vertical tail in its local body axes system are as follows:

$$\begin{bmatrix} u_{\mathrm{VT}} \\ v_{\mathrm{VT}} \\ w_{\mathrm{VT}} \end{bmatrix} = \begin{bmatrix} u_{\mathrm{B}} \\ v_{\mathrm{B}} \\ w_{\mathrm{B}} \end{bmatrix} + \begin{bmatrix} \mathrm{STA}_{\mathrm{VT}} - \mathrm{STA}_{\mathrm{CG}} \\ \mathrm{BL}_{\mathrm{VT}} - \mathrm{BL}_{\mathrm{CG}} \\ \mathrm{WL}_{\mathrm{VT}} - \mathrm{WL}_{\mathrm{CG}} \end{bmatrix} \times \begin{bmatrix} p_{\mathrm{B}} \\ q_{\mathrm{B}} \\ r_{\mathrm{B}} \end{bmatrix} + \begin{bmatrix} 0 \\ k_{v_{\mathrm{tr}}} w_{i_{\mathrm{tr}}} \\ w_{i\mathrm{VT}} \end{bmatrix} \tag{2.75}$$

where $w_{i\mathrm{VT}}$ is the interference velocity from the main rotor to the vertical tail. The term $w_{i_{\mathrm{tr}}}$ accounts for the blockage effect due to the tail rotor. The angle of attack and sideslip angle of the vertical tail are calculated in consideration of different orientations between the horizontal tail and vertical tail:

$$\alpha_{\mathrm{HT}} = \tan^{-1}\left(\frac{v_{\mathrm{VT}}}{u_{\mathrm{VT}}}\right) + i_{\mathrm{VT}} \tag{2.76}$$

$$\beta_{\mathrm{HT}} = \sin^{-1}\left(\frac{w_{\mathrm{VT}}}{V_{\mathrm{VT}}}\right) \tag{2.77}$$

where

$$V_{\mathrm{VT}} = \sqrt{u_{\mathrm{VT}}^2 + v_{\mathrm{VT}}^2 + w_{\mathrm{VT}}^2} \tag{2.78}$$

and the angle $i_{\mathrm{HT}}$ is the incidence angle of the vertical tail. The angle $i_{\mathrm{HT}}$ is adjusted to include the angle of attack for zero-lift for cambered airfoil and can be used as a trim variable for movable vertical tail, like a rudder for a fixed-wing aircraft. The remaining calculation of forces and moments for the vertical tail can be directly referred to Sect. 2.3.9.

### 2.3.12 Fuselage

Aerodynamic forces and moments acting on the fuselage can be computed in almost the exact manner as the horizontal tail. However, one difficulty in the case of the fuselage is the determination of the aerodynamic database. While such database for the horizontal tail is still relatively easy to find due to its airfoil shape, it is generally difficult for the fuselage due to its bluff body shape. Conducting a wind-tunnel experiment or a comprehensive computational study are certainly beneficial, although they may sometimes be costly and time-consuming.

An alternative is to estimate equivalent flat-plate areas for all body-x, body-y, and body-z axes and calculate correspondingly forces along those axes. Similar to the horizontal and vertical tails, the velocity components acting on the fuselage in its local body axes system are [97]:

$$\begin{bmatrix} u_{\mathrm{FU}} \\ v_{\mathrm{FU}} \\ w_{\mathrm{FU}} \end{bmatrix} = \begin{bmatrix} u_{\mathrm{B}} \\ v_{\mathrm{B}} \\ w_{\mathrm{B}} \end{bmatrix} + \begin{bmatrix} 0 \\ 0 \\ w_{i_{\mathrm{FU}}} \end{bmatrix} \tag{2.79}$$

where $w_{i_{\mathrm{FU}}}$ is the interference velocity from the main rotor to the fuselage. Note that the effects from angular rates are ignored here due to close distance between fuselage aerodynamic center and body C.G. Let $S_{\mathrm{Ref}}^{\mathrm{front}}$, $S_{\mathrm{Ref}}^{\mathrm{side}}$, and $S_{\mathrm{Ref}}^{\mathrm{top}}$ be the equivalent flat-plate areas along the body-x, body-y, body-z axes, the forces on the fuselage are

$$\begin{bmatrix} X_{\mathrm{FU}} \\ Y_{\mathrm{FU}} \\ Z_{\mathrm{FU}} \end{bmatrix} = \begin{bmatrix} \frac{1}{2}\rho u_{\mathrm{FU}}^2 S_{\mathrm{Ref}}^{\mathrm{front}} \\ \frac{1}{2}\rho v_{\mathrm{FU}}^2 S_{\mathrm{Ref}}^{\mathrm{side}} \\ \frac{1}{2}\rho w_{\mathrm{FU}}^2 S_{\mathrm{Ref}}^{\mathrm{top}} \end{bmatrix} \tag{2.80}$$

The above forces are assumed to act on the fuselage's center of pressure. A suggestion is provided to relate fuselage frontal drag area with the rest: $S_{\mathrm{Ref}}^{\mathrm{side}} = 2.2 S_{\mathrm{Ref}}^{\mathrm{front}}$ and $S_{\mathrm{Ref}}^{\mathrm{top}} = 1.5 S_{\mathrm{Ref}}^{\mathrm{front}}$. Moments due to the aerodynamic forces with respect to the body C.G. are

$$\begin{bmatrix} L_{\mathrm{FU}} \\ M_{\mathrm{FU}} \\ N_{\mathrm{FU}} \end{bmatrix} = - \begin{bmatrix} \mathrm{STA}_{\mathrm{FU}} - \mathrm{STA}_{\mathrm{CG}} \\ \mathrm{BL}_{\mathrm{FU}} - \mathrm{STA}_{\mathrm{CG}} \\ \mathrm{WL}_{\mathrm{FU}} - \mathrm{WL}_{\mathrm{CG}} \end{bmatrix} \times \begin{bmatrix} X_{\mathrm{FU}} \\ Y_{\mathrm{FU}} \\ Z_{\mathrm{FU}} \end{bmatrix} \tag{2.81}$$

### 2.3.13 Aerodynamic Interference

Aerodynamic interference among different modules are essential in rotorcraft modeling. Sometimes the interference introduces power penalties or loss of efficiency. For example, additional vertical loading on the fuselage due to main rotor downwash in hover condition could count for as much as 5% of total weight. In another example, for a tail rotor in pusher configuration, its aerodynamic efficiency decreases due to blackage effect from the vertical tail. The interference can be productive in other cases. For example, a horizontal tail immersed in the main rotor downwash contributes a stabilizing pitch moment for longitudinal static stability. A rudder immersed in a propwash for an autogyro could increase its control effectiveness at low speed.

A typical assumption in dealing with interference is airflow superposition. An additional velocity vector from the source of the interference is superimposed to the resultant module's local velocity vector, thus resulting in changes in local angle of attack and sideslip angle. The superposition method shall be treated as a first-order approximation of the actual interference.

From data presented in [42], an empirical expression for interference velocity $w_i$ due to a rotor was provided in [97]:

$$w_i = k_f v_i \tag{2.82}$$

where $v_i$ is the averaged downwash at the rotor plane. The empirical factor $k_f$ is given as follows:

$$k_f = 1.299 + 0.671\chi - 1.172\chi^2 + 0.351\chi^3 \tag{2.83}$$

where the rotor wake angle $\chi$ is defined as

$$\chi = \tan^{-1}\left(\frac{\mu}{-\lambda}\right) \tag{2.84}$$

For convenience, the graphical relationship between the interference factor and the rotor wake angle is provided in Fig. 2.5. The interference velocity in Sects. 2.3.7–2.3.12 can be computed using (2.82).

### 2.3.14 Rotor Rotational Degree of Freedom

In most rotary-wing flight dynamics modeling, rotor speed is assumed to be constant. This assumption is certainly valid in the presence of a governor module, which is designed to maintain the rotor speed through a feedback system. Nevertheless, for a helicopter operating in auto-rotation mode or an autogyro, the rotor speed is not determined by an engine but rather by the balancing of driving torque and driven

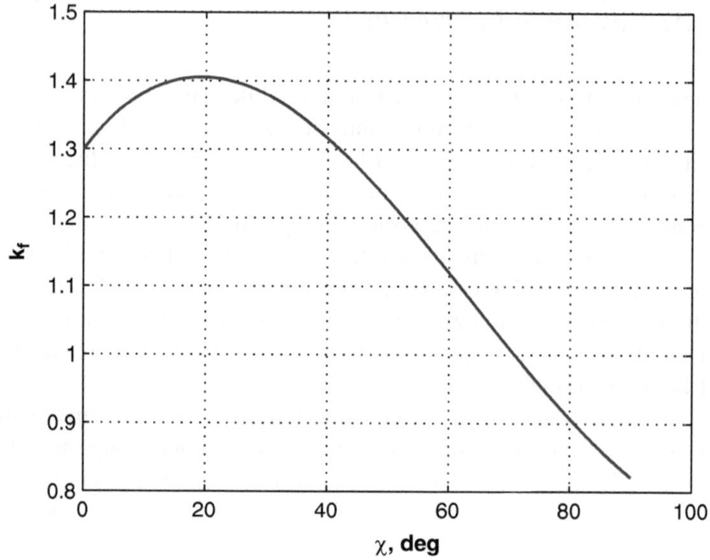


**Fig. 2.5** Interference factor with respect to rotor wake angle

torque at the rotor. In such a case, it is necessary to set up rotor rotational degree of freedom. This equation is described as follows:

$$J\dot{\Omega} = -Q \tag{2.85}$$

where $Q$ is the rotor torque. Rotor rotation inertia $J$ can be approximated using the blade flapping inertia $I_\beta$:

$$J = N_b I_\beta \tag{2.86}$$

where $N_b$ is the number of blades.

### 2.3.15 Flight Control System

In general, the flight control module consists of at least one of the following systems:

- Feedforward mechanical system, including pilot's controls, mechanical linkage, actuation system, swashplate, and control rods. This system can normally be represented by linear equations between the pilot's controls and blade pitch variations. It may also include first order transfer functions for actuators and nonlinear saturation representing control limits.

**Fig. 2.6** Teetering rotor configuration with a fly bar for a model helicopter

- Automatic flight control system (AFCS), including stability augmentation system (SAS) and control augmentation system (CAS). Details of AFCS design will be illustrated in later chapters.
- Fly bar (also known as stabilizer bar) for teetering rotor configuration (see Fig. 2.6).

The fly bar deserves an in-depth discussion as it is a common feature found in the 2-bladed teetering rotor configuration. The reason is that this type of rotor configuration tends to have excessively rapid response to cyclic controls. The installation of fly bar provides lagged rate feedback to slow down the fast response.

The bar consists essentially of a rod with two small airfoils at two ends. Without cyclic controls, the bar maintains its rotational plane parallel to the ground. With cyclic controls, the bar responds in a manner similar to the main rotor blades. There are historically two types of fly bars: Hiller bar and Bell–Hiller bar. For the Hiller bar, cyclic controls are first transmitted from the rotating swashplate to the bar. Subsequently, cyclic pitch variations of the rotor blades are controlled entirely by the tilting of the bar. For the Bell–Hiller bar, there is a mechanical mixer between the bar and the rotor blades. Cyclic controls go to the bar as in the case of the Hiller bar. However, there is another portion of cyclic controls that go directly to the rotor blades. The tilting of the bar due to the cyclic controls goes further to the rotor blades via the mixer. One advantage of the Bell–Hiller bar is that it introduces flexibility in adjusting the delay to the cyclic controls since the rotor blades are now controlled by both fly bar tilting and a portion of cyclic controls.

Let $\bar{\beta}_{1c}$ and $\bar{\beta}_{1s}$ be the longitudinal and lateral first harmonic terms for the fly bar. The fly bar can be formulated using a first order dynamic system [8]:

$$\tau_s \dot{\bar{\beta}}_{1c} = -\bar{\beta}_{1c} - \tau_s q_{\mathrm{B}} + \bar{d}\theta^s_{1s} \tag{2.87}$$

$$\tau_s \dot{\bar{\beta}}_{1s} = -\bar{\beta}_{1s} - \tau_s p_{\mathrm{B}} + \bar{d}\theta^s_{1c} \tag{2.88}$$

where $\tau_s$ and $\bar{d}$ are time constants of the fly bar and control derivative, respectively. While $\bar{d}$ can be determined experimentally, the time constant $\tau_s$ is a function of the fly bar's lock number $\gamma_s$ and rotor speed $\Omega$:

$$\tau_s = \frac{16}{\gamma_s \Omega} \tag{2.89}$$

where lock number for the fly bar can be determined by the following equation:

$$\gamma_s = \frac{\rho a_{\mathrm{bar}} c_{\mathrm{bar}} (r_2^2 - r_1^2)}{I_b^{\mathrm{bar}}} \tag{2.90}$$

where $r_2$ and $r_1$ are the outer and inner radii of the fly bar, respectively. $\theta^s_{1c}$ and $\theta^s_{1s}$ represent longitudinal and lateral cyclic controls at the rotating swashplate. The cyclic controls at the rotor blades are expressed as

$$\theta_{1c} = \theta^s_{1c} + k_{\mathrm{bar}}\bar{\beta}_{1c} \tag{2.91}$$

$$\theta_{1s} = \theta^s_{1s} + k_{\mathrm{bar}}\bar{\beta}_{1s} \tag{2.92}$$

where $k_{\mathrm{bar}}$ is the fly bar bearing determined by the geometry of the mechanic mixer. Experimentally, $k_{\mathrm{bar}}$ is not difficult to measure. Setting both $\theta_{1c}$ and $\theta_{1s}$ to zero followed by tilting the fly bar flapping angle, the value of $k_{\mathrm{bar}}$ is the ratio of measured changing pitch angle along the blade to the fly bar flapping angle.

## 2.4 Helicopter Performance Prediction

Endurance and range are two important performance indices in helicopter flight. Both indices can be directly derived from helicopter power calculation. Total power required for a helicopter comes from three difference sources:

- *Main Rotor Power:* This is the most dominant component in the total power consumption. This rotor power includes induced power, profile power, and parasite power. Induced power is associated with rotor thrust production, which is considered as *useful* power. Profile power is required to turn the rotor in viscous air. Parasite power is needed to overcome the helicopter drag including fuselage and empennage.

- *Tail Rotor Power:* As the tail rotor is used to counter the main rotor torque, its power consumption can be derived based on the torque and its own rotational speed. Typically, the tail rotor's power is approximately 10% of main rotor's power at hover and low speed flight. As the helicopter transits to high speed, the vertical fin may be effective enough to offset an essential portion of the main rotor torque. Thus, the required power by the tail rotor typically drops as the flight speed increases.
- *Ancillary Power:* This is to calculate the power used to drive ancillaries such as hydraulics and generators. The ancillary power also takes into account various power losses presented in the whole system. Typically, ancillary power counts for $5--10\%$ of the total power required.

A formula for computing main rotor power is shown as follows:

$$P = \kappa T V_i + \rho A(\Omega R)^3 \frac{\sigma c_{d0}}{8}(1 + 4.6\mu^2) + DV \tag{2.93}$$

The first component in the power equation is the induced power. An empirical factor $\kappa$ is associated with this component to account for wake tip loss, nonuniform inflow and other losses. A number of values for the factor $\kappa$ are suggested in the literature with a typical value taken to be 1.15. Rotor thrust $T$ can be approximated as $W$ at hover and low speed flight, or $\sqrt{W^2 + D^2}$ for moderate and high speed flight. A complete formula for the calculation of induced velocity $V_i$ is provided in Sect. 2.3. However, this formula is not easy to solve especially in forward flight and a Newton–Raphson algorithm is required. Fortunately, this formula can be simplified with respect to the speed regime:

$$V_i = \begin{cases} \sqrt{\frac{W}{2\rho A}}, & \text{hover} \\ \frac{T}{2\rho A V}, & V > 0.1\Omega R \end{cases} \tag{2.94}$$

The second component in (2.93) is the profile power. A typical value of drag coefficient $c_{d0}$ is taken as 0.0087. However, this value is obtained with ideal testing conditions and for a blade in near perfect form. In reality, a 50% increase of $c_{d0} = 0.0087$ is usually adopted for conservative purposes. The last component in (2.93) is the parasite power, which is mainly contributed from fuselage, empennage, and rotor hub. A detailed description of parasite drag in forward flight can be referred to in [84]. Values for equivalent flat-plate area $S_{\text{Ref}}$ range from $10ft^2$ for small helicopters to $50ft^2$ for large utility helicopters. In the case of UAVs, the values can vary from $0.5ft^2$ to $2ft^2$ depending on the UAV size, exposure of mechanical linkages, and landing systems.

Power consumption due to the tail rotor can be calculated using (2.93) with minor modifications. The parasite power component is dropped off in the calculation, while parameters for main rotor are exchanged with those for tail rotor. A simple way of computing tail rotor power is to take 10% of the main rotor power for hover and 5% high speed.

**Table 2.1** RMAX general data for power prediction

| Rotor diameter | 3115 mm |
|---|---|
| Number of blades | 2 |
| Blade chord | 138 mm |
| Rotor speed | 830 RPM |
| Gross weight | 88 kg |

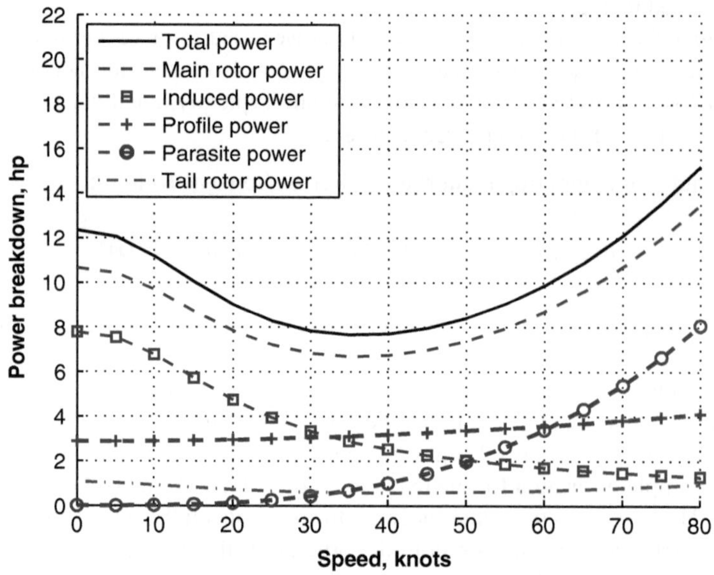


**Fig. 2.7** Power prediction for a Yamaha RMAX helicopter

An example is now provided to illustrate power prediction for a hobby helicopter, i.e., the Yamaha RMAX helicopter. The RMAX is a popular radio-controlled rotary-wing platform. It has been widely used in universities and research institutes as a testbed for a range of activities including nonlinear modeling, flight control, and system identification [91]. General data for the RMAX helicopter is provided in Table 2.1. Other parameters required for power calculation include: $\kappa = 1.15$, $c_{d0} = 0.013$ (50% more than the nominal value of 0.0087 for NACA 23012 airfoil), $S_{\text{ref}} = 1.5 ft^2$, and $\eta_{\text{auxillary}} = 0.9$.

The U-shaped total power curve in Fig. 2.7 illustrates power required for the RMAX in a straight and level flight at sea level. Induced power is the dominant component at hover, but reduces significantly as the speed increases. Profile power gradually increases over the speed to reflect the growing effects from compressibility at the advancing blade and stall and reverse flow at the retreating blade. Parasite power will be dominant at high speed flight as it is proportional to $V^3$. The values for tail rotor power are almost flat when the effect from the vertical fin is not considered here. However, in the presence of a vertical fin, the tail rotor power in general decreases with the increase of flight speed.

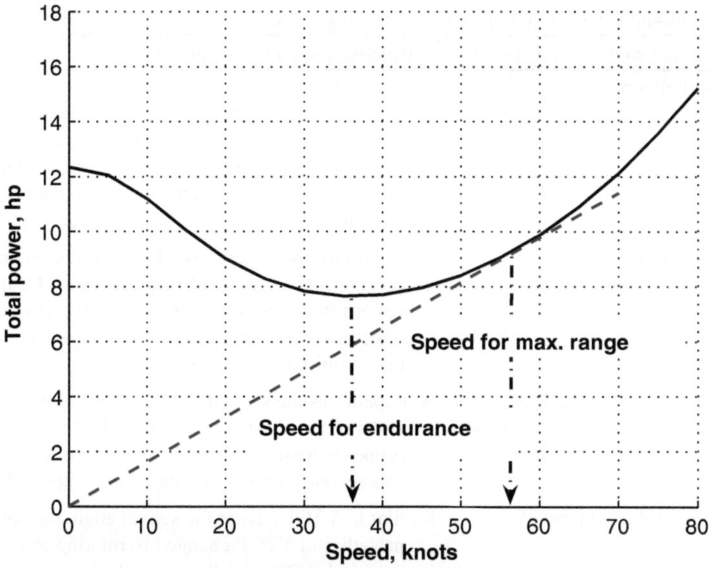


**Fig. 2.8** Determination of endurance and range

Maximum endurance can be achieved when a helicopter operates at a flight speed with the least required power. As observed from Fig. 2.8, this speed can be found at the bottom of the U-shaped power curve (37 knots in Fig. 2.3). Interestingly enough, the endurance speed does not coincide with the minimum drag point, which can be graphically found by drawing the tangent to the U-shaped power curve. A helicopter can achieve the best range at this point as it corresponds to the minimum value of the ratio $P/V$, which can be further interpreted as the best lift-to-drag ratio. In general, endurance and range can not be achieved at one speed and the endurance speed is always slower than the range speed.

## 2.5 Conclusion

As stated in Sect. 2.1, we provided a modeling platform for a general class of rotor-wing vehicles. These rotary-wing vehicles include but are not limited to the compound helicopter, the coaxial helicopter, the helicopter with vectored thrust open propeller (VTOP), the helicopter with vectored thrust ducted propeller (VTDP), autogyro (also known as gyrocopter, autogiro, and gyrodyne), the tilt-rotor aircraft, and the tilt-wing aircraft.

The common feature of these vehicles that they have at least one main rotor. Some have more than one main rotor, like the coaxial helicopter, the tilt-rotor,

**Table 2.2** Modular-based Modeling for rotor-wing vehicles

| Unconventional rotor-wing vehicles | Distinctive features compared with classical helicopters |
|---|---|
| Compound helicopter | Two secondary rotors: one for a tail rotor and another for a pusher propeller; tail rotor used to counter main rotor torque at low speed; pusher propeller used to provide a forward thrust; wing module required to share lift with the main rotor; interference between main rotor and wing needed. |
| Co-axial helicopter | Two main rotors; interference between two main rotors crucial: interference velocity at lower rotor hub due to upper rotor superimposed to the induced velocity for lower rotor and vice versus; no horizontal and vertical tails required. |
| Vectored thrust open propeller | A propeller as the secondary rotor: tail rotor at low speed and a propeller at high speed; swivel angle of open propeller treated as a trim variable; wing required; one less secondary rotor compared to compound helicopter. |
| Vectored thrust ducted propeller | Similar to VTOP except the swivel angle for open propeller in VTOP changed to rotating angle for the vane in VTDP; aerodynamic effect from duct considered. |
| Tilt-rotor aircraft | Two main-rotor modules required; interference between the main rotor and the wing necessary when operating in the helicopter mode. |
| Tilt-wing aircraft | Two main-rotor modules required; FCS related to wing tilting required. |
| Autogyro | Main rotor operating in auto-rotation mode with induced velocity in the steep descent regime; main rotor rotational degree-of-freedom necessary; use secondary rotor module for propeller; interference from propwash to rudder essential for directional effectiveness at low speed. |

and the tilt-wing. Some have more than two rotors, like the compound helicopter. Some instead have an extra propeller module, including the autogyro, the VTOP, and the VTDP. Besides the differences in the rotor system, interference can be critical in some cases. For example, whether there is interference between the rotor and the wing is a distinct feature that separates the tilt-rotor aircraft from the tilt-wing aircraft. For an autogyro, when its rudder is immersed in the propwash, it could have added directional capability. Distinctive features of those rotary-wing vehicles are listed in both Table 2.2.

The key concept in modeling the rotary-wing vehicles is to follow the diagram in Fig. 2.2. For one type of vehicle, its associated modules may be different from those of the classical single main-rotor configuration. However, they can be either special cases of those modules illustrated in Sect. 2.3 (for example, a propeller module), or duplicated modules with minimum differences (for example, two identical rotors on a tilt-rotor aircraft with opposite rotational directions). Once established, those modules produce forces and moments to the general equations of motion. The general equations of motion will update its motion variables based on the forces and moments, and transmit those motion variables back to each individual module.

# Chapter 3
# Stability Analysis for Rotary-Wing Aircraft

## 3.1 Introduction

With the flight dynamic modeling from Chap. 2, the natural next step is to conduct trim analysis. Trim is a state in which force and moment equilibrium are maintained. Trim analysis is the basis of many rotary-wing aircraft studies, including performance and stability analysis, control system design, handling qualities assessment, and software-in-the-loop simulation.

During the linearization process, small perturbation can be applied to the trimmed states to extract the linearized model. All elements in the linearized model can be grouped into four categories: gravity terms, kinematic terms, stability derivative terms, and control derivative terms. Focus is given on a sanity check of these terms. At the end of the chapter, three examples will be provided on the dynamic stability study of the rotary-wing aircraft.

## 3.2 Trim

Trim calculation can be conducted analytically or numerically. A good illustration of an analytical trim calculation can be referred to in [84]. While analytical trim calculation may provide insight on critical forces and moments acting on the rotary-wing aircraft, its application is restricted to simplified analytical equations or special flight conditions in which great simplifications are required.

It may be noticed by readers that the mathematical modeling detailed in Chap. 2 tends to be matrix-centric and can thus be easily implemented in *Matlab*™. As a result, numerical trim calculation can be conducted based on the *Trim* command available in *Matlab*™. A description of this command can be referred in the *Simulink*™ User Manual.

B. Ren et al., *Modeling, Control and Coordination of Helicopter Systems*,
DOI 10.1007/978-1-4614-1563-3_3, © Springer Science+Business Media, LLC 2012

In this section, the procedure for numerical trim calculation is outlined based on the Newton method. The procedure can be implemented in the *Matlab*™ environment, as well as other programming languages like C. Compared to the *Trim* command in *Matlab*™, this set of procedures can provide greater flexibilities in fine-tuning parameters related to convergence and its speed.

Typical difficulties related to the Newton method include the existence of multiple equilibria, computation and singularity of the Jacobian matrix, and closeness of the initial guess to the final converged solution. In the application of rotary-wing aircraft, one possible case for multiple equilibria occurs when the blade sectional angle of attack gets close to the stall region. Here, a single value of lift coefficient may correspond to two angles of attack near to each other. A remedy is thus to smooth out the lift coefficient curve. Generating the Jacobian matrix may not be a big problem as the trim problem only involves a small number of trim variables (typically six) and the matrix can be obtained through numerical perturbation. In terms of the initial guess, one can always start with hover condition and increase flight speed in a comfortably small step. Singularity of the Jacobian matrix may occur at extreme flight conditions like flying at $V_{ne}$ (never exceed speed). For flight conditions close to the boundaries of the flight envelope, professional judgement needs to be made to determine whether the failure to converge is due to numerical reasons or physical causes.

As a general formulation, suppose that the dynamic system can be represented as follows:

$$\dot{X} = f(X, U) \tag{3.1}$$

Here, $X$ can be states from the general six degree-of-freedom equations of motion, blade flapping equations, rotor rotational degree-of-freedom, or dynamics involving flight control system. $U$ are either control variables in general or trim variables in the trim problem.

The procedures using Newton's method are described as follows:

1. Set flight condition $X$.
2. Set trim targets $Y$ which can be a subset of $\dot{X}$.
3. Set trim variables $U$ and associated perturbation $\Delta h$.
4. Set initial conditions for trim variables $U_0$ and run system (3.1) to obtain $Y_0$.
5. Vary one trim variable at a time as $U(i) + \Delta h(i)$ and run the full system to obtain $Y^i$.
6. Form the Jacobian matrix $J$:

$$J(:, i) = \frac{Y^i - Y_0}{\Delta h(i)} \tag{3.2}$$

7. Obtain $J^{-1}$.

**Table 3.1** Setting on trim variables and trim targets

| Trim flight conditions | Trim variables | Trim targets |
|---|---|---|
| Steady level flight or hover | $\theta_c, \theta_{1c}, \theta_{1s}, \theta_{c_{tr}}, \phi, \theta$ | $\dot{u}, \dot{v}, \dot{w}, \dot{p}, \dot{q}, \dot{r}$ |
| Climb or descent | $\theta_{1c}, \theta_{1s}, \theta_{c_{tr}}, \phi, \theta, \gamma_v$ | $\dot{u}, \dot{v}, \dot{w}, \dot{p}, \dot{q}, \dot{r}$ |
| Autorotation | $\theta_c, \theta_{1c}, \theta_{1s}, \theta_{c_{tr}}, \phi, \theta$ $\gamma_v$ | $\dot{u}, \dot{v}, \dot{w}, \dot{p}, \dot{q}, \dot{r}$ $\dot{\Omega}_{\text{rotor}}$ |
| Coordinated turn | $\theta_c, \theta_{1c}, \theta_{1s}, \theta_{c_{tr}}, \phi, \theta$ | $\dot{u}, \dot{v}, \dot{w}, \dot{p}, \dot{q}, \dot{r}$ |
| Longitudinal static stability | $\theta_{1c}, \theta_{1s}, \theta_{c_{tr}}, \phi, \theta, \gamma_v$ | $\dot{u}, \dot{v}, \dot{w}, \dot{p}, \dot{q}, \dot{r}$ |
| Lateral static stability | $\theta_{1c}, \theta_{1s}, \theta_{c_{tr}}, \phi, \theta, \gamma_h$ | $\dot{u}, \dot{v}, \dot{w}, \dot{p}, \dot{q}, \dot{r}$ |

8. Set the update of $U$:

$$U_{\text{new}} = U_{\text{old}} - J^{-1}(Y_{\text{old}} - Y) \tag{3.3}$$

9. Run the full system to obtain $Y_{\text{new}}$.
10. Test whether $||Y_{\text{new}} - Y|| < \epsilon$. If not, go back to Step 8.

Astute readers may realize that the above procedures do not strictly follow the classical Newton method. It is in fact a variant of the Newton method called the Chord method. The advantage of the Chord method is that it only requires us to compute the Jacobian matrix once, thus saving considerable computational cost. However, while the Newton method can guarantee quadratic convergence, the Chord method can only achieve linear convergence.

A collection of trim variables and trim targets at various trimmed flight conditions is provided in Table 3.1 [14]:

In Table 3.1, $\gamma_h$ and $\gamma_v$ represent horizontal and vertical flight path angles, respectively. In some trim conditions, collective pitch $\theta_0$ is not one of the trim variables as it is set by a predefined value.

## 3.3 Linearization

The process of linearization is based on small perturbation theory in which a variable is the sum of its nominal value plus a perturbation. For example, pitch attitude, $\theta$, is written as

$$\theta = \theta_0 + \Delta\theta \tag{3.4}$$

In another example, the total force acting along $x$-axis in body frame, $X$, can be expressed as

$$X = X_0 + \Delta X \tag{3.5}$$

The incremental force $\Delta X$ can be further extended with the following linear approximation:

$$\Delta X = X_u \Delta u + X_w \Delta w + X_q \Delta q + X_{\theta_c} \Delta \theta_c + X_{\theta_{1s}} \Delta \theta_{1s} \tag{3.6}$$

In the above expression, the terms ($X_u$, $X_w$, $X_q$) are called stability derivatives, while the remaining terms ($X_{\theta_c}$, $X_{\theta_{1s}}$) are referred to as control derivatives. Both stability and control derivatives are first order partial derivatives. While these derivatives can also be obtained through analytical investigation [84], they can be computed by numerical perturbation. For example, drag damping $X_u$ is

$$X_u = \frac{\partial X}{\partial u} \cong \frac{X(U_0 + \Delta u) - X(U_0)}{\Delta u} \tag{3.7}$$

The product from the linearization process is a linearized system with respect to a reference flight condition in the following form:

$$\dot{x} = Ax + Bu \tag{3.8}$$

with states $x^{\mathrm{T}} = [u,\ w,\ q,\ \theta,\ v,\ p,\ r,\ \phi]$ and control inputs $u^{\mathrm{T}} = [\theta_c, \theta_{1s}, \theta_{1c}, \theta_{c_{\mathrm{tr}}}]$. Notice that, for simplicity, the symbol $\Delta$ is not explicitly included in the increments of variables. Matrix $A$ is made of four parts: longitudinal dynamic $A_{\mathrm{long}}$, lateral dynamics $A_{\mathrm{lat}}$, longitudinal/lateral coupling $A_{\mathrm{long/lat}}$, and lateral/longitudinal coupling $A_{\mathrm{lat/long}}$:

$$A = \begin{bmatrix} A_{\mathrm{long}} & A_{\mathrm{long/lat}} \\ A_{\mathrm{lat/long}} & A_{\mathrm{lat}} \end{bmatrix} \tag{3.9}$$

General forms of these four sub-matrices are provided:

$$A_{\mathrm{long}} = \begin{bmatrix} \frac{X_u}{m} & \frac{X_w}{m} & \frac{X_q}{m} - W_0 & -g\cos\theta_0 \\ \frac{Z_u}{m} & \frac{Z_w}{m} & \frac{Z_q}{m} + U_0 & -g\cos\phi_0\sin\theta_0 \\ \frac{M_u}{I_{yy}} & \frac{M_w}{I_{yy}} & \frac{M_q}{I_{yy}} & 0 \\ 0 & 0 & \cos\phi_0 & 0 \end{bmatrix} \tag{3.10}$$

$$
A_{\text{long/lat}} = \begin{bmatrix}
\frac{X_v}{m} & \frac{X_p}{m} & \frac{X_r}{m} + V_0 & 0 \\
\frac{Z_v}{m} & \frac{Z_p}{m} - V_0 & \frac{Z_r}{m} & -g \sin\phi_0 \cos\theta_0 \\
\frac{M_v}{I_{yy}} & \frac{M_p}{I_{yy}} & \frac{M_r}{I_{yy}} & 0 \\
0 & 0 & -\sin\phi_0 & 0
\end{bmatrix} \tag{3.11}
$$

$$
A_{\text{lat/long}} = \begin{bmatrix}
\frac{Y_u}{m} & \frac{Y_w}{m} & \frac{Y_q}{m} & -g \sin\phi_0 \sin\theta_0 \\
\frac{I_{zz}L_u + I_{xz}N_u}{\Gamma} & \frac{I_{zz}L_w + I_{xz}N_w}{\Gamma} & \frac{I_{zz}L_q + I_{xz}N_q}{\Gamma} & 0 \\
\frac{I_{xz}L_u + I_{xx}N_u}{\Gamma} & \frac{I_{xz}L_w + I_{xx}N_w}{\Gamma} & \frac{I_{xz}L_q + I_{xx}N_q}{\Gamma} & 0 \\
0 & 0 & \sin\phi_0 \tan\theta_0 & 0
\end{bmatrix} \tag{3.12}
$$

$$
A_{\text{lat}} = \begin{bmatrix}
\frac{Y_v}{m} & \frac{Y_p}{m} + W_0 & \frac{Y_r}{m} - U_0 & g \cos\phi_0 \cos\theta_0 \\
\frac{I_{zz}L_v + I_{xz}N_v}{\Gamma} & \frac{I_{zz}L_p + I_{xz}N_p}{\Gamma} & \frac{I_{zz}L_r + I_{xz}N_r}{\Gamma} & 0 \\
\frac{I_{xz}L_v + I_{xx}N_v}{\Gamma} & \frac{I_{xz}L_p + I_{xs}N_p}{\Gamma} & \frac{I_{xz}L_r + I_{xx}N_r}{\Gamma} & 0 \\
0 & 1 & \cos\phi_0 \tan\theta_0 & 0
\end{bmatrix} \tag{3.13}
$$

All the elements in the above matrices can be obtained through numerical perturbation. However, there are gravitational and inertial terms in the matrix that can be accurately obtained through an analytical study of the equations of motion. For this reason, these gravitational and inertial terms can serve as a good references for the sanity check once $A$ is obtained from the linearization process. For example, the term $-g \cos\theta_0$ in the matrix $A_{\text{long}}$ can be computed with trim pitch attitude $\theta_0$. In another example, the term $\frac{Z_q}{m} + U_0$ in the matrix $A_{\text{long}}$ is dominated by $U_0$ in forward flight.

Finally, the control matrix $B$ is provided:

$$B = \begin{bmatrix} \frac{X_{\theta_c}}{m} & \frac{X_{\theta_{1s}}}{m} & \frac{X_{\theta_{1c}}}{m} & \frac{X_{\theta_{\mathrm{ctr}}}}{m} \\ \frac{Z_{\theta_c}}{m} & \frac{Z_{\theta_{1s}}}{m} & \frac{Z_{\theta_{1c}}}{m} & \frac{Z_{\theta_{\mathrm{ctr}}}}{m} \\ \frac{M_{\theta_c}}{I_{yy}} & \frac{M_{\theta_{1s}}}{I_{yy}} & \frac{M_{\theta_{1c}}}{I_{yy}} & \frac{M_{\theta_{\mathrm{ctr}}}}{I_{yy}} \\ 0 & 0 & 0 & 0 \\ \frac{Y_{\theta_c}}{m} & \frac{Y_{\theta_{1s}}}{m} & \frac{Y_{\theta_{1c}}}{m} & \frac{Y_{\theta_{\mathrm{ctr}}}}{m} \\ \frac{I_{zz}L_{\theta_c}+I_{xz}M_{\theta_c}}{\Gamma} & \frac{I_{zz}L_{\theta_{1s}}+I_{xz}M_{\theta_{1s}}}{\Gamma} & \frac{I_{zz}L_{\theta_{1c}}+I_{xz}M_{\theta_{1c}}}{\Gamma} & \frac{I_{zz}L_{\theta_{\mathrm{ctr}}}+I_{xz}M_{\theta_{0\mathrm{tr}}}}{\Gamma} \\ \frac{I_{xz}L_{\theta_c}+I_{xx}M_{\theta_c}}{\Gamma} & \frac{I_{xz}L_{\theta_{1s}}+I_{xx}M_{\theta_{1s}}}{\Gamma} & \frac{I_{xz}L_{\theta_{1c}}+I_{xx}M_{\theta_{1c}}}{\Gamma} & \frac{I_{xz}L_{\theta_{\mathrm{ctr}}}+I_{xx}M_{\theta_{\mathrm{ctr}}}}{\Gamma} \\ 0 & 0 & 0 & 0 \end{bmatrix} \tag{3.14}$$

Besides the state (3.8), the measurement equation is also required:

$$y = Cx \tag{3.15}$$

The measurement vector $y$ is determined by the available sensors. In the later section, controllability and observability of the resultant system ((3.8) and (3.15)) need to be determined by calculating $M = (B\ AB\ A^2B\ \ldots\ A^{n-1}B)$ and $O = (C\ CA\ CA^2\ \ldots\ CA^{n-1})$ , where $n$ is the number of states.

## *Case Study for Longitudinal Motion of Helicopter at Hover*

In this section, the mathematical model of a single helicopter in hover is linearized. The following Table 3.2 provides the symbols and subscripts to be used.

A helicopter model at hover, where only the longitudinal motion is considered, can be modeled in state space form as

$$\begin{aligned} \dot{x} &= f(x,u) = A(x,u)x + B(x,u)u \\ y &= Cx \end{aligned} \tag{3.16}$$

**Table 3.2** Symbols and subscripts of helicopter model in hover

| Symbols | Meanings |
|---|---|
| $x_1, x_3$ | Forward, vertical displacement |
| $x_2, x_4$ | Forward, vertical velocity |
| $\theta, q$ | Pitch angle and pitch rate |
| $B_1$ | Longitudinal cyclic angle |
| $\theta_o$ | Collective pitch angle |
| $w_c$ | Weight coefficient |
| $x, z$ | Aerodynamic force derivatives |
| $m$ | Aerodynamic pitch moment derivative |
| Subscripts | Meanings |
| $u$ | Aerodynamic derivative due to forward velocity |
| $w$ | Aerodynamic derivative due to vertical velocity |
| $q$ | Aerodynamic derivative due to pitch rate |
| $B_1$ | Aerodynamic derivative due to longitudinal cyclic angle |
| $\theta_o$ | Aerodynamic derivative due to pitch angle |

where $x = [x_1, x_2, x_3, x_4, \theta, q]^T$, $u = [B_1, \theta_o]^T$, $y = [\theta, x_4]^T$ and the state matrices $A$, $B$ and $C$ are

$$A = \begin{bmatrix} 0 & 1 & 0 & 0 & 0 & 0 \\ 0 & x_u & 0 & x_w & -w_c & x_q \\ 0 & 0 & 0 & 1 & 0 & 0 \\ 0 & z_u & 0 & z_w & 0 & z_q \\ 0 & 0 & 0 & 0 & 0 & 1 \\ 0 & m_u & 0 & m_w & 0 & m_q \end{bmatrix}$$

$$B = \begin{bmatrix} 0 & 0 \\ x_{B_1} & x_{\theta_o} \\ 0 & 0 \\ z_{B_1} & z_{\theta_o} \\ 0 & 0 \\ m_{B_1} & m_{\theta_o} \end{bmatrix}$$

$$C = \begin{bmatrix} 0 & 0 & 0 & 0 & 1 & 0 \\ 0 & 0 & 0 & 1 & 0 & 0 \end{bmatrix}$$

In particular, the aerodynamic parameters are calculated according to [9] as

$$x_u = -2.1268\theta_o^2 + 0.0353\theta_o - 0.004$$

$$x_w = -(\theta - B_1)(5.6090\theta_o^2 - 2.7970\theta_o - 0.0621)$$

$$x_q = 2.0074\theta_o - 0.0141$$

$$x_{\theta_o} = -(\theta - B_1)(4.34280\theta_o^2 - 2.4937\theta_o - 0.0354)$$

$$x_{B_1} = 0.081850\theta_o - 0.0035$$

$$
\begin{aligned}
z_u &= 0 \\
z_w &= 5.609\theta_o^2 - 2.797\theta_o - 0.0621 \\
z_q &= 0 \\
z_{\theta_o} &= 4.3428\theta_o^2 - 2.4937\theta_o - 0.0354 \\
z_{B_1} &= 0
\end{aligned}
$$

$$
\begin{aligned}
m_u &= 233.3.64\theta_o^2 + 24.7213\theta_o + 0.4129 \\
m_w &= B_1(606.7919\theta_o^2 - 302.58\theta_o - 6.7138) \\
m_q &= -4.5622\theta_o - 0.06324 \\
m_{\theta_o} &= B_1(469.8083\theta_o^2 - 269.7760\theta_o - 3.835) \\
m_{B_1} &= -88.5473\theta_o - 11.4745
\end{aligned}
$$

$$
w_c = -0.0856
$$

The linearization of the mathematical model in (3.16) may be done by expanding the nonlinear function into a Taylor series about the operating point and neglecting the higher order terms of the expansion [74]. Considering (3.16), the equilibrium point of the system may be found by solving

$$
f(x, u) = 0 \tag{3.17}
$$

The equilibrium points are found to be at $(x_0, u_0) = (0, 0)$. This is followed by doing a Taylor series expansion of (3.16) about the equilibrium point, which results in

$$
\begin{aligned}
\dot{x} &= f(x, u) \\
&= f(x_0, u_0) + \left[ \frac{\partial f}{\partial x_1}(x_1 - x_{01}) + \frac{\partial f}{\partial x_2}(x_2 - x_{02}) + \cdots + \frac{\partial f}{\partial x_6}(x_6 - x_{06}) \right. \\
&\qquad \left. + \frac{\partial f}{\partial u_1}(u_1 - u_{01}) + \frac{\partial f}{\partial u_2}(u_2 - u_{02}) \right] \\
&\quad + \frac{1}{2!} \left[ \frac{\partial^2 f}{\partial x_1^2}(x_1 - x_{01})^2 + \frac{\partial^2 f}{\partial x_2^2}(x_2 - x_{02})^2 + \cdots + \frac{\partial^2 f}{\partial u_2^2}(u_2 - u_{02})^2 \right] + \cdots
\end{aligned} \tag{3.18}
$$

The partial derivatives are evaluated at the operating point, and neglecting the higher order derivatives yield

$$\dot{x} - f(x_0, u_0) = \left[ \frac{\partial f}{\partial x_1}(x_1 - x_{01}) + \frac{\partial f}{\partial x_2}(x_2 - x_{02}) + \cdots + \frac{\partial f}{\partial x_6}(x_6 - x_{06}) \right.$$
$$\left. + \frac{\partial f}{\partial u_1}(u_1 - u_{01}) + \frac{\partial f}{\partial u_2}(u_2 - u_{02}) \right]$$

It is noted that the matrix $C$ in (3.16) is already constant and it therefore does not need to be linearized. The resulting linearized state space representation of the system is given by

$$\begin{aligned} \dot{x} &= Ax + Bu \\ y &= Cx \end{aligned} \tag{3.19}$$

where $A$ and $B$ are constant matrices defined as

$$A = \begin{bmatrix} \frac{\partial f_1}{\partial x_1} & \cdots & \frac{\partial f_1}{\partial x_6} \\ \vdots & \ddots & \vdots \\ \frac{\partial f_6}{\partial x_1} & \cdots & \frac{\partial f_6}{\partial x_6} \end{bmatrix} = \begin{bmatrix} 0 & 1 & 0 & 0 & 0 & 0 \\ 0 & -0.004 & 0 & 0 & -0.856 & -0.0141 \\ 0 & 0 & 0 & 1 & 0 & 0 \\ 0 & 0 & 0 & -0.0621 & 0 & 0 \\ 0 & 0 & 0 & 0 & 0 & 1 \\ 0 & 0.4129 & 0 & 0 & 0 & -0.0632 \end{bmatrix}$$

$$B = \begin{bmatrix} \frac{\partial f_1}{\partial u_1} & \frac{\partial f_1}{\partial u_2} \\ \vdots & \vdots \\ \frac{\partial f_6}{\partial u_1} & \frac{\partial f_6}{\partial u_2} \end{bmatrix} = \begin{bmatrix} 0 & 0 \\ -0.0035 & 0 \\ 0 & 0 \\ 0 & -0.0354 \\ 0 & 0 \\ -11.4745 & 0 \end{bmatrix}$$

Following the linearization of the system, the controllability and observability matrices of the resulting system are given by $M = (B\ AB\ A^2B\ ...\ A^5B)$ and $O = (C\ CA\ CA^2\ ...\ CA^5)^{\mathrm{T}}$ are calculated using MATLAB. Both the matrices are found to have full rank of 6. This implies that the linearized system is both controllable and observable.

## 3.4 Description of Stability and Control Derivatives

In this section, a detailed illustration is provided on key stability and control derivatives. This description focuses on contributing sources, physical interpretation, and typical signs of these derivatives. Part of the description is a variation from [81].

- $X_u$: this term represents drag damping, contributed mainly from fuselage parasite drag and main rotor H-force. By definition, the sign of $X_u$ is negative and the magnitude grows larger as speed increases. As $X_{\text{FU}} \cong -\frac{1}{2}\rho U_0^2 S_{\text{ref}}$, where $S_{\text{ref}}$ is the equivalent flat plate drag area, it follows that:

$$\frac{X_u}{m} \cong \frac{1}{m}\frac{\partial X_{\text{FU}}}{\partial u} = -\frac{\rho U_0 S_{\text{ref}}}{m} \tag{3.20}$$

- $Z_w$: heave damping. At hover, the value of $Z_w$ determines the time constant of vertical response.
- $M_u$: speed stability. At hover, the rotor flaps back in response to a head wind disturbance. As the resultant nose-up moment is in the direction opposing the disturbance, the rotary-wing vehicle is statically stable with $M_u$ being positive. However, an excessive value of $M_u$ may lead to unstable phugoid response and is sensitive to gust. In forward flight, the rotor has a similar contribution while the horizontal stabilizer (if mounted) may have a significant effect to $M_u$ depending on the location of the stabilizer. If the horizontal stabilizer is mounted directly under the main rotor and experiences download from rotor downwash, it enhances $M_u$. If the horizontal stabilizer experiences upload during flight, the effect is destabilizing.
- $M_w$: angle of attack stability. At hover, the value of $M_w$ is close to zero. In forward flight, a positive increase of angle of attack leads to backward tilting of the rotor, creating a nose-up moment to further increase the angle of attack. Thus, the contribution from the rotor to $M_w$ is destabilizing with its corresponding value taking positive sign. A horizontal stabilizer (whether it experiences download or upload) contributes a stabilizing effect, which is the main reason to justify its existence in a pure helicopter configuration. In addition, a forward C.G. location (ahead of rotor shaft) will contribute a stabilizing moment.
- $M_q$: pitch damping. For a rotor with counter clockwise rotation, a positive change in the pitch rate results in a negative roll due to gyroscopic moment. This in turn causes a flap-down for the blade over the nose and a flap-up for the blade over the tail. The resultant nose-down moment opposes the original pitch rate variation, thus the damping effect. In fact, the above explanation can be referred to (2.16):

$$\beta_{1c} = -\frac{16}{\gamma\Omega} q_{\text{H}} \tag{3.21}$$

  Another interesting interpretation of the above equation is that when the fuselage is tilting at a constant pitch rate $q_{\text{H}}$, the time lag between the fuselage and the rotor is $\frac{16}{\gamma\Omega}$ s.
- $L_v$: Dihedral effect. $L_v$ is the lateral counterpart of $M_u$. For a right sideslip, the rotor responds with a left roll. A positive dihedral effect thus takes a negative value. The contributions from both tail rotor and vertical tail depend on their relative location above or below C.G. When above C.G., the effect is stabilizing and *vice versa*. Similar to $M_u$, an excessively large value of $L_v$ may not be preferable. In some helicopter configurations, the vertical tail is placed underneath the tail boom in order to have a moderate value of $L_v$.

- $L_p$: roll damping. $L_p$ is the lateral counterpart of $M_q$. Again, from (2.16), it follows that:

$$\beta_{1s} = -\frac{16}{\gamma\Omega} p_{\mathrm{H}} \tag{3.22}$$

- $N_v$: weathercock stability. The main contributions to $N_v$ are from the tail rotor and vertical tail. Both are stabilizing with $N_v$ taking a positive value.
- $M_p$ and $L_q$: cross-coupling due to roll/pitch rates. Once again, (2.16) can be used:

$$\beta_{1s} = -\frac{q_{\mathrm{H}}}{\Omega} \tag{3.23}$$

$$\beta_{1c} = \frac{p_{\mathrm{H}}}{\Omega} \tag{3.24}$$

Thus, $M_p$ takes a positive value while $L_q$ is negative.
- $Z_{\theta_c}$, $M_{\theta_{1s}}$, $L_{\theta_{1c}}$, and $N_{\theta_{ctr}}$: collective/pitch/roll/yaw control power.

In the subsequent sections, three examples will be provided on linearized dynamics for the rotary-wing aircraft. The first two examples deal with two different hobby helicopters (Yamaha R50 and Copterwork AF25B) at two different flight conditions (hover and forward flight). The last example illustrates a combined system consisting of a helicopter and an underslung load.

## 3.5 Yamaha R50 Helicopter at Hover

Consider the longitudinal linearized model of the Yamaha R50 helicopter [54] at hover:

$$\begin{bmatrix} \dot{u} \\ \dot{w} \\ \dot{q} \\ \dot{\theta} \end{bmatrix} = \begin{bmatrix} -0.0553 & 0.0039 & 1.413 & -32.1731 \\ -0.0027 & -0.5727 & -0.0236 & -0.2358 \\ 0.2373 & 0.002 & -6.9424 & 0 \\ 0 & 0 & 1 & 0 \end{bmatrix} \begin{bmatrix} u \\ w \\ q \\ \theta \end{bmatrix} + \begin{bmatrix} 11.2579 & 0 \\ 0.0698 & -0.199 \\ -38.6267 & 0 \\ 0 & 0 \end{bmatrix} \begin{bmatrix} \delta \\ \delta_\Omega \end{bmatrix} \tag{3.25}$$

where $\delta$ is longitudinal cyclic control and $\delta_\Omega$ denotes the main rotor RPM variation.

It is interesting to compare the state matrix in (3.25) and $A_{\mathrm{long}}$ in (3.10). Although trim roll and pitch attitudes are not provided in [54], their values can be assumed to be small. Therefore, the two gravitational terms, $-g\cos\theta_0$ and $-g\cos\phi_0\sin\theta_0$,

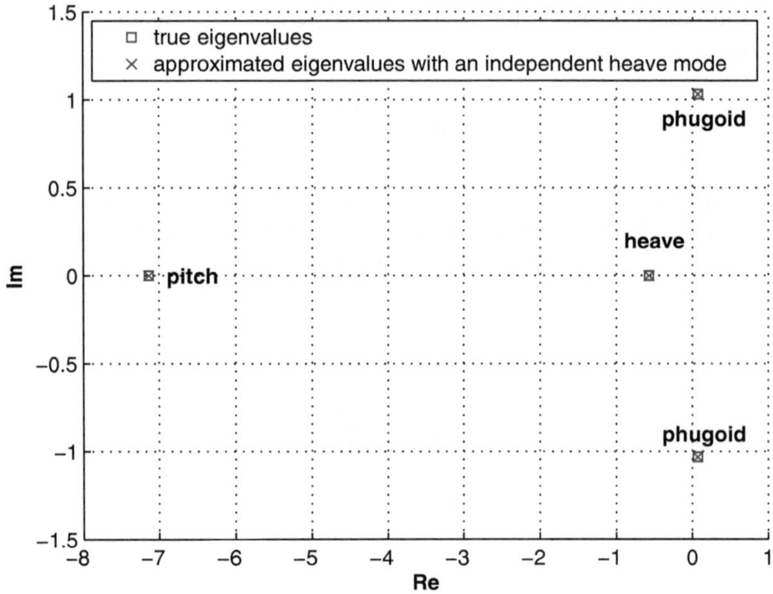


**Fig. 3.1** Eigenvalue plot for Yamaha R50 helicopter at hover

take estimated values of $-32.2\ ft/s^2$ and $0\ ft/s^2$, respectively. By the same token, the kinematic term $\cos\phi_0$ is close to 1. Moreover, some terms are weak at hover, including $X_u$, $X_w$, $Z_u$, $Z_q$, and $M_w$. The term $X_q$ is drag due to pitch rate with the main contribution from main rotor. Following a positive change in pitch rate, the main rotor with a counter-clockwise rotation has a tendency of left roll due to gyroscopic moment, which results in a forward tilting of the rotor disk. Thus, the sign of $X_q$ should be positive. In fact, this process is how the opposing pitch-down moment is generated for the pitch damping $M_q$. Due to hingeless rotor configuration for the R50 helicopter, the value of $M_q$ is fairly large if compared to normal articulated or even teetering rotor configurations. The magnitude of $M_q$ can also take the contribution from the fly bar. The positive sign of $M_u$ has already been discussed in the last section. Thus, the state matrix in (3.25) appears to be reasonable from the above sanity check.

From the given linearized dynamics in (3.25), a distribution of eigenvalues can be obtained (see Fig. 3.1). An approximation can be made by further assuming that the heave dynamics are independent from the remaining dynamics. From the figure, it can be seen that the differences in eigenvalues from both approximated dynamics and full dynamics are indiscernible. This is not surprising as the rotor is aerodynamically symmetric at hover condition. It also shows that the heave mode at hover can be characterized as a first-order dynamic system with time constant at 1.75 s.

Apart from heave mode, the remaining eigenvalues include one pair of complex conjugates, often known as the phugoid mode, and one at real axis, known as the pitch mode. While the pitch mode is stable and aperiodic, the phugoid mode here is mildly unstable and oscillatory.

Going back to matrix $A_{\text{long}}$, it reduces to the following form when taking out the heave dynamics [81]:

$$A_{\text{long,reduced}} = \begin{bmatrix} \frac{X_u}{m} & \frac{X_q}{m} & -g \\ \frac{M_u}{I_{yy}} & \frac{M_q}{I_{yy}} & 0 \\ 0 & 1 & 0 \end{bmatrix} \tag{3.26}$$

The corresponding characteristic equation is

$$s^3 - \left(\frac{X_u}{m} + \frac{M_q}{I_{yy}}\right) s^2 + \left(\frac{X_u}{m}\frac{M_q}{I_{yy}} - \frac{M_u}{I_{yy}}\frac{X_q}{m}\right) s + \frac{M_u}{I_{yy}} g = 0 \tag{3.27}$$

When $M_u = 0$, the resulting characteristic equation will have three roots on the real axis: 0, $X_u/m$, and $M_q/I_{yy}$. When the value of $M_u/I_{yy}$ is increased from 0 to 0.2373, it can be found that the root close to $M_q/I_{yy}$ hardly moves, while the root locus corresponding to 0 and $X_u/m$ first merges together on the real axis before becoming a complex pair and moving towards the right half of the plane. The result is shown in Fig. 3.2.

Figure 3.2 indicates that the eigenvalue associated with the pitch mode is mainly determined by $M_q$, and an increase in value of $M_u$ may result in an unstable phugoid mode.

It is also interesting to discuss the effect of $M_q$. The results from $M_q$ variation can be seen in Fig. 3.3. There are three values of $M_q$: original value, half of the original value, and double of the original value. From the figure, it is clear that increasing pitch damping is beneficial to the system stability.

As a side note, the lateral linearized dynamics is a mirror image of longitudinal counterpart due to the main rotor's aerodynamic symmetry at hover. Similar to longitudinal dynamics, there are three modes in lateral dynamics: yaw mode, roll mode, and dutch roll model. The correspondence between the two dynamics are: heave mode vs. yaw mode, pitch mode vs. roll mode, and phugoid mode vs. dutch roll mode.

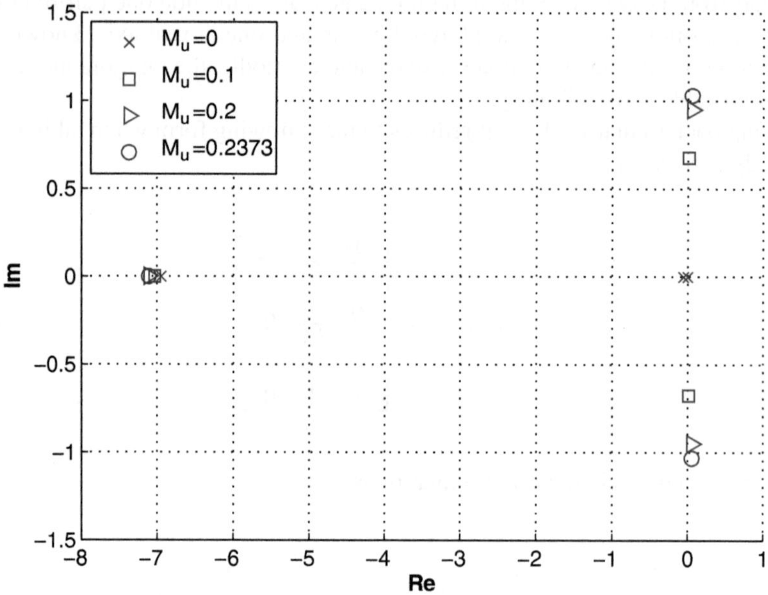


**Fig. 3.2** Influence of speed stability $M_u$

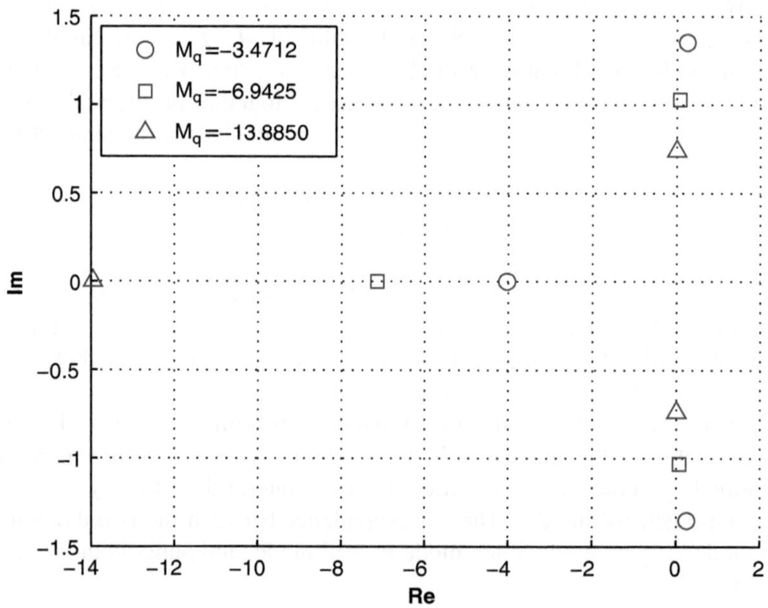


**Fig. 3.3** Influence of pitch damping $M_q$

## 3.6 Copterworks AF25B Helicopter in Forward Flight

Due to aerodynamic asymmetry on the main rotor in forward flight and dynamic coupling from the flapping blades and constantly spinning rotor, longitudinal and lateral dynamics may not always be sufficiently decoupled. This is added complexity when compared to the analysis of fixed-wing aircraft.

In this section, a coupled linearized model for a Copterworks AF25B radio control helicopter is analyzed (see Fig. 3.4). The condition is straight level flight at 40 knots. The gross weight of the helicopter is 30 kg. Trimmed roll and pitch attitudes are 1.43° and −10.2°, respectively. Velocities along the body $x$-axis and $z$-axis are $U_0 = 66.4915$ and $W_0 = -11.9628$. The linearized model is shown in (3.28):

$$A = \begin{bmatrix} -0.1634 & 0.0676 & 14.1039 & -31.6911 & -0.0029 & -1.8475 & -1.4723 & 0.0003 \\ -0.0792 & -0.9827 & 73.9560 & 5.6832 & 0.0314 & 2.4586 & -0.4810 & -0.7882 \\ 0.0208 & 0.1940 & -24.1599 & -0.0014 & 0.1118 & 41.7746 & -0.0236 & -0.0127 \\ 0 & 0 & 0.9997 & 0 & 0 & 0 & -0.0249 & 0.0001 \\ -0.0014 & -0.0046 & -1.6321 & 0.1414 & -0.0804 & -14.0134 & -66.2104 & 31.6828 \\ 0.3740 & 0.2245 & -106.5635 & 0.0010 & -0.3005 & -67.5186 & -0.1599 & 0.0049 \\ 0.0272 & 0.0304 & -0.2059 & 0 & 0.1869 & 0.6160 & -0.8763 & 0 \\ 0 & 0 & -0.0045 & -0.0001 & 0 & 1.0000 & -0.1793 & -0.0001 \end{bmatrix} \tag{3.28}$$

Once again, a sanity check of the state matrix is in order. In the follow-up discussion, entry (n,m) stands for $n^{\text{th}}$ row and $m^{\text{th}}$ column.

1. Entry (1,1): drag damping $X_u$. In the nonlinear modeling, equivalent flat-plate drag area is assumed to be $2ft^2$. Based on (3.20), $X_u$ is estimated at −0.1577, a value close to −0.1634 from the linearization.
2. Entry (1,3): $X_q - W_0$. This term is dominated by vertical speed along the body $z$-axis $-W_0$.
3. Entries (1,4), (2,4), (2,8), (5,4), (5,8): gravitational terms. The calculated values based on corresponding entries in (3.8) are −31.6843, 5.6999, −0.7908, 0.1423, and 31.6813, sufficiently close to corresponding entries in the state matrix.

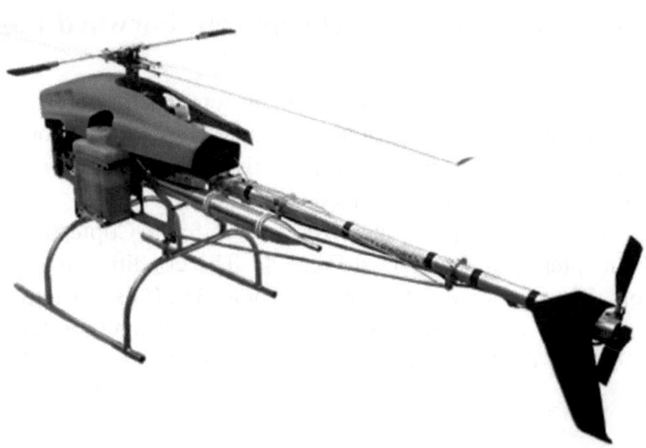

**Fig. 3.4** AF25B radio control helicopter from Copterworks(Courtesy of Copterworks Inc.)

4. Entry (2,2): heave damping $Z_w$. At hover, $Z_w = -0.4$ for AF25B. The magnitude of $Z_w$ tends to increase with higher speed [78].
5. Entry (2,3): $Z_q + U_0$. This term is dominated by forward speed along the body $x$-axis $U_0$.
6. Entry (3,1): speed stability $M_u$. In reference to $M_u$ for Lynx helicopter (see Fig. 4.15 of [78]), the value of $M_u$ is reduced with the increase of speed from hover to moderate speed. The value of $M_u$ is 0.0825 at hover for the AF25B, decreasing to 0.0208 at 40 knots. This comparison is reasonable since both the AF25B and the Lynx have hingeless rotor configurations.
7. Entry (3,2): angle of attack stability $M_w$. The positive value here indicates the destabilizing effect of $M_w$. This is because the AF25B has no horizontal stabilizer to produce opposing pitching moment from a variation of vertical velocity.
8. Entries (3,3) and (6,6): pitch damping term $M_q$ and roll damping term $L_p$. Both values are negative. Compared to the value in entry (3,3), the one in entry (6,6) is almost three times larger. This is due to roll moment of inertia $I_x$ which is three times less than the pitch moment of inertia $I_y$.
9. Entries (3,6) and (6,3): cross-coupling $M_p$ and $L_q$. The signs are positive for $M_p$ and negative for $L_q$, which conform to the analysis in Sect. 3.4.
10. Entries (4,3), (4,7), (8,3), (8,7): kinematic terms. The calculated values based on corresponding entries in (3.8) are 0.9997, –0.0250, –0.0045, and –0.1799, which are in good agreement with corresponding entries in the state matrix.
11. Entry (6,5): dihedral effect $L_v$. Here, the negative value has a stabilizing effect, which mainly comes from the main rotor. However, it should be noted that the effect from the vertical fin is destabilizing as the fin is mounted below the C.G..
12. Entry (7,5): weathercock effect $N_v$. As expected, the value of $N_v$ in the state matrix is positive with main contributions from the tail rotor and vertical fin.

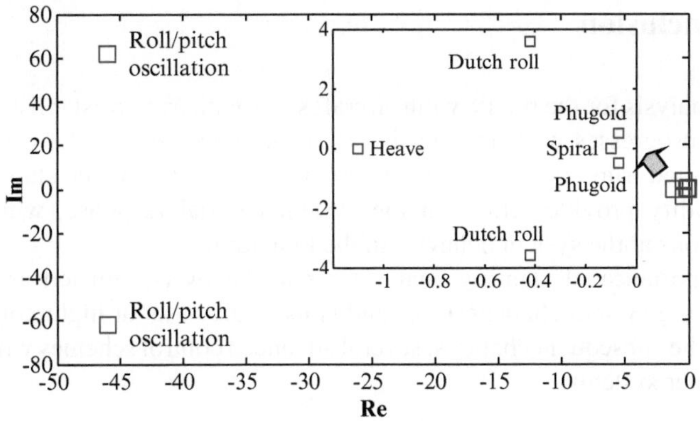


**Fig. 3.5** Eigenvalue distribution for the AF25B at 40 knots

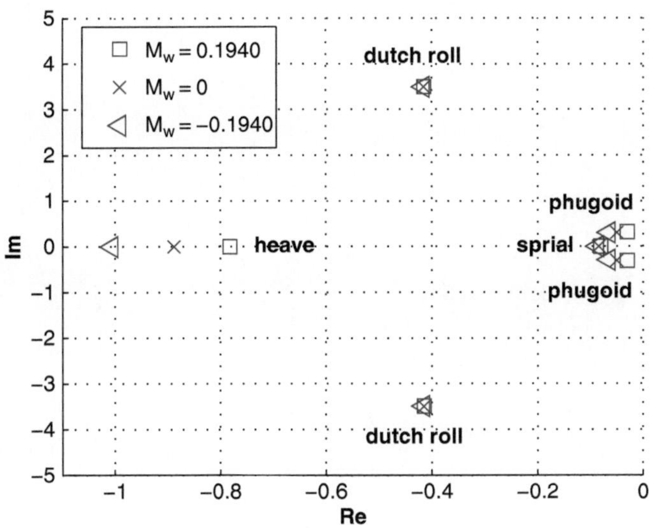


**Fig. 3.6** Variation of $M_w$ on the eigenvalue distribution

Following the sanity check on the state matrix, the corresponding eigenvalues are plotted in Fig. 3.5. Far away to the imaginary axis in the left half of the complex plane is the roll/pitch oscillation mode. Mainly due to the hingeless rotor configuration and its associated large hub moment, the dynamics of this mode are rapid. Other modes are all close to the imaginary axis and stable, including heave, spiral, dutch roll, and phugoid modes. In particular, the phugoid mode is the closest to the imaginary axis, indicating minimum stability margin. An improvement on $M_w$ is preferable (see Fig. 3.6), with the corresponding means ranging from the addition of horizontal stabilizer to forward C.G. (ahead of rotor thrust vector).

## 3.7 Conclusion

Stability analysis for the rotary-wing aircraft starts with trim to establish steady state condition, progresses with linearization to obtain linearized model with respect to the established trim condition, and centers on static and dynamic stability studies. Static stability provides clues on the system's initial response, while dynamic stability looks at the system behavior in the long term.

The importance of stability analysis cannot be over-estimated as it provides insights to the system characteristics and is itself the basis of flight control system design. In the subsequent chapters, several advanced control schemes will be offered for helicopter systems.

# Chapter 4
# Altitude Control of Helicopters with Unknown Dynamics

## 4.1 Introduction

The linearized models of Chap. 3 are useful not only for stability analysis, but also for control design, by gain scheduling the models at different operating points. This traditional technique has been successfully implemented in a wide variety of applications. However, it requires extensive modeling, which is costly and time-consuming, and the models are highly specific to a particular helicopter system. There is a need for controllers that can operate with minimal model information, handle nonlinearities over the entire flight regime, and are portable across different helicopter systems. In this chapter, we address this need by presenting a robust adaptive neural network (NN) control for helicopters.

Helicopter control design is challenging because helicopters are inherently unstable without closed loop control, differing from many classes of mechanical systems that are naturally passive or dissipative. Unrestrained helicopter motion is governed by underactuated configuration, i.e., the number of control inputs is less than the number of degrees of freedom to be stabilized, which makes it difficult to apply the conventional robotics approach for controlling Euler–Lagrange systems. In addition, helicopter dynamics are highly nonlinear and strongly coupled, such that disturbances along a single degree of freedom can easily propagate to the other degrees of freedom and lead to loss of performance or even destabilization.

Increasing effort has been made towards control design that guarantees stability for helicopter systems. Apart from the above-mentioned traditional linear control, many nonlinear techniques have been proposed, ranging from feedback linearization to model reference adaptive control and dynamic inversion. Dynamic sliding mode control was proposed for helicopter vertical regulation in [93]. Output tracking with nonhyperbolic and near nonhyperbolic internal dynamics in helicopter hover control was discussed in [19]. In [55], approximate input–output linearization was employed to obtain a dynamically linearizable helicopter system without zero dynamics, and output tracking was achieved. In [16], a high-bandwidth $H\infty$ loop shaping control was designed and tested for a robotic helicopter. Internal


DOI 10.1007/978-1-4614-1563-3_4, 

model-based control was applied to the nonlinear motion control of a helicopter in [41]. In [104], model-based control was applied to the altitude and yaw angle tracking of a Lagrangian helicopter model.

Since helicopter control applications are characterized by unknown aerodynamical disturbances, they are generally difficult to model accurately. The presence of modeling errors, in the form of parametric and functional uncertainties, unmodeled dynamics, and disturbances from the environment, is a common problem. In this context, model-based control, such as the aforementioned schemes, tend to be susceptible to uncertainties and disturbances that cause performance degradation. How to handle model uncertainties and disturbances is one of the important issues in the control of helicopters.

Owing to the universal approximation capabilities, learning and adaptation, and parallel distributed structures of NNs, the feasibility of applying NNs to model unknown functions in dynamic systems has been demonstrated in several studies [23, 30, 32, 60, 61, 73]. As such, several flight control approaches using NNs have been proposed. Among of them, approximate dynamic inversion with augmented neural networks was proposed to handle unmodeled dynamics in [38, 53, 87], while neural dynamic programming was shown to be effective for tracking and trimming control of helicopters in [21]. During the adaptive trajectory control of an autonomous helicopter in [43] and [51], the method of pseudocontrol hedging (PCH) was used to protect the adaptation process from actuator limits and dynamics.

In this chapter, we propose an adaptive NN control for helicopters in vertical flight, which can be represented by single-input-single-output (SISO) models to yield useful results, because the coupling between longitudinal and lateral- directional equations in this flight regime is weak [10]. While the proposed controller handles vertical flight, other flight regimes can be handled by other control modules. Motivated by results in the NN control of nonlinear systems [30], we utilize Lyapunov-based techniques to design a robust adaptive NN control for helicopters with guaranteed stability. Although a nonaffine system can be rendered affine by adding an integrator to the control input, thus allowing many control methods for affine nonlinear system to be used, the disadvantage of this approach is that the dimension of the system is increased, and control efforts are not direct and immediate either [30]. Subsequently, effective control for the system may not be achieved. In this chapter, we focus on control design for the nonaffine system directly, without adding any integrators to the input.

Differing from the approaches in [38, 54], which were based on approximate dynamic inversion with augmented NNs, we utilize the Mean Value Theorem and the Implicit Function Theorem as mathematical tools to handle the nonaffine nonlinearities in the helicopter dynamics, based on the pioneering work of [31]. While the NNs in [38, 54] compensate for inaccuracy of the inversion model, those in our proposed scheme approximate the ideal feedback control law directly. In cases where reasonably accurate knowledge of the dynamic inversion model is available, the method of [38, 54] has been shown to provide an effective solution to the problem. However, the construction of the dynamic inversion for a nonaffine system may not be an easy task in general. For such cases, our approach offers a feasible means of tackling the problem, since a priori knowledge of the inversion is not required.

## 4.2 Problem Formulation and Preliminaries

Consider the class of SISO helicopter systems described by the following differential equations nonaffine in the control:

$$\begin{aligned} \dot{x} &= f(x, u) \\ y &= h(x) \end{aligned} \tag{4.1}$$

where $x \in R^n$ are the states of the system; $u, y \in R$ denote the input and output respectively; and $f : R^n \times R \to R^n$ is an unknown function.

The control objective is output tracking of a desired reference trajectory such that the tracking error converges to a neighborhood of zero, i.e., $|y(t) - y_d(t)| \leq \delta$, where $\delta > 0$. At the same time, all closed loop signals are to be kept bounded. The reference trajectory $y_d(t)$ is generated by the following reference model:

$$\begin{aligned} \dot{\xi}_{di} &= \xi_{di+1}, \quad 1 \leq i \leq \rho - 1, \\ \dot{\xi}_{d\rho} &= f_d(\xi_d), \\ y_d &= \xi_{d1}, \end{aligned} \tag{4.2}$$

where $\rho \geq 2$ is a constant index; $\xi_d = [\xi_{d1}, \xi_{d2}, \ldots, \xi_{d\rho}]^{\mathrm{T}} \in R^\rho$ are the states of the reference system; $y_d \in R$ is the system output; and $f_d : R^\rho \to R$ is a known function.

**Assumption 4.1.** The reference trajectory $y_d(t)$ and its $\rho$ derivatives remain bounded, i.e., $\xi_d \in \Omega_d \subset R^\rho$, $\forall t \geq 0$, where $\rho$ is the relative degree of (4.1).

*Remark 4.1.* The SISO representation considered in this chapter is valid for simple operations involving the regulation or tracking of single degree of freedom, such as altitude tracking or pitch regulation, among others.

*Remark 4.2.* System (4.1) is a general description of the nonlinear helicopter dynamics, for which the control input is not necessarily affine. This is a realistic representation for helicopters, due to the fact that the inputs are not torques or forces, which would yield an affine representation. Rather, the inputs are position variables implicit in aerodynamical forces or torques, resulting in a nonaffine form. Due to the lack of mathematical tools, the control design for such systems is very challenging. Note that control-affine nonlinear systems as well as linear systems are special cases of (4.1). As such, by designing a controller for (4.1), we cover the other systems as well.

**Assumption 4.2.** System (4.1) is input–output linearizable with strong relative degree $\rho < n$.

Define $\phi_j(x) = L_f^{j-1}h$ for $j = 1, ..., \rho$, where $L_f h$ denotes the Lie derivative of the function $h(x)$ with respect to the vector field $f(x,u)$. Due to Assumption 4.2, it was shown in [39] that there exist other $n-\rho$ functions $\phi_{\rho+1}, ..., \phi_n$ independent of $u$, such that the mapping $\Phi(x) = [\phi_1(x), \phi_2(x), ..., \phi_n(x)]^{\mathrm{T}}$ has a Jacobian matrix which is nonsingular for all $x \in \Omega_x$. Therefore, $\Phi(x)$ is a diffeomorphism on $\Omega_x$. By setting $\xi = [\phi_1(x), \phi_2(x), ..., \phi_\rho(x)]^{\mathrm{T}}$ and $\eta = [\phi_{\rho+1}(x), \phi_{\rho+2}(x), ..., \phi_n(x)]^{\mathrm{T}}$, system (4.1) can be expressed in the normal form :

$$\begin{aligned}
\dot{\eta} &= q(\eta,\xi) \\
\dot{\xi}_j &= \xi_{j+1}, \quad j = 1, ..., \rho - 1 \\
\dot{\xi}_\rho &= b(\xi,\eta,u) \\
y &= \xi_1
\end{aligned} \tag{4.3}$$

where $b(\xi,\eta,u) = L_f^{j-1}h$; $q(\xi,\eta) = [L_f\phi_{\rho+1}(x), L_f\phi_{\rho+2}(x), ..., L_f\phi_n(x)]^{\mathrm{T}}$; $x = \Phi^{-1}(\xi, n)$, for $(\xi,\eta,u) \in \bar{U} := \{(\xi,\eta,u)|(\xi,\eta) \in \Phi(\Omega_x); u \in \Omega_u\}$.

**Assumption 4.3.** The zero dynamics of system (4.3), given by $\dot{\eta} = q(0,\eta)$ are exponentially stable. In addition, the function $q(\xi,\eta)$ is Lipschitz in $\xi$, i.e.,

$$\|q(\xi,\eta) - q(0,\eta)\| \le a_\xi\|\xi\| + a_q, \quad \forall(\xi,\eta) \in \Phi(\Omega_x) \tag{4.4}$$

Under Assumption 4.3, by the converse Lyapunov theorem, there exists a Lyapunov function $V_0(\eta)$ which satisfies the following inequalities:

$$\gamma_1\|\eta\|^2 \le V_0(\eta) \le \gamma_2\|\eta\|^2 \tag{4.5}$$

$$\frac{\partial V_0}{\partial \eta}q(0,\eta) \le -\lambda_a\|\eta\|^2 \tag{4.6}$$

$$\left\|\frac{\partial V_0}{\partial \eta}\right\| \le \lambda_b\|\eta\| \tag{4.7}$$

where $\gamma_1$, $\gamma_2$, $\lambda_a$, and $\lambda_b$ are positive constants.

For ease of notation, define $g(x,u) := \frac{\partial b(\xi,\eta,u)}{\partial u}$. The following two assumptions specify some conditions on the unknown function $g(x,u)$.

**Assumption 4.4.** There exist smooth functions $\bar{g}(\xi,\eta)$ and a positive constant $\underline{g} > 0$, such that $\bar{g}(\xi,\eta) \ge |g(\xi,\eta,u)| \ge \underline{g} > 0$ holds for all $(\xi,\eta,u) \in \bar{U}$. Without loss of generality, it is further assumed that the sign of $g(\xi,\eta,u)$ is positive for all $(\xi,\eta,u) \in \bar{U}$.

**Assumption 4.5.** There exist a positive function $g_0(\xi,\eta)$ such that $|\dot{g}(\xi,\eta,u)| \le 2g(\xi,\eta,u)g_0(\xi,\eta)$, $\forall(\xi,\eta,u) \in \bar{U}$.

*Remark 4.3.* Assumption 4.4 implies that partial derivative $g(\xi, \eta, u)$ has a fixed sign. In addition, it means that the Taylor series linearization is controllable so one could always linearize and design a linear controller, if possible. This assumption is standard and necessary as otherwise, the system is not controllable.

Based on Assumption 4.4, the following lemma is given to assert the existence of an implicit desired function, which will be used in the design of the NN controller.

**Lemma 4.4 (Implicit Function Theorem).** *[33] For a continuously differentiable function $b(\xi, \eta, u) : R^n \times R \to R$ satisfying Assumption 4.4, there exists a continuous (smooth) function $u^* = u(\xi, \eta)$ such that $b(\xi, \eta, u^*) = 0$.*

**Lemma 4.5 (Mean Value Theorem).** *[1] Assume that $f(x, y) : R^n \times R \to R$ has a derivative (finite or infinite) at each point of an open set $R^n \times (a, b)$, and assume also that it is continuous at both endpoints $y = a$ and $y = b$. Then there is a point $\xi \in (a, b)$ such that $f(x, b) - f(x, a) = f'(x, \xi)(b - a)$.*

*Remark 4.6.* It should be emphasized that the Mean Value Theorem gives an equality condition and is different from Taylor series expansion, which only gives an approximation when the higher order terms are truncated.

*Remark 4.7.* The combination of the Implicit Function Theorem, the Mean Value Theorem, and NNs is instrumental to solving the control problem for generalized nonaffine helicopter systems described by (4.1). While the Implicit Function Theorem asserts the existence of a desired control, it does not provide any means of constructing it. NNs are thus employed for this purpose. On the other hand, the Mean Value Theorem expresses the nonaffine function into a form where the actual and desired inputs are linearly matched, facilitating the design of adaptive NN control via certainty equivalence and Lyapunov-based techniques. These will be elaborated in detail in the subsequent developments.

Lastly, we present the following definition and Lemma, which are important for stability and performance analysis.

**Definition 4.8.** The solution of (4.1) is Semi-Globally Uniformly Ultimately Bounded (SGUUB) if, for any compact set $\Omega_0$, there exists an $S > 0$ and $T(S, X(t_0))$ such that $\|X(t)\| \leq S$ for all $X(t_0) \in \Omega_0$ and $t \geq t_0 + T$.

**Lemma 4.9.** *[34] Suppose that there exists a $C^1$ continuous and positive definite Lyapunov function $V(x)$ satisfying*

$$\gamma_1(\|x\|) \leq V(x) \leq \gamma_2(\|x\|), \tag{4.8}$$

*such that*

$$\dot{V}(x) \leq -c_1 V(x) + c_2, \tag{4.9}$$

*where $\gamma_1, \gamma_2 : R^n \to R$ are class $K_\infty$ functions and $c_1$, $c_2$ are positive constants, then the solution $x(t)$ is SGUUB.*

## 4.3 Function Approximation with Neural Networks

Due to the existence of model uncertainties in practice, we introduce a NN here to approximate and compensate for them using the good function approximation capability of NN. In particular, two types of NN will be discussed, i.e., the radial basis function neural network (RBFNN), which is linearly parameterized; and the multilayer neural network (MNN), which is nonlinearly parameterized.

### *4.3.1 Function Approximation with RBFNN*

The RBFNN can be used to approximate the continuous function $f(Z): R^m \to R$ as follows:

$$f(Z) = W^{\mathrm{T}} S(Z) + \varepsilon(Z) \tag{4.10}$$

where the input vector $Z \in \Omega_Z \subset R^m$; weight vector $W = [w_1, w_2, ..., w_l]^{\mathrm{T}} \in R^l$, the NNs node number $l > 1$; $S(Z) = [s_1(Z), ..., s_l(Z)]^{\mathrm{T}}$, with $s_i(Z)$ being chosen as the commonly used Gaussian functions, which have the form:

$$s_i(Z) = \exp\left[\frac{-(Z - \mu_i)^{\mathrm{T}}(Z - \mu_i)}{\eta_i^2}\right], \; i = 1, 2, ..., l$$

where $\mu_i = [\mu_{i1}, \mu_{i2}, ..., \mu_{im}]^{\mathrm{T}}$ is the center of the receptive field and $\eta_i$ is the width of the Gaussian function; and $\varepsilon(Z)$ is the approximation error which is bounded over the compact set $\Omega_Z$, i.e., $|\varepsilon(Z)| \leq \bar{\varepsilon}$, $\forall Z \in \Omega_Z$ where $\bar{\varepsilon} > 0$ is an unknown constant.

It has been proven that RBFNN (4.10) can approximate any continuous function $f(Z)$ over a compact set $\Omega_Z \subset R_m$ to arbitrarily any degree of accuracy as

$$f(Z) = W^{*\mathrm{T}} S(Z) + \varepsilon^*(Z), \quad \forall Z \in \Omega_Z \subset R^m \tag{4.11}$$

where $W^*$ is ideal constant weights, and $\varepsilon^*(Z)$ is the approximation error for the special case where $W = W^*$.

**Assumption 4.6.** On the compact set $\Omega_Z$, the ideal NN weights $W^*$ is bounded by

$$\|W^*\| \leq w_m \tag{4.12}$$

The ideal weight vector $W^*$ is defined as the value of $W$ that minimizes $|\varepsilon(Z)|$ for all $Z \in \Omega_Z \subset R^m$:

$$W^* = \arg\min_{W}\{\sup_{Z\in\Omega_Z} |f(Z) - W^{\mathrm{T}} S(Z)|\}$$

In general, the ideal weights $W^*$ are unknown and need to be estimated in control design. Let $\hat{W}$ be the estimates of $W^*$, and the weight estimation errors $\tilde{W} = \hat{W} - W^*$.

### 4.3.2 Function Approximation with MNN

The other popular type of NN, nonlinearly parameterized MNN, is used to approximated the continuous function $f(Z) : R^m \to R$ as follows:

$$f(Z) = W^{\mathrm{T}} S(V^{\mathrm{T}} Z) + \varepsilon(Z)$$

where the vector $Z = [z_1, z_2, ..., z_m, 1]^{\mathrm{T}} \in \Omega_Z \subset R^{m+1}$ are the input variables to the NNs; $S(\cdot) \in R^l$ is a vector of known continuous basis functions, with $l$ denoting the number of neural nodes; $W \in R^l$ and $V \in R^{(m+1)\times l}$ are adaptable weights; and $\varepsilon(Z)$ is the approximation error which is bounded over the compact set $\Omega_Z$, i.e., $|\varepsilon(Z)| \le \bar{\varepsilon}$, $\forall Z \in \Omega_Z$ where $\bar{\varepsilon} > 0$ is an unknown constant.

According to the universal approximation property [27], MNNs can smoothly approximate any continuous function $f(Z)$ over a compact set $\Omega_Z \subset R^{m+1}$ to arbitrarily any degree of accuracy as that

$$f(Z) = W^{*\mathrm{T}} S(V^{*\mathrm{T}} Z) + \varepsilon^*(Z), \quad \forall Z \in \Omega_Z \subset R^{m+1}$$

where $W^*$ and $V^*$ are the ideal constant weights, and $\varepsilon^*(Z)$ is the approximation error for the special case where $W = W^*$ and $V = V^*$. The ideal weights $W^*$ and $V^*$ are defined as the values of $W$ and $V$ that minimize $|\varepsilon(Z)|$ for all $Z \in \Omega_Z \subset R^{m+1}$, i.e.,

$$(W^*, V^*) := \arg\min_{(W,V)}\{\sup_{Z\in\Omega_Z} |f(Z) - W^{\mathrm{T}} S(V^{\mathrm{T}} Z)|\}$$

**Assumption 4.7.** On the compact set $\Omega_Z$, the ideal NN weights $W^*$, $V^*$ are bounded by

$$\|W^*\| \le w_m, \quad \|V^*\|_F \le v_m$$

In general, the ideal weights $W^*$ and $V^*$ are unknown and need to be estimated in control design. Let $\hat{W}$ and $\hat{V}$ be the estimates of $W^*$ and $V^*$, respectively, and the weight estimation errors $\tilde{W} = \hat{W} - W^*$ and $\tilde{V} = \hat{V} - V^*$.

**Lemma 4.10.** *[30] Using $f_{mnn} = \hat{W}^{\mathrm{T}} S(\hat{V}^{\mathrm{T}} Z)$ to approximate the ideal function $f(Z)$, its approximation error can be expressed as*

$$\hat{W}^{\mathrm{T}} S(\hat{V}^{\mathrm{T}} Z) - W^{*\mathrm{T}} S(V^{*\mathrm{T}} Z) = \tilde{W}^{\mathrm{T}}(\hat{S} - \hat{S}'\hat{V}^{\mathrm{T}} Z) + \hat{W}^{\mathrm{T}}\hat{S}'\tilde{V}^{\mathrm{T}} Z + d_u$$

where $\hat{S} = S(\hat{V}^{\mathrm{T}} Z)$, $\hat{S}' = \mathrm{diag}\,\{\hat{s}'_1, \hat{s}'_2, ..., \hat{s}'_l\}$ with

$$\hat{s}'_i = s'(\hat{v}_i^{\mathrm{T}} Z) = \frac{\mathrm{d}[s(z_a)]}{\mathrm{d}z_a}\Big|_{z_a = \hat{v}_i^{\mathrm{T}} Z}$$

and the residual term $d_u$ is bounded by

$$|d_u| \le \|V^*\|_F \|Z\hat{W}^{\mathrm{T}}\hat{S}'\|_F + \|W^*\|\|\hat{S}'\hat{V}^{\mathrm{T}} Z\| + |W^*|_1$$

Throughout this chapter, we employ sigmoidal functions as basis functions for the MNN, which are defined by

$$s_i(z_a) = \frac{1}{1 + \mathrm{e}^{-\mu z_a}}, \quad i = 1, 2, ..., l \tag{4.13}$$

where $\mu > 0$ is a design constant.

## 4.4 Adaptive NN Control Design

We employ backstepping for the $\xi$ subsystem, and then make use of the exponential stability of the zero dynamics to show that the overall closed-loop system is stable and that output tracking is achieved. The control design is performed first for the full state case and subsequently for the output feedback case with high gain observers.

### 4.4.1 *Full State Feedback Control*

**Step 1:** Let $z_1(t) = \xi_1(t) - y_d(t)$ and $z_2(t) = \xi_2(t) - \alpha_1(t)$, where $\alpha_1(t)$ is a virtual control function to be determined. Define quadratic function $V_1 = \frac{1}{2}z_1^2$. Choosing the virtual control $\alpha_1$ as

$$\alpha_1 = -k_1 z_1 + \dot{y}_d, \tag{4.14}$$

we can show that

$$\dot{V}_1 = -k_1 z_1^2 + z_1 z_2, \tag{4.15}$$

where the term $z_1 z_2$ will be canceled in the subsequent step.

**Step i (i = 2, ..., $\rho - 1$):** Let $z_i(t) = \xi_i(t) - \alpha_{i-1}(t)$, where $\bar{\xi}_i := [\xi_1, ..., \xi_i]^T$ and $\alpha_i(t)$ is a virtual control function to be determined. Define quadratic function $V_i = V_{i-1} + \frac{1}{2} z_i^2$. Choose the virtual control $\alpha_i$ as

$$\alpha_i = -k_i z_i - z_{i-1} + \dot{\alpha}_{i-1}, \tag{4.16}$$

where the derivative can be written as

$$\dot{\alpha}_{i-1} = \frac{\partial \alpha_{i-1}}{\partial \bar{\xi}} \dot{\bar{\xi}} + \sum_{k=0}^{i-1} \frac{\partial \alpha_{i-1}}{\partial y_d^{(k)}} y_d^{(k+1)}. \tag{4.17}$$

It can be shown that

$$\dot{V}_i = -\sum_{j=1}^{i} k_j z_j^2 + z_i z_{i+1}, \tag{4.18}$$

where the term $z_i z_{i+1}$ will be canceled in the subsequent step.

**Step $\rho$:** This is the final step where the actual control law $u$ will be designed.

From Assumption 2, we know that $g(\xi, \eta, u) \geq d > 0$ for all $(\xi, \eta, u) \in R^{n+1}$. Define $\nu$ as

$$\nu = -\dot{\alpha}_{\rho-1} + g_0(\xi, \eta) z_\rho, \tag{4.19}$$

where $k_n > 0$ is a constant. It is clear that $\nu$ is a function of $\xi$, $\eta$, $y_d$, $y_d^{(1)}$,..., $y_d^{(\rho)}$. Considering the fact that $\frac{\partial \nu}{\partial u} = 0$, the following inequality holds

$$\frac{\partial [b(\xi, \eta, u) + \nu]}{\partial u} \geq d > 0. \tag{4.20}$$

According to Lemma 1, for every value of $\xi$, $\eta$ and $\nu$, there exists a smooth ideal control input $u \in R$ such that

$$b(\xi, \eta, u^*) + \nu = 0, \tag{4.21}$$

Using the Mean Value Theorem in Lemma 4.5, there exists $(0 < \lambda < 1)$ such that

$$b(\xi, \eta, u) = b(\xi, \eta, u^*) + g_\lambda (u - u^*), \tag{4.22}$$

where $g_\lambda := g(\xi, \eta, u_\lambda)$. Combining (4.19)–(4.22) yields

$$\dot{z}_\rho = -g_0(\xi, \eta) z_\rho + g_\lambda (u - u^*). \tag{4.23}$$

We employ a robust MNN controller of the form:

$$u = u_{nn} + u_b, \tag{4.24}$$

where

$$u_{nn} = \hat{W}^{\mathrm{T}} S(\hat{V}^{\mathrm{T}} Z), \tag{4.25}$$

$$u_b = -k_\rho z_\rho - z_{\rho-1} - k_b \left( \left\| Z \hat{W}^{\mathrm{T}} \hat{S}' \right\|_F^2 + \left\| \hat{S}' \hat{V}^{\mathrm{T}} Z \right\|^2 \right) z_\rho. \tag{4.26}$$

The component $u_{nn}$ is an MNN that approximates $u^*(\xi, \eta)$, which can be expressed as

$$u^* = W^{*\mathrm{T}} S(V^{*\mathrm{T}} Z) + \varepsilon, \tag{4.27}$$

where $Z = [\xi, \eta, z_\rho, \dot{\alpha}_{\rho-1}]^{\mathrm{T}} \in \Omega \subset R^{n+2}$; $W^*$ denotes the vector of ideal constant weights, and $|\varepsilon| \le \bar{\varepsilon}$ is the approximation error with constant $\bar{\varepsilon} > 0$. As detailed in [30], the component $u_b$ ensures robustness to the approximation error of the MNN.

Consider the Lyapunov function candidate

$$V_\rho = V_{\rho-1} + \frac{1}{2 g_\lambda} z_\rho^2 + \frac{1}{2} \tilde{W}^{\mathrm{T}} \Gamma_W^{-1} \tilde{W} + \frac{1}{2} \mathrm{tr} \{ \tilde{V}^{\mathrm{T}} \Gamma_V^{-1} \tilde{V} \}, \tag{4.28}$$

where $\tilde{W} := \hat{W} - W^*$ and $\tilde{V} := \hat{V} - V^*$. The derivative of $V_\rho$ is

$$\begin{aligned}
\dot{V}_\rho &= -\sum_{j=1}^{\rho-1} k_j z_j^2 + z_{\rho-1} z_\rho + \frac{z_\rho \dot{z}_\rho}{g_\lambda} - \frac{\dot{g}_\lambda z_\rho^2}{2 g_\lambda^2} + \tilde{W}^{\mathrm{T}} \Gamma_W^{-1} \dot{\hat{W}} + \mathrm{tr} \left\{ \tilde{V}^{\mathrm{T}} \Gamma_V^{-1} \dot{\hat{V}} \right\} \\
&= -\sum_{j=1}^{\rho-1} k_j z_j^2 + z_{\rho-1} z_\rho - \frac{g_0}{g_\lambda} z_\rho^2 - \frac{\dot{g}_\lambda}{2 g_\lambda^2} z_\rho^2 + z_\rho (u - u^* - \varepsilon) + \tilde{W}^{\mathrm{T}} \Gamma_W^{-1} \dot{\hat{W}} \\
&\quad + \mathrm{tr} \left\{ \tilde{V}^{\mathrm{T}} \Gamma_V^{-1} \dot{\hat{V}} \right\} \\
&= -\sum_{j=1}^{\rho-1} k_j z_j^2 + z_{\rho-1} z_\rho - \left( g_0 + \frac{\dot{g}_\lambda}{2 g_\lambda} \right) \frac{z_\rho^2}{g_\lambda} + \tilde{W}^{\mathrm{T}} \left[ (\hat{S} - \hat{S}' \hat{V}^{\mathrm{T}} Z) z_\rho \right. \\
&\quad \left. + \Gamma_W^{-1} \dot{\hat{W}} \right] + \mathrm{tr} \left\{ \tilde{V}^{\mathrm{T}} [Z \hat{W}^{\mathrm{T}} \hat{S}' z_\rho \} + \Gamma_V^{-1} \dot{\hat{V}} ] \right\} + z_\rho (d_u + u_b) \\
&\quad - z_\rho \varepsilon
\end{aligned} \tag{4.29}$$

Consider the following adaptation laws

$$\dot{\hat{W}} = -\Gamma_W \left[ (\hat{S} - \hat{S}'\hat{V}^{\mathrm{T}} Z) z_\rho + \sigma_W \hat{W} \right] \tag{4.30}$$

$$\dot{\hat{V}} = -\Gamma_V \left[ Z \hat{W}^{\mathrm{T}} \hat{S}' z_\rho + \sigma_V \hat{V} \right] \tag{4.31}$$

where $\Gamma_W = \Gamma_W^{\mathrm{T}} > 0$, $\Gamma_V = \Gamma_V^{\mathrm{T}} > 0$, $\sigma_W > 0$ and $\sigma_V > 0$ are constant design parameters. Then, substituting (4.26), (4.30), and (4.31) into (4.29) yields

$$\begin{aligned}
\dot{V}_\rho = & -\sum_{j=1}^{\rho-1} k_j z_j^2 + z_{\rho-1} z_\rho - \left( g_0 + \frac{\dot{g}_\lambda}{2 g_\lambda} \right) \frac{z_\rho^2}{g_\lambda} - \sigma_W \tilde{W}^{\mathrm{T}} \hat{W} - \sigma_V \mathrm{tr}\{\tilde{V}^{\mathrm{T}} \hat{V}\} \\
& - z_\rho \varepsilon + z_\rho (d_u + u_b) \\
\le & -\sum_{j=1}^{\rho-1} k_j z_j^2 - \left( g_\lambda k_\rho + g_0 + \frac{\dot{g}_\lambda}{2 g_\lambda} \right) \frac{z_\rho^2}{g_\lambda} - \sigma_W \tilde{W}^{\mathrm{T}} \hat{W} - \sigma_V \mathrm{tr}\{\tilde{V}^{\mathrm{T}} \hat{V}\} \\
& - z_\rho \varepsilon + |z_\rho| \left( \|W^*\| \|\hat{S}' \hat{V}^{\mathrm{T}} Z\| + \|V^*\|_F \|Z \hat{W}^{\mathrm{T}} \hat{S}'\|_F + \|W^*\| \right) \\
& - k_b \left( \|S' \hat{V}^{\mathrm{T}} Z\|^2 + \|Z \hat{W}^{\mathrm{T}} S'\|_F^2 \right) z_\rho^2 \\
\le & -\sum_{j=1}^{\rho-1} k_j z_j^2 - \left( g_\lambda (k_\rho - 1) + g_0 + \frac{\dot{g}_\lambda}{2 g_\lambda} \right) \frac{z_\rho^2}{g_\lambda} - \sigma_W \tilde{W}^{\mathrm{T}} \hat{W} - \sigma_V \mathrm{tr}\{\tilde{V}^{\mathrm{T}} \hat{V}\} \\
& + \|W^*\|^2 + \frac{1}{2} \|V^*\|_F^2 - \left( k_b - \frac{1}{2} \right) \left( \|\hat{S}' \hat{V}^{\mathrm{T}} Z\|^2 + \|Z \hat{W}^{\mathrm{T}} \hat{S}'\|_F^2 \right) z_\rho^2 + \frac{1}{2} \bar{\varepsilon}^2
\end{aligned} \tag{4.32}$$

By completion of squares, the following inequalities hold

$$-\sigma_W \tilde{W}^{\mathrm{T}} \hat{W} \le \frac{\sigma_W}{2} (-\|\tilde{W}\|^2 + \|W^*\|^2) \tag{4.33}$$

$$-\sigma_V \mathrm{tr}\{\tilde{V}^{\mathrm{T}} \hat{V}\} \le \frac{\sigma_V}{2} (-\|\tilde{V}\|_F^2 + \|V^*\|_F^2) \tag{4.34}$$

Substituting (4.33) and (4.34) into (4.32) yields the following

$$\begin{aligned}
\dot{V}_\rho \le & -\sum_{j=1}^{\rho-1} k_j z_j^2 - \left( g_\lambda (k_\rho - 1) + g_0 + \frac{\dot{g}_\lambda}{2 g_\lambda} \right) \frac{z_\rho^2}{g_\lambda} - \left( k_b - \frac{1}{2} \right) \left( \left\| \hat{S}' \hat{V}^{\mathrm{T}} Z \right\|^2 \right. \\
& \left. + \left\| Z \hat{W}^{\mathrm{T}} \hat{S}' \right\|_F^2 \right) z_\rho^2 - \frac{\sigma_W}{2} \|\tilde{W}\|^2 - \frac{\sigma_V}{2} \|\tilde{V}\|_F^2 + \frac{1}{2} \bar{\varepsilon}^2 + \frac{\sigma_W + 2}{2} \|W^*\|^2 \\
& + \frac{\sigma_V + 1}{2} \|V^*\|_F^2
\end{aligned} \tag{4.35}$$

From Assumption 4.5, we know that $\left(g_0 + \frac{\dot{g}_\lambda}{2g_\lambda}\right) \geq 0$. Hence, by choosing the control parameters $k_\rho$ and $k_b$ as follows

$$k_\rho > 1, \quad k_b > \frac{1}{2} \tag{4.36}$$

the second and third right-hand-side (RHS) terms of (4.35) are strictly negative, thus leading to the following simplification

$$\begin{aligned}\dot{V}_\rho &\leq -\sum_{j=1}^{\rho-1} k_j z_j^2 - (k_\rho - 1)z_\rho^2 - \frac{\sigma_W}{2}\|\tilde{W}\|^2 - \frac{\sigma_V}{2}\|\tilde{V}\|_F^2 \\ &\quad + \frac{1}{2}\bar{\varepsilon}^2 + \frac{\sigma_W + 2}{2}\|W^*\|^2 + \frac{\sigma_V + 1}{2}\|V^*\|_F^2 \\ &\leq -c_1 V_\rho + c_2\end{aligned} \tag{4.37}$$

where

$$c_1 = \min\left\{2k_1,\ 2k_2, ...,\ 2k_{\rho-1},\ 2\underline{g}(k_\rho - 1),\ \frac{\sigma_W}{\lambda_{\max}(\Gamma_W^{-1})},\ \frac{\sigma_V}{\lambda_{\max}(\Gamma_W^{-1})}\right\} \tag{4.38}$$

$$c_2 = \frac{1}{2}\bar{\varepsilon}^2 + \frac{\sigma_W + 2}{2}\|W^*\|^2 + \frac{\sigma_V + 1}{2}\|V^*\|_F^2 \tag{4.39}$$

The following lemma is useful for stability analysis of the internal dynamics.

**Lemma 4.11.** *Given that Assumptions 4.1 and 4.3 are satisfied, there exist positive constants $a_1$, $a_2$ and $T_0$ such that the trajectories $\eta(t)$ of the internal dynamics satisfy*

$$\|\eta(t)\| \leq a_1(\|z(t)\| + \|\xi_d(t)\|) + a_2 \ \ \forall\, t > T_0. \tag{4.40}$$

*Proof.* The proof is very similar to that given in [35]. For completeness, it is shown here. According to Assumption 4.3, there exists a Lyapunov function $V_0(\eta)$. Differentiating $V_0(\eta)$ along (4.3) yields

$$\begin{aligned}\dot{V}_0(\eta) &= \frac{\partial V_0}{\partial \eta} q(\xi, \eta) \\ &= \frac{\partial V_0}{\partial \eta} q(0, \eta) + \frac{\partial V_0}{\partial \eta}(q(\xi, \eta) - q(0, \eta)).\end{aligned} \tag{4.41}$$

Noting (4.4)–(4.7), (4.41) can be written as

$$\dot{V}_0 \leq -\lambda_a \|\eta\|^2 + \lambda_b a_\xi \|\eta\| \|\xi\| + \lambda_b a_q \|\eta\|. \tag{4.42}$$

From the fact that $\|\xi\| \leq \|z\| + \|\xi_d\|$, where $\xi_d := [y_d,\ \alpha_1,\ \alpha_2, ..., \alpha_{\rho-1}]^{\mathrm{T}}$, we obtain the following:

$$\dot{V}_0(\eta) \leq \lambda_a \|\eta\|^2 + \lambda_b \|\eta\| (a_\xi \|\xi_d\| + a_q + a_\xi \|z\|). \tag{4.43}$$

Therefore, $\dot{V}_0(\eta) \leq 0$ whenever

$$\|\eta\| \geq \frac{\lambda_b}{\lambda_a} (a_\xi \|\xi_d\| + a_q + a_\xi \|z\|). \tag{4.44}$$

By letting

$$a_1 = \frac{\lambda_b a_\xi}{\lambda_a}, \quad a_2 = \frac{\lambda_b a_q}{\lambda_a}, \tag{4.45}$$

we conclude that there exist $T_0$ such that (4.40) holds. □

We summarize our results for the full-state feedback case in the following theorem.

**Theorem 4.12.** *Consider the SISO helicopter dynamics (4.1) satisfying Assumptions 4.1–4.5, with control law (4.24) and adaptation laws (4.30)–(4.31). For initial conditions $\xi(0)$, $\eta(0)$, $\tilde{W}(0)$, $\tilde{V}(0)$ belonging to any compact set $\Omega_0$, all closed loop signals are SGUUB, and the tracking error $z_1 = y - y_d$ converges to the compact set*

$$\Omega_{z_1} := \left\{ z_1 \in R \,\middle|\, \|z_1\| \leq \sqrt{\frac{2c_2}{c_1}} \right\}, \tag{4.46}$$

*where $c_1$ and $c_2$ are constants defined in (4.38) and (4.39), respectively.*

*Proof.* According to Lemma 1.1–1.2 in [34], we know from (4.37) that $z$, $\tilde{W}$ and $\tilde{V}$ are bounded within the compact sets for all $t > 0$:

$$\Omega_z = \left\{ z \in R^\rho \,\middle|\, \|z\| \leq \sqrt{2\left(V_\rho(0) + \frac{c_2}{c_1}\right)} \right\}, \tag{4.47}$$

$$\Omega_W = \left\{ \tilde{W} \in R^l \,\middle|\, \|\tilde{W}\| \leq \sqrt{\frac{2\left(V_\rho(0) + \frac{c_2}{c_1}\right)}{\lambda_{\min}\left(\Gamma_W^{-1}\right)}} \right\}, \tag{4.48}$$

$$\Omega_V = \left\{ \tilde{V} \in R^{(m+1)\times l} \,\middle|\, \|\tilde{V}\|_F \leq \sqrt{\frac{2\left(V_\rho(0) + \frac{c_2}{c_1}\right)}{\lambda_{\min}\left(\Gamma_V^{-1}\right)}} \right\}. \tag{4.49}$$

Since $W^*$ and $V^*$ are bounded, it is obvious that $\hat{W}$ and $\hat{V}$ are also bounded.

From (4.14), (4.16), and the fact that $z$, $y_d$, $y_d^{(1)}, \ldots, y_d^{(\rho)}$ are bounded, we know that the virtual controls $\alpha_i, i = 1, 2, \ldots, \rho$ are bounded. Hence, there exists a constant $a_3 > 0$ such that $\|\xi_d\| \leq a_3$.

From Lemma 4.11, it can be seen that $\eta$ is bounded if both $z$ and $\xi_d$ are bounded. As a result, we can conclude that the states of the internal dynamics will converge to the compact set

$$\Omega_\eta := \left\{ \eta \in R^{n-\rho} \,\middle|\, \|\eta\| \leq a_1 \left( \sqrt{\frac{2c_2}{c_1}} + a_3 \right) + a_2 \right\}, \tag{4.50}$$

where $a_1$ and $a_2$ are defined in (4.45). Since the control signal $u(t)$ is a function of the weights $\hat{W}(t)$, $\hat{V}(t)$ and the states $\xi(t)$, $\eta(t)$, we know that it is also bounded. Therefore, we have shown that all the closed loop signals are SGUUB.

To show that the tracking error $z_1 = y - y_d$ converges to the compact set $\Omega_{z_1}$, we multiply (4.37) by $e^{-c_1 t}$ and integrate over [0,t] to obtain that

$$|z(t)| \leq \sqrt{2\left(V_\rho(0) + \frac{c_2}{c_1}\right) e^{-c_1 t} + 2\frac{c_2}{c_1}}, \tag{4.51}$$

from which it is easy to see that $|z_1(t)| \leq \sqrt{\frac{2c_2}{c_1}}$ as $t \to \infty$. □

*Remark 4.13.* From Theorem 4.12, we know that the size of the steady state compact set $\Omega_{z1}$, to which the tracking error converges, is governed by the constants $c_1$ and $c_2$, which in turn depend on the control and NN parameters. It follows that by appropriate tuning of the parameters, the guaranteed upper bound for the steady state tracking error can be reduced. For instance, increasing the control gains $k_1, \ldots, k_\rho$ increases $c_1$ accordingly, and leads to a reduction in the size of $\Omega_{z_1}$.

*Remark 4.14.* Although the theoretical results in this chapter are obtained under Assumption 4.4 that $g_\lambda(\cdot) > 0$, there is no loss of generality. For the case of $g_\lambda(\cdot) < 0$, the Lyapunov function candidate (4.28) can be changed to

$$V_\rho = V_{\rho-1} - \frac{1}{2g_\lambda} z_\rho^2 + \frac{1}{2}\tilde{W}^{\mathrm{T}} \Gamma_W^{-1} \tilde{W} + \frac{1}{2}\mathrm{tr}\left\{\tilde{V}^{\mathrm{T}} \Gamma_V^{-1} \tilde{V}\right\}, \tag{4.52}$$

for which a correspondingly stable controller can be constructed as $u = -u_{nn} - u_b$, where $u_{nn}$ and $u_b$ are defined in (4.25) and (4.26) respectively.

### 4.4.2 Output Feedback Control

In the previous section, we have considered the case where full state measurement is possible, that is, $\eta$ and $\xi$ are all available. In this section, we tackle the output feedback problem, where only $\eta$ and $\xi_1$ are available, by utilizing high gain observers.

**Lemma 4.15.** *[7, 35] Consider the following linear system:*

$$\begin{aligned} \epsilon\dot{\pi}_i &= \pi_{i+1}, \; i = 1, 2, \ldots, \rho - 1, \\ \epsilon\dot{\pi}_\rho &= -\gamma_1\pi_\rho - \gamma_2\pi_{\rho-1} - \cdots - \gamma_{\rho-1}\pi_2 - \pi_1 + \xi_1(t), \end{aligned} \tag{4.53}$$

*where $\epsilon$ is a small positive constant and the parameters $\gamma_1$ to $\gamma_{\rho-1}$ are chosen such that the polynomial $s^\rho + \gamma_1 s^{\rho-1} + \cdot + \gamma_{\rho-1}s + 1$ is Hurwitz. Suppose the states $\xi$ belong to a compact set, so that $|\xi_k| < Y_k$, then the following property holds:*

$$\tilde{\xi}_k := \frac{\pi_k}{\epsilon^{k-1}} - \xi_k = -\epsilon\zeta^{(k)}, \; k = 1, 2, \ldots, \rho, \tag{4.54}$$

*where $\zeta := \pi_\rho + \gamma_1\pi_{\rho-1} + \cdots + \gamma_{\rho-1}\pi_1$ and $\zeta^{(k)}$ denotes the $k$-th derivative of $\zeta$. Furthermore, there exist positive constants $h_k$ and $t^*$ such that $|\tilde{\xi}_k| \le \epsilon h_k$ is satisfied for $t > t^*$.*

*Proof.* Differentiating the last equation of (4.53) with respect to time and substituting into $\frac{\pi_2}{\epsilon} - \dot{\chi}$ yields

$$\frac{\pi_2}{\epsilon} - \dot{\chi} = \frac{\pi_2}{\epsilon} - \epsilon\ddot{\pi}_n - \gamma_1\dot{\pi}_n - \gamma_2\dot{\pi}_{n-1} - \cdots - \gamma_{n-1}\dot{\pi}_2 - \dot{\pi}$$

Noting from (4.53) that $\dot{\pi}_i = \epsilon\ddot{\pi}_{i-1}$ for $i = 2, 3, \ldots, n$ leads to

$$\begin{aligned} \frac{\pi_2}{\epsilon} - \dot{\chi} &= -\epsilon(\ddot{\pi}_n + \gamma_1\ddot{\pi}_{n-1} + \gamma_2\ddot{\pi}_{n-2} + \cdots + \gamma_{n-1}\ddot{\pi}_1) \\ &= -\epsilon\ddot{\zeta} \end{aligned}$$

By repeatedly differentiating the above and utilizing (4.53), we arrive at (4.54).

To show that $|\tilde{y}(t)| \le \epsilon h_k$ for $t > t^*$, we first note that the derivatives of the vector $[\pi_1 \; \pi_2 \; \ldots \; \pi_\rho]^{\mathrm{T}}$ can be computed as follows

$$\begin{aligned} \pi^{(j)}(t) = \frac{1}{\epsilon^j}A^j e^{\frac{At}{\epsilon}} \big[&\pi(0) + A^{-1}by(0) + \epsilon A^{-2}by^{(1)}(0) \\ &+ \cdots + \epsilon^{j-1}A^{-j}by^{(j-1)}(0)\big] \\ + \frac{1}{\epsilon}\mathrm{e}^{\frac{At}{\epsilon}} &\int_0^t \mathrm{e}^{\frac{-A\tau}{\epsilon}} by^{(j)}(\tau)\, \mathrm{d}\tau \end{aligned} \tag{4.55}$$

for $j = 1, 2, ..., \rho$, where

$$A = \begin{bmatrix} 0 & 1 & 0 & \cdots & 0 \\ 0 & 0 & 1 & \cdots & 0 \\ \cdots & \cdots & \cdots & \cdots & \cdots \\ 0 & 0 & 0 & \cdots & 1 \\ -1 & \gamma_1 & \gamma_2 & \cdots & \gamma_{\rho-1} \end{bmatrix}, \quad b = \begin{bmatrix} 0 \\ 0 \\ \vdots \\ 0 \\ 1 \end{bmatrix} \tag{4.56}$$

Since $\xi$ belongs to a compact set, and $u$ is bounded, we know that $|y^{(j)}| \leq Y_j$. Therefore, there exists a constant $t^* > 0$ such that for $t > t^*$, $|\pi^{(j)}(t)| \leq D_j$, where $D_j$ is a positive constant independent of $\epsilon$. This leads to $|\zeta^{(j)}| \leq h_j := BD_j$, where $B$ is the normof $[1\ \gamma_1\ \gamma_2\ ...\ \gamma_{\rho-1}]^{\mathrm{T}}$. □

*Remark 4.16.* Note that $\frac{\pi_k}{\epsilon^{k-1}}$ converges to a neighborhood of $\xi_k$, provided that $y$ and its derivatives up to the $\rho$-th order are bounded. Hence, $\frac{\pi_{k+1}}{\epsilon^k}$ is a suitable observer to estimate the $k$th order output derivative.

To prevent peaking [52], saturation functions are employed on the observer signals whenever they are outside the domain of interest:

$$\pi_i^s = \bar{\pi}_i \operatorname{sat}\left(\frac{\pi_i}{\bar{\pi}_i}\right), \quad \bar{\pi}_i \geq \max_{(z,\tilde{W},\tilde{V})\in\Omega} (\pi_i), \quad \operatorname{sat}(a) = \begin{cases} -1 & \text{for } a < -1 \\ a & \text{for } |a| \leq 1 \\ 1 & \text{for } a > 1 \end{cases} \tag{4.57}$$

for $i = 1, 2, ..., \rho$, where $\tilde{\xi} = [\tilde{\xi}_1, ..., \tilde{\xi}_\rho]^{\mathrm{T}}$, and the compact set $\Omega := \Omega_z \times \Omega_W \times \Omega_V$, where $\Omega_z$, $\Omega_W$, and $\Omega_V$ are defined in (4.47), (4.48), and (4.49) respectively, denotes the domain of interest.

Now, we revisit the control law (4.24)–(4.26) and adaptation laws (4.30)–(4.31) for the full-state feedback case. Via the certainty equivalence approach, we modify them by replacing the unavailable quantities $z_i$ and $Z$ with their estimates, $\hat{z}_i := \frac{\pi_i^s}{\epsilon^{i-1}} - \hat{\alpha}_{i-1}$ and $\hat{Z} := [\xi_1, \frac{\pi_2^s}{\epsilon}, ..., \frac{\pi_\rho^s}{\epsilon^{\rho-1}}, \eta, \hat{z}_\rho, \dot{\hat{\alpha}}_{\rho-1}]^{\mathrm{T}}$ respectively, for $i = 2, ..., \rho$. Therefore, the control laws are given by

$$\hat{\alpha}_1 = -k_1 z_1 + \dot{y}_d,$$

$$\hat{\alpha}_i = -k_i \hat{z}_i - \hat{z}_{i-1} + \dot{\hat{\alpha}}_{i-1}, \tag{4.58}$$

$$u_{nn} = \hat{W}^{\mathrm{T}} S(\hat{V}^{\mathrm{T}} \hat{Z}), \tag{4.59}$$

$$u_b = -k_\rho \hat{z}_\rho - \hat{z}_{\rho-1} - k_b \left( \left\| \hat{Z} \hat{W}^{\mathrm{T}} \hat{S}_o' \right\|_F^2 + \left( \left\| \hat{S}_o \right\| + \left\| \hat{S}_o' \hat{V}^{\mathrm{T}} \hat{Z} \right\| \right)^2 \right) \hat{z}_\rho \tag{4.60}$$

Due to the fact that the actual NN is in terms of $\hat{Z}$ while the ideal NN is in terms of $Z$, the following Lemma is needed.

**Lemma 4.17.** *[35] The error between the actual and ideal NN output can be written as*

$$\hat{W}^{\mathrm{T}} S(\hat{V}^{\mathrm{T}}\hat{Z}) - W^{*\mathrm{T}} S(V^{*\mathrm{T}} Z) = \tilde{W}^{\mathrm{T}}(\hat{S}_o - \hat{S}_o' \hat{V}^{\mathrm{T}}\hat{Z}) + \hat{W}^{\mathrm{T}}\hat{S}_o'\tilde{V}^{\mathrm{T}}\hat{Z} + d_u \tag{4.61}$$

*where* $\hat{S}_o := S(\hat{V}^{\mathrm{T}}\hat{Z})$; $\hat{S}_o' :=\mathrm{diag}\{\hat{s}_{o1}', ..., \hat{s}_{ol}'\}$ *with*

$$\hat{s}_{oi}' = s'(\hat{v}_i^{\mathrm{T}}\hat{Z}) = \frac{\mathrm{d}s(z_a)}{\mathrm{d}z_a}\Big|_{z_a=\hat{v}_i^{\mathrm{T}}\hat{Z}}, \quad i = 1, 2, ..., l, \tag{4.62}$$

*and the residual term* $d_u$ *is bounded by*

$$d_u \le \|W^*\| \left( \left\|\hat{S}_o'\hat{V}^{\mathrm{T}}\hat{Z}\right\| + \left\|\hat{S}_o\right\| \right) + \|V^*\|_F \left\|\hat{Z}\hat{W}^{\mathrm{T}}\hat{S}_o'\right\|_F \tag{4.63}$$

Accordingly, the adaptation laws are designed as

$$\dot{\hat{W}} = -\Gamma_W \left[ (\hat{S}_o - \hat{S}_o'\hat{V}^{\mathrm{T}}\hat{Z})\hat{z}_\rho + \sigma_W \hat{W} \right] \tag{4.64}$$

$$\dot{\hat{V}} = -\Gamma_V \left[ \hat{Z}\hat{W}^{\mathrm{T}}\hat{S}_o'\hat{z}_\rho + \sigma_V \hat{V} \right] \tag{4.65}$$

Using the backstepping procedure similar to Sect. 4.4.1, and substituting (4.59), (4.64), and (4.65) into the derivative of $V_\rho$ along the closed loop trajectories, it can be shown that

$$\begin{aligned}
\dot{V}_\rho \le & -\sum_{j=1}^{\rho-1} k_j z_j^2 - \left( g_0 + \frac{\dot{g}_\lambda}{2g_\lambda} \right) \frac{z_\rho^2}{g_\lambda} - \sum_{j=2}^{\rho-1} k_j z_j \tilde{z}_j + \sum_{j=2}^{\rho-1} z_j (\dot{\hat{\alpha}}_{j-1} - \dot{\alpha}_{j-1}) \\
& + \sum_{j=3}^{\rho-1} z_j \tilde{z}_{j-1} - z_\rho \varepsilon - \tilde{W}^{\mathrm{T}} \left[ (\hat{S}_o - \hat{S}_o'\hat{V}^{\mathrm{T}}\hat{Z})\tilde{z}_\rho + \sigma_W \hat{W} \right] \\
& - \mathrm{tr}\left\{ \tilde{V}^{\mathrm{T}} \left[ \hat{Z}\hat{W}^{\mathrm{T}}\hat{S}_o'\tilde{z}_\rho + \sigma_V \hat{V} \right] \right\} + |z_\rho| \left[ \|W^*\| \left( \left\|\hat{S}_o'\hat{V}^{\mathrm{T}}\hat{Z}\right\| + \left\|\hat{S}_o\right\| \right) \right. \\
& \left. + \|V^*\|_F \left\|\hat{Z}\hat{W}^{\mathrm{T}}\hat{S}_o'\right\|_F \right] + z_\rho u_b
\end{aligned} \tag{4.66}$$

From the inequalities in (4.33) and (4.34), we know that

$$
\begin{aligned}
\dot{V}_\rho \le & -\sum_{j=1}^{\rho-1} k_j z_j^2 - \left(g_0 + \frac{\dot{g}_\lambda}{2g_\lambda}\right)\frac{z_\rho^2}{g_\lambda} - \sum_{j=2}^{\rho-1} k_j z_j \tilde{z}_j + \sum_{j=2}^{\rho-1} z_j(\dot{\hat{\alpha}}_{j-1} - \dot{\alpha}_{j-1}) \\
& + \sum_{j=3}^{\rho-1} z_j \tilde{z}_{j-1} - z_\rho \varepsilon - \tilde{W}^{\mathrm{T}}(\hat{S}_o - \hat{S}_o'\hat{V}^{\mathrm{T}}\hat{Z})\tilde{z}_\rho - \hat{W}^{\mathrm{T}}\hat{S}_o'\tilde{V}^{\mathrm{T}}\hat{Z}\tilde{z}_\rho \\
& + \frac{\sigma_W}{2}(-\|\tilde{W}\|^2 + \|W^*\|^2) + \frac{\sigma_V}{2}(-\|\tilde{V}\|_F^2 + \|V^*\|_F^2) + \frac{1}{2}\|W^*\|^2 \\
& + \frac{1}{2}\|V^*\|_F^2 + \frac{1}{2}z_\rho^2\left[\left(\left\|\hat{S}_o'\hat{V}^{\mathrm{T}}\hat{Z}\right\| + \left\|\hat{S}_o\right\|\right)^2 + \left\|\hat{Z}\hat{W}^{\mathrm{T}}\hat{S}_o'\right\|_F^2\right] + z_\rho u_b
\end{aligned}
\tag{4.67}
$$

Substituting the bounding control (4.60) into (4.66) yields

$$
\begin{aligned}
\dot{V}_\rho \le & -\sum_{j=1}^{\rho-1} k_j z_j^2 - \left(k_\rho - \frac{1}{2}\right)z_\rho^2 - \left(g_0 + \frac{\dot{g}_\lambda}{2g_\lambda}\right)\frac{z_\rho^2}{g_\lambda} - \frac{\sigma_W}{2}\|\tilde{W}\|^2 - \frac{\sigma_V}{2}\|\tilde{V}\|_F^2 \\
& - \tilde{W}^{\mathrm{T}}(\hat{S}_o - \hat{S}_o'\hat{V}^{\mathrm{T}}\hat{Z})\tilde{z}_\rho - \hat{W}^{\mathrm{T}}\hat{S}_o'\tilde{V}^{\mathrm{T}}\hat{Z}\tilde{z}_\rho - \sum_{j=2}^{\rho} k_j z_j \tilde{z}_j + \sum_{j=2}^{\rho} z_j(\dot{\hat{\alpha}}_{j-1} \\
& - \dot{\alpha}_{j-1}) + \sum_{j=3}^{\rho} z_j \tilde{z}_{j-1} + \frac{\sigma_W + 1}{2}\|W^*\|^2 + \frac{\sigma_V + 1}{2}\|V^*\|_F^2 + \frac{1}{2}\bar{\varepsilon}^2 \\
& - \left(k_b - \frac{1}{2}\right)\left[\left(\left\|\hat{S}_o'\hat{V}^{\mathrm{T}}\hat{Z}\right\| + \left\|\hat{S}_o\right\|\right)^2 + \left\|\hat{Z}\hat{W}^{\mathrm{T}}\hat{S}_o'\right\|_F^2\right](z_\rho^2 + z_\rho\tilde{z}_\rho)
\end{aligned}
\tag{4.68}
$$

The following lemma is useful for handling the terms containing the estimation errors $\tilde{z_j}$, for $j = 1, 2, ..., \rho$.

**Lemma 4.18.** *There exist positive constants $F_i$ and $G_i$, which are independent of $\epsilon$, such that, for $t > t^*$, $i = 1, 2, ..., \rho - 1$, the estimates $\hat{\alpha}_i$ and $\hat{z}_i$ satisfy the following inequalities:*

$$
|\dot{\hat{\alpha}}_i - \dot{\alpha}_i| \le \epsilon F_i, \quad |\tilde{z}_i| := |\hat{z}_i - z_i| \le \epsilon G_i. \tag{4.69}
$$

*Proof.* Starting from $i = 1$, we know that

$$\dot{\hat{\alpha}}_1 - \dot{\alpha}_1 = -k_1(\dot{\hat{z}}_1 - \dot{z}_1) = -k_1\tilde{\xi}_2. \quad (4.70)$$

Subsequently, it is easy to obtain that

$$\begin{aligned}\dot{\hat{\alpha}}_2 - \dot{\alpha}_2 &= -k_2(\tilde{\xi}_3 - (\dot{\hat{\alpha}}_1 - \dot{\alpha}_1)) - \tilde{\xi}_2 - k_1\tilde{\xi}_3 \\ &= -(1 + k_1k_2)\tilde{\xi}_2 - (k_1 + k_2)\tilde{\xi}_3 =: a_2^{\mathrm{T}}\Psi_2 \end{aligned} \quad (4.71)$$

where $a_2 := [-(1 + k_1k_2) \ -(k_1 + k_2)]^{\mathrm{T}}$ and $\Psi_2 := [\tilde{\xi}_2 \ \tilde{\xi}_3]^{\mathrm{T}}$. Suppose that

$$\begin{aligned}\dot{\hat{\alpha}}_{i-2} - \dot{\alpha}_{i-2} &= a_{i-2}^{\mathrm{T}}\Psi_{i-2}, \\ \dot{\hat{\alpha}}_{i-1} - \dot{\alpha}_{i-1} &= a_{i-1}^{\mathrm{T}}\Psi_{i-1}, \end{aligned} \quad (4.72)$$

where $\Psi_j = [\tilde{\xi}_2 \ldots \tilde{\xi}_{j+1}]^{\mathrm{T}}$ and $a_j$ is a vector of constants, for $j = i-2, i-1$. Then, by induction, it can be shown that

$$\begin{aligned}\dot{\hat{\alpha}}_i - \dot{\alpha}_i &= -k_i\left(\tilde{\xi}_{i+1} - (\dot{\hat{\alpha}}_{i-1} - \dot{\alpha}_{i-1})\right) - \left(\tilde{\xi}_i - (\dot{\hat{\alpha}}_{i-2} - \dot{\alpha}_{i-2})\right) + \ddot{\hat{\alpha}}_{i-1} - \ddot{\alpha}_{i-1}, \\ &= -k_i\left(\tilde{\xi}_{i+1} - a_{i-1}^{\mathrm{T}}\Psi_{i-1}\right) - \left(\tilde{\xi}_i - a_{i-2}^{\mathrm{T}}\Psi_{i-2}\right) + a_{i-1}^{\mathrm{T}}\dot{\Psi}_{i-1} = a_i^{\mathrm{T}}\Psi_i \end{aligned} \quad (4.73)$$

for $i = 3, \ldots, \rho - 1$. From Lemma 4.15, we know that $\|\Psi_i\| \le \epsilon H_i$, where $H_i := \left\|[h_2, \ldots, h_{i+1}]^{\mathrm{T}}\right\|$, which leads to the fact that $|\dot{\hat{\alpha}}_i - \dot{\alpha}_i| \le \epsilon\|a_i\|H_i =: \epsilon F_i$, and thus (4.69) is proven.

To prove (4.69), note that

$$\begin{aligned}\tilde{z}_i &= \tilde{\xi}_i - (\hat{\alpha}_{i-1} - \alpha_{i-1}) \\ &= \tilde{\xi}_i - (-k_{i-1}\tilde{z}_{i-1} - \tilde{z}_{i-2} + \dot{\hat{\alpha}}_{i-1} - \dot{\alpha}_{i-1}) \end{aligned} \quad (4.74)$$

By following a similar inductive procedure, starting from $\tilde{z}_1 = \tilde{\xi}_1$ and $\tilde{z}_2 = \tilde{\xi}_2 - (\hat{\alpha}_1 - \alpha_1) = \tilde{\xi}_2$, it can be shown that $\tilde{z}_i = a_{zi}^{\mathrm{T}}\Psi_i$, where $a_{zi}$ is a constant vector. Using the property in (4.69), it is straightforward to see that $|\tilde{z}_i| \le \epsilon\|a_{zi}\|H_i =: \epsilon G_i$. The proof is now complete. □

Using Lemma 4.18, it is clear that, for $t > t^*$, the following inequalities hold:

$$-\tilde{W}^{\mathrm{T}}(\hat{S}_o - \hat{S}_o'\hat{V}^{\mathrm{T}}\hat{Z})\tilde{z}_\rho \le \frac{\epsilon}{2}\|\tilde{W}\|^2 + \frac{\epsilon}{2}\left(\|\hat{S}_o\| + \|\hat{S}_o'\hat{V}^{\mathrm{T}}\hat{Z}\|\right)^2 G_\rho^2, \quad (4.75)$$

$$-\hat{W}^{\mathrm{T}}\hat{S}_o'\tilde{V}^{\mathrm{T}}\hat{Z}\tilde{z}_\rho \le \frac{\epsilon}{2}\|\tilde{V}\|_F^2 + \frac{\epsilon}{2}\left\|\hat{Z}\hat{W}^{\mathrm{T}}\hat{S}_o'\right\|_F^2 G_\rho^2, \quad (4.76)$$

$$-\sum_{j=2}^{\rho} k_j z_j \tilde{z}_j \le \sum_{j=2}^{\rho} \frac{k_j}{2}\left(z_j^2 + \epsilon^2 G_j^2\right), \tag{4.77}$$

$$\sum_{j=2}^{\rho-1} z_j(\hat{\dot{\alpha}}_{j-1} - \dot{\alpha}_{j-1}) \le \sum_{j=2}^{\rho} \frac{1}{2}\left(z_j^2 + \epsilon^2 F_j^2\right), \tag{4.78}$$

$$\sum_{j=3}^{\rho} z_j \tilde{z}_{j-1} \le \sum_{j=3}^{\rho} \frac{1}{2}\left(z_j^2 + \epsilon^2 G_{j-1}^2\right). \tag{4.79}$$

By substitution of the inequalities (4.75)–(4.79) into (4.68), it is straightforward to obtain the following expression:

$$\begin{aligned}
\dot{V}_\rho \le & -\sum_{j=1}^{\rho-1} k_j z_j^2 - \left(k_\rho - \frac{1}{2}\right) z_\rho^2 - \left(g_0 + \frac{\dot{g}_\lambda}{2g_\lambda}\right)\frac{z_\rho^2}{g_\lambda} - \frac{(\sigma_W - \epsilon)}{2}\|\tilde{W}\|^2 \\
& - \frac{(\sigma_V - \epsilon)}{2}\|\tilde{V}\|_F^2 + \left[\left(\|\hat{S}_o\| + \|\hat{S}_o'\hat{V}^{\mathrm{T}}\hat{Z}\|\right)^2 + \left\|\hat{Z}\hat{W}^{\mathrm{T}}\hat{S}_o'\right\|_F^2\right]\frac{\epsilon}{2}G_\rho^2 \\
& + \sum_{j=2}^{\rho}\frac{k_j}{2}(z_j^2 + \epsilon^2 G_j^2) + \sum_{j=2}^{\rho}\frac{1}{2}(z_j^2 + \epsilon^2 F_j^2) + \sum_{j=3}^{\rho}\frac{1}{2}(z_j^2 + \epsilon^2 G_{j-1}^2) \\
& + \frac{\sigma_W + 1}{2}\|W^*\|^2 + \frac{\sigma_V + 1}{2}\|V^*\|_F^2 + \frac{1}{2}\bar{\varepsilon}^2 \\
& - \frac{1}{2}\left(k_b - \frac{1}{2}\right)\left[\left(\left\|\hat{S}_o'\hat{V}^{\mathrm{T}}\hat{Z}\right\| + \left\|\hat{S}_o\right\|\right)^2 + \left\|\hat{Z}\hat{W}^{\mathrm{T}}\hat{S}_o'\right\|_F^2\right]\left(z_\rho^2 - \epsilon^2 G_\rho^2\right)
\end{aligned} \tag{4.80}$$

The RHS terms can be rearranged into a more convenient form for analysis:

$$\begin{aligned}
\dot{V}_\rho \le & -k_1 z_1^2 - \frac{1}{2}(k_2 - 1)z_2^2 - \sum_{j=3}^{\rho-1}\frac{1}{2}(k_j - 2)z_j^2 - \frac{1}{2}(k_\rho - 3)z_\rho^2 - \left(g_0 + \frac{\dot{g}_\lambda}{2g_\lambda}\right)\frac{z_\rho^2}{g_\lambda} \\
& - \frac{(\sigma_W - \epsilon)}{2}\|\tilde{W}\|^2 - \frac{(\sigma_V - \epsilon)}{2}\|\tilde{V}\|_F^2 + \sum_{j=2}^{\rho}\frac{k_j}{2}\epsilon^2 G_j^2 + \sum_{j=2}^{\rho}\frac{1}{2}\epsilon^2 F_j^2 \\
& + \sum_{j=3}^{\rho}\frac{1}{2}\epsilon^2 G_{j-1}^2 + \frac{\sigma_W + 1}{2}\|W^*\|^2 + \frac{\sigma_V + 1}{2}\|V^*\|_F^2 + \frac{1}{2}\bar{\varepsilon}^2 \\
& - \frac{1}{2}\left(k_b - \frac{1}{2}\right)\left[\left(\left\|\hat{S}_o'\hat{V}^{\mathrm{T}}\hat{Z}\right\| + \left\|\hat{S}_o\right\|\right)^2 + \left\|\hat{Z}\hat{W}^{\mathrm{T}}\hat{S}_o'\right\|_F^2\right]\Bigg(z_\rho^2 \\
& - \left(\epsilon^2 + \frac{2\epsilon}{2k_b - 1}\right) G_\rho^2\Bigg).
\end{aligned} \tag{4.81}$$

Finally, by appropriately choosing the control parameters $k_\rho$ and $k_b$ as follows:

$$k_2 > 1, \quad k_3, ..., k_{\rho-1} > 2, \quad k_\rho > 3, \quad k_b > \frac{1}{2}, \quad \sigma_W, \sigma_V > \epsilon, \tag{4.82}$$

it can be shown that

$$\dot{V}_\rho \le -c_1 V_\rho + c_2 - K\left(z_\rho^2 - \epsilon c_3\right), \tag{4.83}$$

where

$$c_1 := \min\left\{2k_1,\ (k_2-1),\ (k_3-2), ...,\ (k_{\rho-1}-2),\ \underline{g}(k_\rho-3),\ \frac{(\sigma_W-\epsilon)}{\lambda_{\max}(\Gamma_W^{-1})},\ \frac{(\sigma_V-\epsilon)}{\lambda_{\max}(\Gamma_W^{-1})}\right\}, \tag{4.84}$$

$$c_2 := \frac{1}{2}\bar{\varepsilon}^2 + \frac{\sigma_W+1}{2}\|W^*\|^2 + \frac{\sigma_V+1}{2}\|V^*\|_F^2 + \frac{1}{2}\epsilon^2\left(\sum_{j=2}^{\rho} F_j^2 + \sum_{j=2}^{\rho-1}(k_j+1)G_j^2 + G_\rho^2\right), \tag{4.85}$$

$$c_3 := \left(\epsilon + \frac{2}{2k_b-1}\right)G_\rho^2, \tag{4.86}$$

$$K := \frac{1}{2}\left(k_b - \frac{1}{2}\right)\left[\left(\left\|\hat{S}_o'\hat{V}^{\mathrm{T}}\hat{Z}\right\| + \left\|\hat{S}_o\right\|\right)^2 + \left\|\hat{Z}\hat{W}^{\mathrm{T}}\hat{S}_o'\right\|_F^2\right]. \tag{4.87}$$

It can be shown that

$$\dot{V}_\rho(t) \le -c_{o1}V_\rho(t) + c_{o2}, \quad t \ge t^*$$
$$c_{o1} := \min\left\{2\lambda_{\min}(K_1),\ \frac{\lambda_{\min}\left(K_2 - \frac{3}{2}I\right)}{\lambda_{\max}(M)},\ \frac{\min_{i=1,2,3}\{\frac{\sigma_i}{4}\|\tilde{\theta}_i\|^2\}}{\lambda_{\max}(\Gamma^{-1})}\right\} \tag{4.88}$$

$$c_{o2} := \sum_{i=1}^{3}\left(1 + \frac{\sigma_i}{2}\right)\|\theta_i^*\|^2 + \frac{1}{2}\|\varepsilon\|^2 + 2\lambda_{\max}(\bar{K}_2\bar{K}_2^{\mathrm{T}} + \Lambda)\epsilon\kappa_2 \tag{4.89}$$

To ensure that $\rho > 0$, the control gains $K_1$ and $K_2$ are chosen to satisfy the following conditions:

$$\lambda_{\min}(K_1) > 0, \quad \lambda_{\min}\left(K_2 - \frac{3}{2}I\right) > 0. \tag{4.90}$$

We are ready to summarize our results for the output feedback case under the following theorem.

**Theorem 4.19.** *Consider the helicopter dynamics (4.1) under Assumptions 4.1–4.5, with output feedback control laws (4.59)–(4.60), adaptation laws (4.64)–(4.65), and high gain observer (4.53) which is turned on at time $t^*$ in advance. For initial conditions $\xi(0)$, $\eta(0)$, $\tilde{W}(0)$, $\tilde{V}(0)$ starting in any compact set $\Omega_0$, all closed loop signals are SGUUB, and the tracking error $z_1$ converges to the steady state compact set:*

$$\Omega_{z_1} := \left\{ z_1 \in R \,\middle|\, |z_1| \le \sqrt{\frac{2\bar{c}_2}{c_1}} \right\}, \tag{4.91}$$

*where $\bar{c}_2 := c_2 + c_3\bar{K}$, and $c_1$ is as defined in (4.84).*

*Proof.* We consider the following two cases for the stability analysis:

Case 1: $|z_\rho| > \sqrt{\epsilon c_3}$

For this case, the last term of (4.83) is negative, thus yielding

$$\dot{V}_\rho \le -c_1 V_\rho + c_2, \tag{4.92}$$

which straightforwardly implies that all closed loop signals are SGUUB, according to Lemma 4.9. However, when $|z_\rho| \le \sqrt{\epsilon c_3}$, the last term of (4.83) may not be negative, leading to a more complicated analysis, as shown in the subsequent case.

Case 2: $|z_\rho| \le \sqrt{\epsilon c_3}$

For this case, we want to show that, as a result of $z_\rho$ being bounded, the function $K$ in (4.87) is also bounded, for which (4.83) can be expressed in the form of (4.9), convenient for establishing SGUUB property. To this end, note that the derivative of $V_{\rho-1}$ is given by

$$\dot{V}_{\rho-1} \le -\sum_{i=1}^{\rho-1} k_i z_i^2 - \sum_{i=2}^{\rho-1} k_i z_i \tilde{z}_i + \sum_{i=2}^{\rho-1} z_i(\dot{\hat{\alpha}}_{i-1} - \dot{\alpha}_{i-1}) + \sum_{i=3}^{\rho-1} z_i \tilde{z}_{i-1} + z_{\rho-1} z_\rho \tag{4.93}$$

According to Lemma 4.18, we can show that

$$\begin{aligned}
\dot{V}_{\rho-1} &\le -\sum_{i=1}^{\rho-1} k_i z_i^2 + \sum_{i=2}^{\rho-1} k_i |z_i| \epsilon G_i + \sum_{i=2}^{\rho-1} |z_i| \epsilon F_i + \sum_{i=3}^{\rho-1} |z_i| \epsilon G_{i-1} + |z_{\rho-1}| \sqrt{\epsilon c_3} \\
&\le -k_1 z_1^2 - \frac{1}{2}(k_2 - 1) z_2^2 - \sum_{i=3}^{\rho-2} \frac{1}{2}(k_i - 2) z_3^2 - \frac{1}{2}(k_\rho - 3) z_{\rho-1}^2
\end{aligned}$$

$$
\begin{aligned}
&+\frac{1}{2}\epsilon c_3 + \frac{1}{2}\sum_{i=3}^{\rho-1}\epsilon^2 G_{i-1}^2 + \frac{1}{2}\sum_{i=2}^{\rho-1}\epsilon^2 F_i^2 + \frac{1}{2}\sum_{i=2}^{\rho-1} k_i \epsilon^2 G_i^2 \\
&\le -c_4 V_{\rho-1} + c_5,
\end{aligned}
\tag{4.94}
$$

where the positive constants $c_4$ and $c_5$ are defined by

$$c_4 := \min\{2k_1,\ k_2 - 1,\ k_3 - 2,\ ...,\ k_{\rho-1} - 2,\ k_\rho - 3\}, \tag{4.95}$$

$$c_5 := \frac{\epsilon}{2}\left[c_3 + \epsilon\left(\sum_{i=2}^{\rho-1} F_i^2 + \sum_{i=2}^{\rho-2}(1+k_i)G_i^2 + k_{\rho-1}G_{\rho-1}^2\right)\right]. \tag{4.96}$$

This implies that $z(t)$ satisfies the inequality

$$\|z(t)\| \le \sqrt{2\left(V_{\rho-1}(0) + \frac{c_5}{c_4}\right) + \epsilon c_3} =: \bar{z}\,. \tag{4.97}$$

According to Lemma 4.11, it follows from the boundedness of $z(t)$ that the internal states $\eta(t)$ are also bounded, i.e.,

$$\|\eta(t)\| \le a_1(\bar{z} + \bar{\xi}_d) + a_2 =: \bar{\eta}, \tag{4.98}$$

where $\|\xi_d(t)\| \le \bar{\xi}_d$ for constant $\bar{\xi}_d > 0$, based on Assumption 4.1. Thus, the vector of NN inputs $\hat{Z}$ is also bounded as follows

$$\left\|\hat{Z}\right\| \le \left\|\left[\xi_{1d} + \bar{z}, \frac{\bar{\pi}_2}{\epsilon}, ..., \frac{\bar{\pi}_\rho}{\epsilon^{\rho-1}}, \bar{\eta}, \sqrt{c_3} + \epsilon G_\rho, \bar{\dot{\alpha}}_{\rho-1} + \epsilon F_{\rho-1}\right]^{\mathrm{T}}\right\| =: \bar{\hat{Z}} \tag{4.99}$$

where the constant $\bar{\dot{\alpha}}_{\rho-1} > 0$ is an upper bound for $\dot{\alpha}_{\rho-1}\left(z, \xi_{1d}, \xi_{1d}^{(1)}, ..., \xi_{1d}^{(\rho)}\right)$. Exploiting the properties of sigmoidal NNs [30], it can be shown that

$$\left(\left\|\hat{S}_o'\hat{V}^{\mathrm{T}}\hat{Z}\right\| + \left\|\hat{S}_o\right\|_F\right) \le 1.224\sqrt{l}. \tag{4.100}$$

As a result, from the adaptation law (4.64), the dynamics of the neural weights $\hat{W} = [\hat{W}_1, ..., \hat{W}_i, ..., \hat{W}_l]^{\mathrm{T}}$ can be shown to satisfy the inequality

$$\dot{\hat{W}} \le -\lambda_{\min}(\Gamma_W)\sigma_W \hat{W} + 1.224\sqrt{l}\lambda_{\min}(\Gamma_W)(\sqrt{\epsilon c_3} + \epsilon G_\rho), \tag{4.101}$$

which results in

$$\|\hat{W}(t)\| \le \|\hat{W}(0)\| + \frac{1.224\sqrt{l}(\sqrt{\epsilon c_3} + \epsilon G_\rho)}{\sigma_W} =: \bar{\hat{W}}, \tag{4.102}$$

where $\bar{\hat{W}}$ is a positive constant. Accordingly, from (4.87), and the fact that $\|S_o'\|_F \leq 0.25\mu\sqrt{l}$, we can show that $K$ is bounded as follows:

$$K \leq \frac{l}{2}\left(k_b - \frac{1}{2}\right)\left[1.498 + 0.0625\left(\bar{\hat{Z}}\bar{\hat{W}}\mu\right)^2\right] =: \bar{K}, \tag{4.103}$$

where $\bar{K}$ is a positive constant. From (4.83) and (4.103), we obtain that

$$\dot{V}_\rho \leq -c_1 V_\rho + \bar{c}_2, \tag{4.104}$$

where $\bar{c}_2 := c_2 + \epsilon c_3 \bar{K}$ is a positive constant.

Having obtained (4.104) for Case 2, we can compare it with (4.92) of Case 1 to see that (4.104) describes a larger compact set in which the closed loop signals remain, by virtue of the fact that $\bar{c}_2 \geq c_2$. Hence, the performance bounds can be analyzed from (4.104), as a conservative approach. A nice property is that as $\epsilon$ diminishes to zero, we have $\bar{c}_2 \to c_2$, and the performance can be analyzed from (4.92) instead, albeit conservatively.

Based on (4.104), we can directly invoke Lemma 4.9 to conclude SGUUB for all closed loop signals. Since it is straightforward to prove that the tracking error $z_1 = \xi_1 - \xi_{1d}$ converges to the compact set $\Omega_{z_1}$, by following the steps outlined in the proof of Theorem 4.12, we have omitted the proof. □

*Remark 4.20.* It follows from Theorem 4.19 that the size of the steady state compact set $\Omega_{z1}$, to which the tracking error converges, depends on the ratio $\frac{\bar{c}_2}{c_1}$, which contain tunable parameters. Thus, we can reduce the size of $\Omega_{z_1}$ by appropriately choosing the parameters. For instance, by choosing the control gains $k_1, ..., k_\rho$ large and the observer parameter $\epsilon$ sufficiently small, the ratio $\frac{\bar{c}_2}{c_1}$ can be decreased, to the effect that $\Omega_{z_1}$ diminishes.

## 4.5 Simulation Study

In Sect. 4.2, we have considered a general representation of helicopters as nonaffine nonlinear systems. Although it would be more realistic to perform simulations on nonaffine helicopter models, an accurate model is difficult to obtain. Since the class of nonaffine systems include linear systems as special cases, we shall apply our proposed adaptive NN control for general nonlinear systems to linear and nonlinear affine helicopter models, which are available in the literature.

In particular, we will first investigate the effectiveness of the NN controller on a linear helicopter model for two tasks: altitude tracking and pitch tracking. For the altitude tracking task, we will compare our results with that of [54], while for the pitch tracking study, we will compare with [38]. This will be followed by a study for the case of a nonlinear helicopter model for vertical flight.

For all cases, we use a multi-layer NN as detailed in Sect. 4.3.2, with identical neuronal activation functions for the hidden layer described by

$$s_i(z_a) = \frac{1}{1 + e^{-\mu z_a}}, \quad i = 1, 2, ..., l, \tag{4.105}$$

so that $S(a) = [s_1(a_1), s_2(a_2), ..., s_l(a_l)]^T$. The control law is given by (4.24) and the adaptation laws by (4.30)–(4.31).

## 4.5.1 *Linear Models*

In this section, we consider two linearized helicopter models for altitude and pitch tracking. Since data on linear models is quite rich in the literature, it is useful to employ them in a study of the effectiveness of the proposed nonlinear NN controller. As linear systems are a special subclass of nonlinear nonaffine systems, the NN controller, which is designed for the latter, can be applied on linear systems without any complications.

### 4.5.1.1 Altitude Tracking

To this end, consider the linearized altitude model of the Yamaha R50 helicopter as detailed in [54] with the longitudinal cyclic input $\delta$ set to zero:

$$\begin{bmatrix} \dot{u} \\ \dot{q} \\ \dot{\theta} \\ \dot{\beta} \\ \dot{\omega} \\ \dot{h} \end{bmatrix} = \begin{bmatrix} X_u & X_q & X_\theta & X_\beta & X_\omega & 0 \\ M_u & M_q & 0 & M_\beta & M_\omega & 0 \\ 0 & 1 & 0 & 0 & 0 & 0 \\ B_u & -1 & 0 & B_\beta & 0 & 0 \\ Z_u & Z_q & Z_\theta & Z_\beta & Z_\omega & 0 \\ 0 & 0 & 0 & 0 & -1 & 0 \end{bmatrix} \begin{bmatrix} u \\ q \\ \theta \\ \beta \\ \omega \\ h \end{bmatrix} + \begin{bmatrix} 0 \\ 0 \\ 0 \\ 0 \\ Z_\Omega \\ 0 \end{bmatrix} \delta_\Omega, \tag{4.106}$$

The control parameters are set to be $k_1 = 2.0$, $k_2 = 8.5$, and $k_b = 1.0$, while the NN parameters are $\mu = 1$, $\Gamma_W = 5I$, $\Gamma_V = 50I$, and $\sigma_W = \sigma_V = 5$. For the high gain observer, we choose $\epsilon = 0.01$, $\gamma_1 = 2$, and $\pi_2 = 0.08$. The lower and upper saturation limits of the control are 393 rpm and 1,348 rpm respectively. The initial conditions are $x(0) = [10,\ 0,\ 0,\ 0,\ 0,\ 0]^T$, $\hat{W} = 0$, and $\hat{V} = 0$.

To compare our controller with that of [54], we consider the tracking of the altitude $h$ according to a desired trajectory $h_d(t)$ defined by

$$\begin{bmatrix} \dot{h}_d \\ \ddot{h}_d \end{bmatrix} = \begin{bmatrix} 0 & 1 \\ -2.25 & -2.4 \end{bmatrix} \begin{bmatrix} h_d \\ \dot{h}_d \end{bmatrix} + \begin{bmatrix} 0 \\ 2.25 \end{bmatrix} h_{\text{ref}},$$

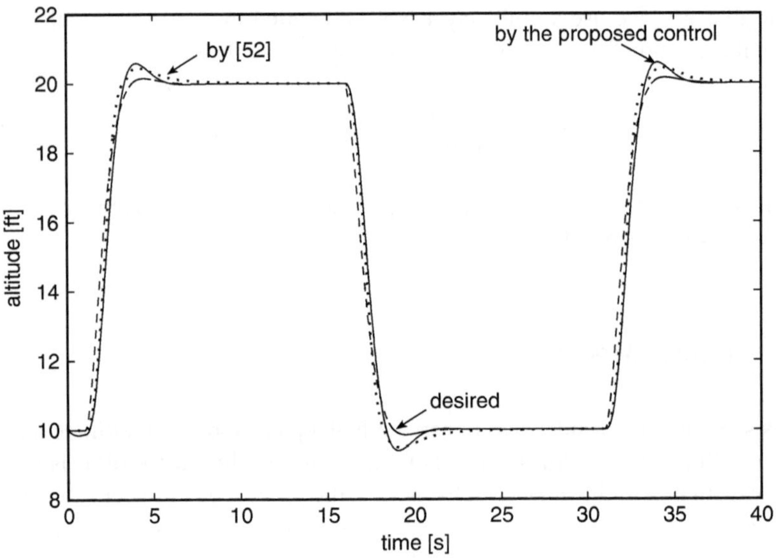


**Fig. 4.1** Comparison of altitude tracking performance between the proposed controller and that of [54]

$$h_{\text{ref}}(t) = \begin{cases} 0 \text{ if } 0 \leq t < 1 \\ 10 \text{ if } 1 \leq t < 16 \\ 0 \text{ if } 16 \leq t < 31 \\ 10 \text{ if } t \geq 31 \end{cases}. \tag{4.107}$$

Note that in our comparison, both controllers are simulated without engine dynamics.

It can be seen in Fig. 4.1 that the tracking performance under the proposed control is reasonably good, with the altitude signal tracking the desired trajectory closely. The comparison shows that the performances under the two different controls are similar. From Fig. 4.2, it is clear that the control signals and neural weights are well-behaved. The control of [54] exhibits more fluctuations, and the neural weights evolve to significantly larger amplitudes. Although this does not set out any comparative advantages, it does demonstrate different mechanisms at work in the two control schemes.

The effect of the parameter $\epsilon$ on tracking errors and observer errors are shown in Fig. 4.3, where it is seen that as $\epsilon$ diminishes, the tracking error under the full-state feedback control is recovered, and faster convergence of the observer is achieved.

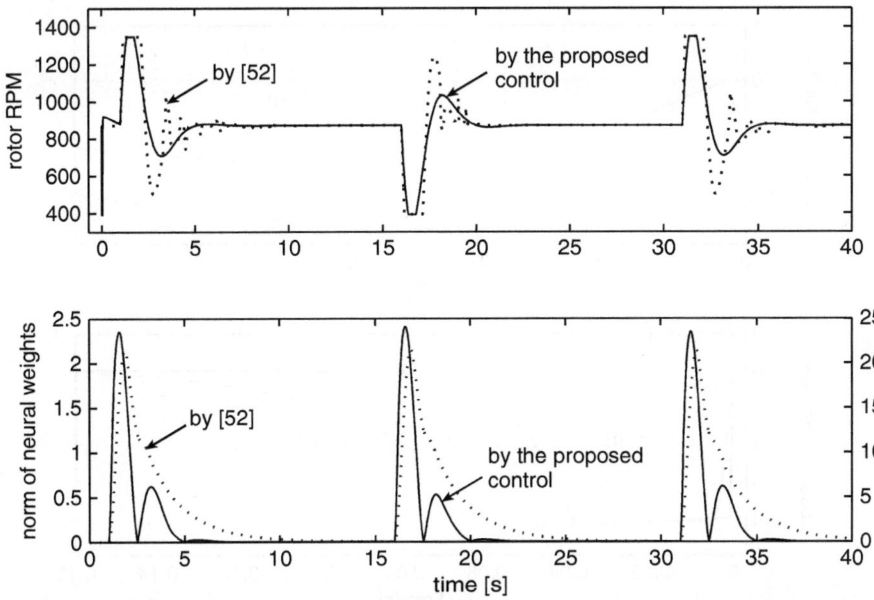


**Fig. 4.2** Comparison of rotor RPM input and norm of neural weights, given by $\sqrt{\|\hat{W}^{\mathrm{T}}\|^2 + \|\hat{V}\|_F^2}$, between the proposed control and that of [54]. For the bottom graph, the left scale corresponds to the proposed control while the right scale corresponds to that of [54]

#### 4.5.1.2 Pitch Tracking

To this end, consider the linearized model of the Yamaha R50 helicopter with actuator dynamics as detailed in [38].

The control parameters are set to be $k_1 = 0.8$, $k_2 = 0.8$, $k_3 = 1.5$ and $k_b = 1.0$, while the NN parameters are $\mu = 0.01$, $\Gamma_W = 5I$, $\Gamma_V = 50I$, $\sigma_W = 5$, and $\sigma_V = 1$. For the high gain observer, we choose $\epsilon = 1 \times 10^{-4}$, $\gamma_1 = 3$, $\gamma_2 = 3$, $\bar{\pi}_2 = 4 \times 10^{-5}$, and $\bar{\pi}_3 = 4 \times 10^{-8}$. The initial conditions are $x(0) = 0$, $\hat{W} = 0$, and $\hat{V} = 0$.

*Remark 4.21.* Although it is stated in (4.82) that $k_2 > 1$ and $k_3 > 3$, those conditions obtained from theoretical analysis are somewhat conservative for this example, although they guarantee a stable, working controller. We found that the lower control gains chosen here are sufficiently good for obtaining satisfactory tracking performance.

To compare our controller with that of [38], we consider the tracking of the pitch angle $\theta$ according to a desired trajectory $\theta_d(t)$ defined by

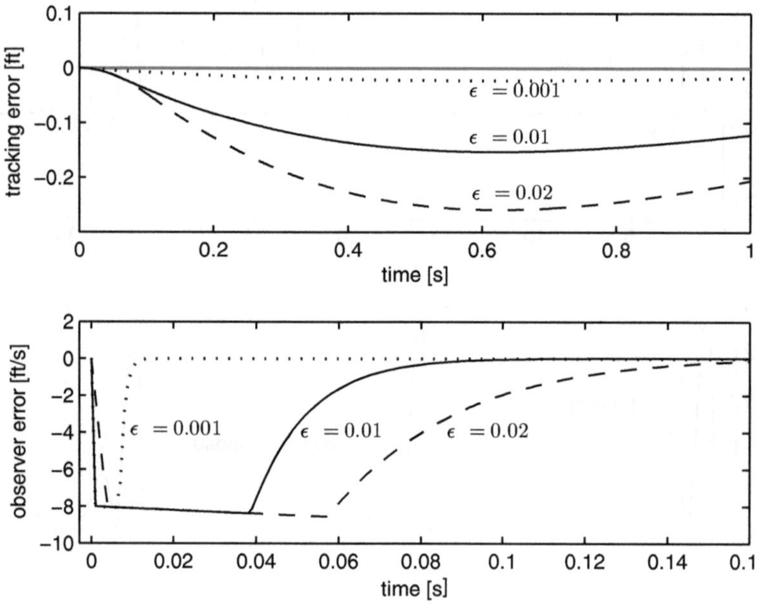


**Fig. 4.3** Effect of $\epsilon$ on tracking errors and observer errors

$$\begin{bmatrix} \dot{\theta}_d \\ \ddot{\theta}_d \\ \dddot{\theta}_d \end{bmatrix} = \begin{bmatrix} 0 & 1 & 0 \\ 0 & 0 & 1 \\ -2,000 & -420 & -36 \end{bmatrix} \begin{bmatrix} \theta_d \\ \theta_d \\ \theta_d \end{bmatrix} + \begin{bmatrix} 0 \\ 0 \\ 2,000 \end{bmatrix} \theta_{\mathrm{ref}},$$

$$\theta_{\mathrm{ref}}(t) = \begin{cases} 0 & \text{if } 0 \le t < 0.5 \\ 0.1 & \text{if } 0.5 \le t < 2 \\ 0 & \text{if } 2 \le t < 3 \\ -0.1 & \text{if } 3 \le t < 4.5 \\ 0 & \text{if } t \ge 4.5 \end{cases} . \tag{4.108}$$

It can be seen in Fig. 4.4 that the tracking performance of our controller is comparable with that of [38] for the pitch tracking task. From Fig. 4.5, we see that the control input and neural weights are bounded. Similar to the results in Sect. 4.5.1.1, the control of [38] exhibits more fluctuations, and the neural weights are larger, illustrating the different mechanisms at work in the two control schemes.

## 4.5.2 *Nonlinear Model*

The previous section considered linearized models which are relatively easy to control, but tend to be more suited for local operation within a neighborhood

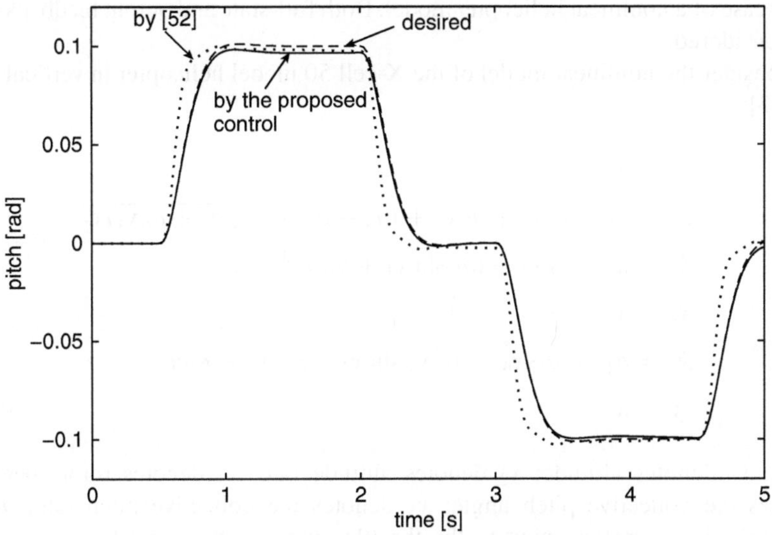


**Fig. 4.4** Comparison of pitch tracking performance between the proposed control and that of [38]

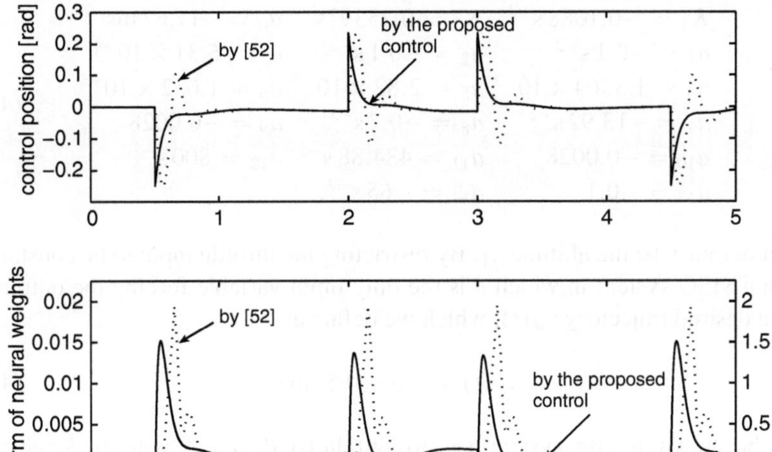


**Fig. 4.5** Comparison of control input and norm of neural weights, given by $\sqrt{\|\hat{W}^{\mathrm{T}}\|^2 + \|\hat{V}\|_F^2}$, between the proposed control and that of [38]. For the bottom graph, the *left scale* corresponds to the proposed control while the *right scale* corresponds to that of [38]

of the operating point, for which the linearized model is valid. Nevertheless, it was demonstrated that the proposed adaptive NN control is effective for the linear models considered, which constitute a subclass of general non-affine systems considered in the control design. In this section, we extend the investigations further to the case of a nonlinear helicopter model. Both full-state and output feedback cases are considered.

Consider the nonlinear model of the X-cell 50 model helicopter in vertical flight [48, 93]:

$$
\begin{aligned}
\dot{x}_1 &= x_2 \\
\dot{x}_2 &= a_0 + a_1 x_2 + a_2 x_2^2 + (a_3 + a_4 x_4 - \sqrt{a_5 + a_6 x_4})x_3^2 \\
\dot{x}_3 &= a_7 + a_8 x_3 + (a_9 \sin x_4 + a_{10})x_3^2 + a_{\text{th}} \\
\dot{x}_4 &= x_5 \\
\dot{x}_5 &= a_{11} + a_{12} x_4 + a_{13} x_3^2 \sin x_4 + a_{14} x_5 - K_2 u \\
y &= x_1,
\end{aligned}
\tag{4.109}
$$

where $x_1$ denotes altitude; $x_2$ denotes altitude rate; $x_3$ denotes rotor speed; $x_4$ denotes the collective pitch angle; $x_5$ denotes the collective pitch rate; $a_{\text{th}} = 111.69\text{s}^{-2}$ is a constant input to the throttle; and $u$ is the input to the collective servomechanisms. The parameters are

$$
\begin{array}{lll}
K_1 = -0.1088\,\text{s}^{-2} & K_2 = 0.25397\text{s}^{-2} & a_0 = -17.67\,\text{ms}^{-2} \\
a_1 = -0.1\,\text{s}^{-2} & a_2 = -0.1\,\text{s}^{-2} & a_3 = 5.31 \times 10^{-4} \\
a_4 = 1.5364 \times 10^{-2} & a_5 = 2.82 \times 10^{-7} & a_6 = 1.632 \times 10^{-5} \\
a_7 = -13.92\,\text{s}^{-2} & a_8 = -0.7\,\text{s}^{-2} & a_9 = -0.0028 \\
a_{10} = -0.0028 & a_{11} = 434.88\,\text{s}^{-2} & a_{12} = 800\,\text{s}^{-2} \\
a_{13} = -0.1 & a_{14} = -65\,\text{s}^{-2}. &
\end{array}
\tag{4.110}
$$

Let output $y$ be the altitude $x_1$. By restricting the throttle input to be constant, we obtain a SISO system in which $u$ is the only input variable forcing the output $y$ to track a desired trajectory $y_d(t)$, which we define as

$$
y_d(t) = 5.5 - 0.5 \sin t. \tag{4.111}
$$

It can be shown that the system has strong relative degree 4, with the $\xi$ subsystem given by

$$
\begin{aligned}
\dot{\xi}_1 &= \xi_2 = x_2 \\
\dot{\xi}_2 &= \xi_3 = a_0 + a_1 x_2 + a_2 x_2^2 + (a_3 + a_4 x_4 - \sqrt{a_5 + a_6 x_4})x_3^2 \\
\dot{\xi}_3 &= \xi_4 = (a_1 + 2a_2\xi_2)\xi_3 + a_3 x_3^2 + \left(a_4 - \frac{a_6}{2\sqrt{a_5 + a_6 x_4}}\right) x_3^2 x_5 \\
&\quad + 2(a_3 + a_4 x_4 - \sqrt{a_5 + a_6 x_4})x_3(a_8 x_3 + (a_9 \sin x_4 + a_{10})x_3^2) \\
\dot{\xi}_4 &= b(x) + g(x)u_2,
\end{aligned}
\tag{4.112}
$$

where

$$g(x) = -K_2 x_3^2 \left( a_4 - \frac{a_6}{2\sqrt{a_5 + a_6 x_4}} \right). \tag{4.113}$$

The derivation of $b(x)$ in (4.112) is omitted, and we proceed to verify that the system indeed satisfies the assumptions supposed in the control design. Assumptions 4.1 and 4.2 are obviously satisfied from (4.108) and (4.112) respectively.

To verify Assumption 4.4, we first note, from a practical standpoint, that the collective pitch angle, $x_4$, is restricted within a range, typically from 0 to 0.44 rad [83]. It can be verified that the bracketed terms in (4.113) are virtually constant: they take values in the range $[1.4, 1.5] \times 10^{-3}$. Thus, the control coefficient $g(x)$ in (4.113) is always negative. Together with the fact that rotor speed $x_3$ is nonzero during flight, it can be concluded that there does not exist any control singularities or zero crossings of $g(x)$. Therefore, the first part of Assumption 4.4 is satisfied.

*Remark 4.22.* Although the second part of the assumption, that $g(x) > 0$, does not correspond to this example, there is no loss of applicability of the theoretical results, as explained in Remark 4.14. The control is still valid under a simple change of sign, i.e., $u = -u_{nn} - u_b$.

Lastly, it is not difficult to verify the existence of a function

$$g_0(x) = 2\left( |a_8| + |a_9 \sin x_4 + a_{10}||x_3| + \frac{\frac{a_6}{4(a_5+a_6x_4)^{1.5}}}{2\left(a_4 - \frac{a_6}{2\sqrt{a_5+a_6x_4}}\right)} \right) > 0,$$
$$\forall x_3 \in R^+, \quad x_4 \in [0, 0.44], \tag{4.114}$$

which fulfils Assumption 4.5 for the case of $g(x) < 0$. Note that this function need not be known; we only need to show its existence.

The control parameters are chosen as $k_1 = 2.0$, $k_2 = 3.0$, $k_2 = 4.5$, $k_4 = 5.5$, $k_b = 0.6$, while the NN parameters are $\mu = 0.01$, $\Gamma_W = 50I$, $\Gamma_V = 20.4I$, $\sigma_V = 0.055$ and $\sigma_W = 0.05$. For the high gain observer, we choose $\epsilon = 5 \times 10^{-4}$, $\gamma_1 = 4$, $\gamma_2 = 6$, $\gamma_3 = 4$, $\bar{\pi}_2 = 5 \times 10^{-4}$, $\bar{\pi}_3 = 5 \times 10^{-8}$, and $\bar{\pi}_4 = 1 \times 10^{-11}$. The saturation limits of the control are $\pm 400$ mrad. The initial conditions are $x(0) = [5.2,\ 0,\ 95.36,\ 0,\ 0]^{\mathrm{T}}$, $\hat{W} = 0$, and $\hat{V} = 0$.

From Fig. 4.6, it can be seen that good tracking performance is achieved by the proposed adaptive NN control. The tracking performance for the full-state and output feedback cases are similar for the choice of $\epsilon$ made. The initial error is efficiently reduced and the altitude trajectory lies in close proximity of the desired sinusoidal trajectory. We compare the performance of the NN controller with a linear PD controller

$$u_{pd} = K_p(y - y_d) + K_d(\dot{y} - \dot{y}_d), \tag{4.115}$$

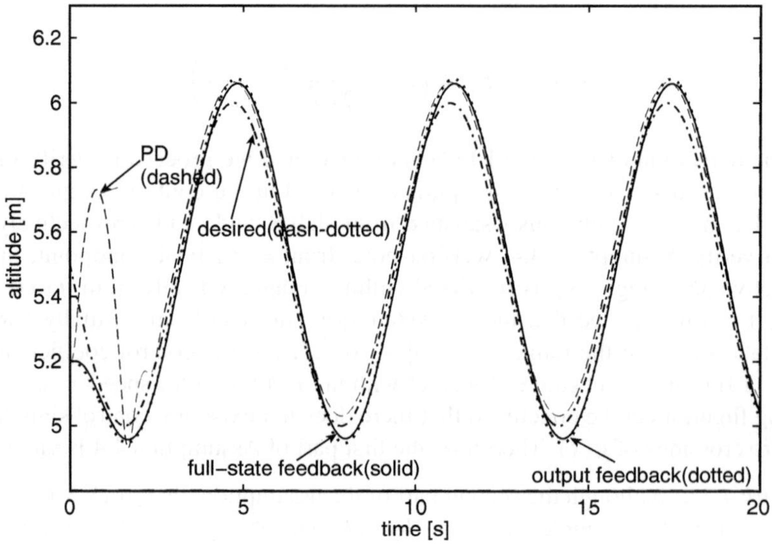


**Fig. 4.6** Comparison of tracking performance between adaptive NN and PD control for nonlinear helicopter model

where $K_p = 5{,}000$ rad and $K_d = 500$ rad s are chosen so that the tracking errors are reasonably small and the control magnitude is constrained to $|u_{pd}| \leq 400$ mrad. Although steady state errors are comparable between PD and NN control, the PD control gives poorer transient performance as it attempts to compensate for the initial error, due to the inability of the linear PD control to adequately compensate for the effects of nonlinearity and coupling. Clearly, a dynamic model compensator is essential to achieve better performance.

The boundedness of the control input and the neural weights, for full-state and output feedback NN control, as well as the PD control, are shown in Fig. 4.7. The size of input signal under PD control is much larger than that under NN control, as seen by the fact that the PD control signal initially fluctuates between the saturation limits. This can be explained by the fact that a large PD control gain is required to compensate for nonlinearities, thus amplifying the control effort greatly when the initial error is large.

In Fig. 4.8, it is shown that the rotor speed and collective pitch angle, for both full-state and output feedback NN control, are bounded. In particular, it is confirmed that the collective pitch angle remains in the region $[0, 0.44]$ rad as restricted in practical operations.

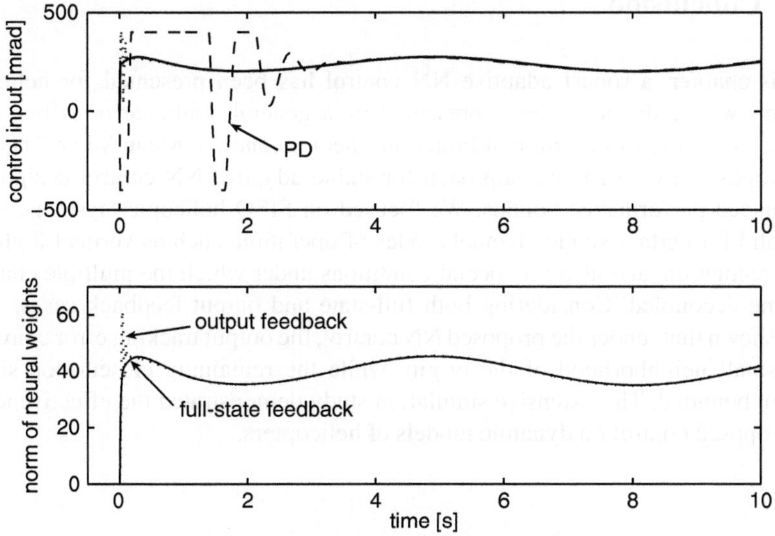


**Fig. 4.7** *Top*: Control signals under adaptive NN and PD control. *Bottom*: Norm of neural weights, given by $\sqrt{\|\hat{W}^{\mathrm{T}}\|^2 + \|\hat{V}\|_F^2}$, under full-state and output feedback NN control

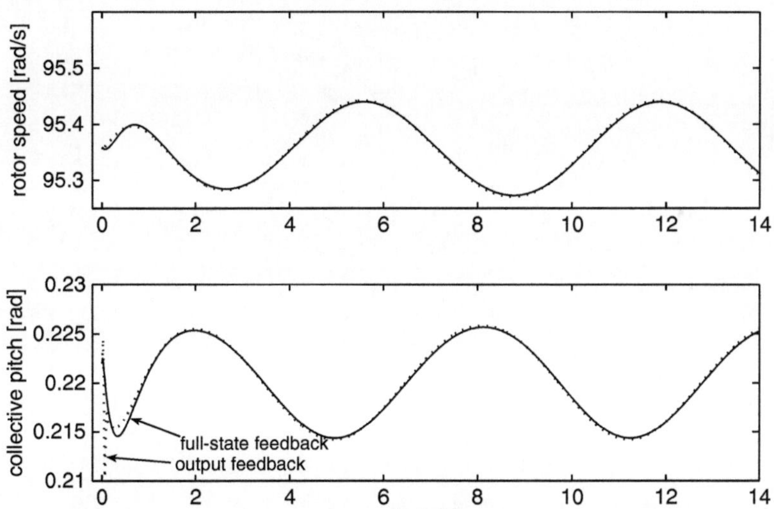


**Fig. 4.8** *Top*: Speed of rotor. *Bottom*: Collective pitch angle

## 4.6 Conclusion

In this chapter, a robust adaptive NN control has been presented for helicopter systems whose dynamics are represented by a general nonlinear nonaffine form. Based on the use of the Implicit Function Theorem and the Mean Value Theorem, we proposed a constructive approach for stable adaptive NN control design with guaranteed performance bounds. We focused on SISO helicopter systems, which are valid for certain single-channel modes of operation, such as vertical flight and pitch regulation, and also for special conditions under which the multiple channels become decoupled. Considering both full-state and output feedback cases, it has been shown that, under the proposed NN control, the output tracking error converges to a small neighborhood of the origin, while the remaining closed-loop signals remain bounded. The extensive simulation study demonstrated the effectiveness of the proposed control on dynamic models of helicopters.

# Chapter 5
# Altitude and Yaw Control of Helicopters with Uncertain Dynamics

## 5.1 Introduction

In Chap. 4, a robust adaptive neural network (NN) control is presented for helicopters in vertical flight, with dynamics in single-input single-output (SISO) nonlinear nonaffine form. By limiting the scope to the vertical flight regime, SISO models can be used to yield useful results, since the coupling between longitudinal and lateral-directional equations in this flight regime is weak [84]. While the proposed controller handles vertical flight, other flight regimes can be handled by other control modules. Evidently, SISO control designs have limited practical use, and many more investigations are needed in the control of multi-input multi-output (MIMO) helicopter dynamics for generality in applications.

Practical helicopter motion governed by a MIMO model has an underactuated configuration, i.e., the number of control inputs is less than the number of degrees of freedom to be stabilized, which makes it difficult to apply the conventional robotics approach for controlling Euler–Lagrange systems. Thus, some flight control techniques need to be further developed for the nonlinear MIMO helicopter dynamics. In [104], model-based control was applied to an autonomous scale MIMO model helicopter mounted in a 2-degree-of-freedom (2DOF) platform. Since helicopter control applications are characterized by unknown aerodynamical disturbances, they are generally difficult to model accurately. The presence of modeling errors, in the form of parametric and functional uncertainties, and unmodeled dynamics and disturbances from the environment, is a common problem. In this context, model-based control, such as the aforementioned schemes, tends to be susceptible to uncertainties and disturbances that cause performance degradation.

In this chapter, altitude and yaw angle tracking are considered for a scale MIMO model helicopter [104] in the presence of model uncertainties, which may be caused by unmodeled dynamics, sensor errors or aerodynamical disturbances from the environment. To deal with the presence of model uncertainties, approximation-based techniques using a NN have been proposed. In particular, two commonly used NNs, namely the multilayer neural network (MNN) and the radial basis

B. Ren et al., *Modeling, Control and Coordination of Helicopter Systems*,
DOI 10.1007/978-1-4614-1563-3_5, © Springer Science+Business Media, LLC 2012

function neural network (RBFNN) are adopted in control design and stability analysis. Based on Lyapunov synthesis, the proposed adaptive NN control ensures that both the altitude and the yaw angle track the given bounded reference signals to a small neighborhood of zero, and guarantees the Semi-Globally Uniformly Ultimate Boundedness (SGUUB) of all the closed-loop signals at the same time. The effectiveness of the proposed control is illustrated through extensive simulations. Compared with the model-based control used in [104], approximation-based control using NN, proposed in this chapter, can accommodate the presence of model uncertainties, reduce the dependence on accurate model building, and thus lead to the tracking performance improvement.

## 5.2 Problem Formulation and Preliminaries

We consider the VARIO scale model helicopter [104], however, the functions and parameters involved in the model are unknown. For clarity, we restate the helicopter dynamics here, which are described by Lagrangian formulation in the following:

$$D(q)\ddot{q} + C(q,\dot{q})\dot{q} + F(\dot{q}) + G(q) + \Delta(q,\dot{q}) = B(\dot{q})\tau \tag{5.1}$$

where $q, \dot{q}$, and $\ddot{q}$ are referred as the vectors of generalized coordinates, generalized velocities, and generalized accelerations, respectively. In particular, $q = [q_1, q_2, q_3]^{\mathrm{T}} = [z, \phi, \gamma]^{\mathrm{T}}$ with $z$ as the attitude ($z > 0$ downwards), $\phi$ as the yaw angle, and $\gamma$ as the main rotor azimuth angle; $\dot{q} = [\dot{q}_1, \dot{q}_2, \dot{q}_3]^{\mathrm{T}} = [\dot{z}, \dot{\phi}, \dot{\gamma}]^{\mathrm{T}}$ with $\dot{z}$ as the vertical velocity, $\dot{\phi}$ as the yaw rate, and $\dot{\gamma}$ as the main rotor angular velocity; $\ddot{q} = [\ddot{q}_1, \ddot{q}_2, \ddot{q}_3]^{\mathrm{T}} = [\ddot{z}, \ddot{\phi}, \ddot{\gamma}]^{\mathrm{T}}$ with $\ddot{z}$ as the vertical acceleration, $\ddot{\phi}$ as the yaw acceleration, and $\ddot{\gamma}$ as the main rotor angular acceleration; $D(q) \in R^{3\times 3}$ is the inertia matrix; $C(q,\dot{q})\dot{q} \in R^3$ is the vector of Coriolis and centrifugal forces; $F(\dot{q}) \in R^3$ is the vector of friction forces; $G(q) \in R^3$ is the vector of gravitational forces; $\Delta(q,\dot{q}) \in R^3$ is the vector of the model uncertainties, which may be caused by unmodeled dynamics, sensor errors or aerodynamical disturbances from the environment; $B(\dot{q}) \in R^{3\times 2}$ is the matrix of control coefficients; and the control inputs $\tau = [\tau_1, \tau_2]^{\mathrm{T}} \in R^2$ are the main and tail rotor collectives (swash plate displacements), respectively. By exploiting the physical properties of the helicopter, e.g., how the control inputs are distributed to the helicopter dynamics, or the coupling relationship between the states, better performance can be achieved. To this end, we assume partial knowledge of the structure of the dynamics [104], although the functions and parameters involved are unknown:

$$D(q) = \begin{bmatrix} d_{11} & 0 & 0 \\ 0 & d_{22}(q_3) & d_{23} \\ 0 & d_{23} & d_{33} \end{bmatrix} \quad C(q,\dot{q}) = \begin{bmatrix} 0 & 0 & 0 \\ 0 & c_{22}(q_3,\dot{q}_3) & c_{23}(q_3,\dot{q}_2) \\ 0 & c_{32}(q_3,\dot{q}_2) & 0 \end{bmatrix}$$

$$F(\dot{q}) = \begin{bmatrix} f_1(\dot{q}_3) \\ 0 \\ f_3(\dot{q}_3) \end{bmatrix} \quad G(q) = \begin{bmatrix} g_1 \\ 0 \\ g_3 \end{bmatrix} \quad \Delta(q,\dot{q}) = \begin{bmatrix} \Delta_1(q,\dot{q}) \\ \Delta_2(q,\dot{q}) \\ \Delta_3(q,\dot{q}) \end{bmatrix}$$

$$B(\dot{q}) = \begin{bmatrix} b_{11}(\dot{q}_3) & 0 \\ 0 & b_{22}(\dot{q}_3) \\ b_{31}(\dot{q}_3) & 0 \end{bmatrix} \tag{5.2}$$

where $d_{11}$, $d_{23}$, $d_{33}$, $g_1$, $g_3$ are unknown constants, $d_{22}(q_3)$, $c_{22}(q_3, \dot{q}_3)$, $c_{23}(q_3, \dot{q}_2)$, $c_{32}(q_3, \dot{q}_2)$, $f_1(\dot{q}_3)$, $f_3(\dot{q}_3)$, $b_{11}(\dot{q}_3)$, $b_{22}(\dot{q}_3)$, $b_{31}(\dot{q}_3)$, $\Delta_1(q, \dot{q})$, $\Delta_2(q, \dot{q})$ and $\Delta_3(q, \dot{q})$ are unknown functions.

To facilitate control design in Sect. 5.3, the following assumptions are in order:

**Assumption 5.1.** The terms $d_{11}$ and $\frac{d_{22}(q_3)d_{33} - d_{23}^2}{2d_{33}}$ are positive.

**Assumption 5.2.** The following equation $\dot{d}_{22}(q_3) - 2c_{22}(q_3, \dot{q}_3) = 0$ holds.

*Remark 5.1.* It is easy to know that the helicopter model in (5.1) with the parameters given in [104], which will be used in the subsequent simulation section, satisfies both Assumptions 5.1 and 5.2.

**Assumption 5.3.** The signs of $b_{11}(\dot{q}_3)$ and $b_{22}(\dot{q}_3)$ are known. Without losing generality, assume that $b_{11}(\dot{q}_3)$ is positive and $b_{22}(\dot{q}_3)$ is negative. There exist positive constants $\underline{b}_{11}$ and $\underline{b}_{22}$, such that $0 \leq \underline{b}_{11} \leq |b_{11}(\dot{q}_3)|$ and $0 \leq \underline{b}_{22} \leq |b_{22}(\dot{q}_3)|$.

*Remark 5.2.* In this section, the vertical flight mode after take-off is considered. From physical analysis, to lift the helicopter up for flight operation, $|\dot{q}_3|$ has to be larger than some certain positive value (e.g., $c_0$) to overcome the gravity. It is noted that in the specific helicopter model given in (5.1), $b_{11}(\dot{q}_3) = 3.411\dot{q}_3^2 \geq 3.411c_0^2 > 0$. Therefore, there always exist some positive constants $\underline{b}_{11}$ such that $0 \leq \underline{b}_{11} \leq |b_{11}(\dot{q}_3)|$ during the vertical flight mode. Similar analysis can be applied to $b_{22}(\dot{q}_3)$ as in Assumption 5.3.

**Assumption 5.4.** There exist positive constants $\underline{d}_{22}$ and $\bar{d}_{22}$, such that $\underline{d}_{22} \leq |d_{22}(q_3)| \leq \bar{d}_{22}$.

*Remark 5.3.* Assumption 5.4 is reasonable due to $d_{22}(q_3) = 0.4305 + 0.0003 \cos^2(-4.143q_3)$ in the specific helicopter model given in (5.1), which will be used in the subsequent simulation section.

The control objective is to ensure that the tracking errors for the altitude $q_1(t)$ and yaw angle $q_2(t)$ from their respective desired trajectories $q_{1d}(t)$ and $q_{2d}(t)$, are driven to a small neighborhood of zero, i.e., $|q_i(t) - q_{id}(t)| \leq \epsilon_i$, where $\epsilon_i > 0$, $i = 1, 2$; at the same time, the main rotor angular velocity $\dot{q}_3(t)$ is stable.

**Assumption 5.5.** The desired trajectories $q_{1d}(t)$ and $q_{2d}(t)$ and their time derivatives up to the 3rd order are continuously differentiable and bounded for all $t \geq 0$.

The following technical lemma is required in the subsequent control design and stability analysis.

**Lemma 5.4.** *For $a, b \in R^+$, the following inequality holds*

$$\frac{ab}{a+b} \leq a \tag{5.3}$$

## 5.3 Control Design

In this section, we will design an adaptive neural control to accommodate the presence of uncertainties in the dynamics (5.1), appearing in the functions $D(q)$, $C(q,\dot{q})$, $F(\dot{q})$, $G(q)$, $\Delta(q,\dot{q})$ and $B(\dot{q})$. After some simple manipulations on (5.1) and (5.2), we can obtain three subsystems: $q_1$-subsystem (5.4), $q_2$-subsystem (5.5) and $q_3$-subsystem (5.6) as follows:

$$d_{11}\ddot{q}_1 + f_1(\dot{q}_3) + g_1 + \Delta_1(q,\dot{q}) = b_{11}(\dot{q}_3)\tau_1 \tag{5.4}$$

$$\begin{aligned} &\frac{d_{22}(q_3)d_{33} - d_{23}^2}{d_{33}}\ddot{q}_2 + c_{22}(q_3,\dot{q}_3)\dot{q}_2 + c_{23}(q_3,\dot{q}_2)\dot{q}_3 + \Delta_2(q,\dot{q}) + \frac{d_{23}}{d_{33}}(b_{31}(\dot{q}_3)\tau_1 \\ &-c_{32}(q_3,\dot{q}_2)\dot{q}_2 - f_3(\dot{q}) - g_3 - \Delta_3(q,\dot{q})) = b_{22}(\dot{q}_3)\tau_2 \end{aligned} \tag{5.5}$$

$$\begin{aligned} &\frac{d_{22}(q_3)d_{33} - d_{23}^2}{d_{22}(q_3)}\ddot{q}_3 + c_{32}(q_3,\dot{q}_2)\dot{q}_2 + f_3(\dot{q}_3) + g_3 + \Delta_3(q,\dot{q}) \\ &+\frac{d_{23}}{d_{22}(q_3)}(b_{22}(\dot{q}_3)\tau_2 - c_{22}(q_3,\dot{q}_3)\dot{q}_2 - c_{23}(q_3,\dot{q}_2)\dot{q}_3 - \Delta_2(q,\dot{q})) = b_{31}(\dot{q}_3)\tau_1 \end{aligned} \tag{5.6}$$

In the following, we will analyze and design a control for each subsystem. For clarity, define the tracking errors and the filtered tracking errors as

$$e_i = q_i - q_{id}, \quad r_i = \dot{e}_i + \lambda_i e_i \tag{5.7}$$

where $\lambda_i$ is a positive number, $i = 1, 2$. Then, the boundedness of $r_i$ guarantees the boundedness of $e_i$ and $\dot{e}_i$ [10, 71–94]. To study the stability of $e_i$ and $\dot{e}_i$, we only need to study the properties of $r_i$. In addition, the following computable signals are defined:

$$\dot{q}_{ir} = \dot{q}_{id} - \lambda_i e_i, \quad \ddot{q}_{ir} = \ddot{q}_{id} - \lambda_i \dot{e}_i$$

### 5.3.1 RBFNN-Based Control

In this section, we will investigate the RBFNN based control design by Lyapunov synthesis to achieve the control objective. Regarding to the obtained three subsystems (5.4)–(5.6), our control design consists of three steps: First, we will design control $\tau_1$ based on the $q_1$-subsystem (5.4); Second, design $\tau_2$ based on the $q_2$-subsystem (5.5) and $\tau_1$; finally, analyze the stability of the internal dynamics of $q_3$-subsystem (5.6).

□ *$q_1$-subsystem*

Since $\dot{q}_1 = \dot{q}_{1r} + r_1$, $\ddot{q}_1 = \ddot{q}_{1r} + \dot{r}_1$, (5.4) becomes

$$d_{11}\dot{r}_1 = b_{11}(\dot{q}_3)\tau_1 - f_{S1,1} \tag{5.8}$$

where

$$f_{S1,1} = d_{11}\ddot{q}_{1r} + f_1(\dot{q}_3) + g_1 + \Delta_1(q,\dot{q}) \tag{5.9}$$

is an unknown continuous function, which is approximated by RBFNN to arbitrarily any accuracy as

$$f_{S1,1} = W_1^{*\mathrm{T}} S_1(Z_1) + \varepsilon_1(Z_1) \tag{5.10}$$

where the input vector $Z_1 = [q_1,\ \dot{q}_1,\ q_2,\ \dot{q}_2,\ q_3,\ \dot{q}_3,\ \dot{q}_{1d},\ \ddot{q}_{1d}]^{\mathrm{T}} \in \Omega_{Z1} \subset R^8$; $\varepsilon_1(Z_1)$ is the approximation error satisfying $|\varepsilon_1(Z_1)| \le \bar{\varepsilon}_1$, where $\bar{\varepsilon}_1$ is a positive constant; $W_1^*$ are ideal constant weights satisfying $\|W_1^*\| \le w_{1m}$, where $w_{1m}$ is a positive constant; and $S_1(Z_1)$ are the basis functions. By using $\hat{W}_1$ to approximate $W_1^*$, the error between the actual and the ideal RBFNNs can be expressed as

$$\hat{W}_1^{\mathrm{T}} S_1(Z_1) - W_1^{*\mathrm{T}} S_1(Z_1) = \tilde{W}_1^{\mathrm{T}} S_1(Z_1) \tag{5.11}$$

where $\tilde{W}_1 = \hat{W}_1 - W_1^*$.

Consider the following Lyapunov function candidate

$$V_1 = \frac{1}{2} d_{11} r_1^2 + \frac{1}{2} \tilde{W}_1^{\mathrm{T}} \Gamma_1^{-1} \tilde{W}_1 \tag{5.12}$$

The time derivative of (5.12) along (5.8) and (5.10) is given by

$$\begin{aligned} \dot{V}_1 &= d_{11} r_1 \dot{r}_1 + \tilde{W}_1^{\mathrm{T}} \Gamma_1^{-1} \dot{\hat{W}}_1 \\ &= r_1 \left[ b_{11}(\dot{q}_3)\tau_1 - W_1^{*\mathrm{T}} S_1(Z_1) - \varepsilon_1(Z_1) \right] + \tilde{W}_1^{\mathrm{T}} \Gamma_1^{-1} \dot{\hat{W}}_1 \end{aligned} \tag{5.13}$$

As $W_1^*$ is a constant vector, we know that $\dot{\tilde{W}}_1 = \dot{\hat{W}}_1$. Therefore, (5.13) becomes

$$\dot{V}_1 = r_1 \left[b_{11}(\dot{q}_3)\tau_1 - W_1^{*\mathrm{T}} S_1(Z_1) - \varepsilon_1(Z_1)\right] + \tilde{W}_1^{\mathrm{T}} \Gamma_1^{-1} \dot{\hat{W}}_1 \tag{5.14}$$

Consider the following RBFNN-based control law and RBFNN weight adaptation law:

$$\tau_1 = -k_1 r_1 - \frac{r_1 \left(\hat{W}_1^{\mathrm{T}} S_1(Z_1)\right)^2}{\underline{b}_{11}\left(|r_1 \hat{W}_1^{\mathrm{T}} S_1(Z_1)| + \delta_1\right)} \tag{5.15}$$

$$\dot{\hat{W}}_1 = -\Gamma_1 \left[S_1(Z_1) r_1 + \sigma_1 \hat{W}_1\right] \tag{5.16}$$

where $k_1 > 0$, $\delta_1 > 0$, $\Gamma_1 = \Gamma_1^{\mathrm{T}} > 0$, and $\sigma_1 > 0$.

*Remark 5.5.* The above $\sigma$-modification adaptation law (5.16) can be replaced by $e$-modification adaptation law like $\dot{\hat{W}}_1 = -\Gamma_1 \left[S_1(Z_1) r_1 + \sigma_1 |r_1| \hat{W}_1\right]$ easily. The control design based on $\sigma$-modification adaptation law in this chapter can be extended to the case based on $e$-modification adaptation law without any difficulty.

Substituting (5.15) and (5.16) into (5.14), we have

$$\dot{V}_1 = -k_1 b_{11}(\dot{q}_3) r_1^2 - \frac{b_{11}(\dot{q}_3)}{\underline{b}_{11}} \frac{r_1^2 \left(\hat{W}_1^{\mathrm{T}} S_1(Z_1)\right)^2}{|r_1 \hat{W}_1^{\mathrm{T}} S_1(Z_1)| + \delta_1} - r_1 W_1^{*\mathrm{T}} S_1(Z_1) - r_1 \varepsilon_1(Z_1)$$
$$-r_1 \tilde{W}_1^{\mathrm{T}} S_1(Z_1) - \sigma_1 \tilde{W}_1^{\mathrm{T}} \hat{W}_1 \tag{5.17}$$

According to Assumption 5.3 and (5.11), we can rewrite (5.17) as

$$\begin{aligned} \dot{V}_1 &\leq -k_1 \underline{b}_{11} r_1^2 - \frac{r_1^2 \left(\hat{W}_1^{\mathrm{T}} S_1(Z_1)\right)^2}{|r_1 \hat{W}_1^{\mathrm{T}} S_1(Z_1)| + \delta_1} - r_1 \hat{W}_1^{\mathrm{T}} S_1(Z_1) - r_1 \varepsilon_1(Z_1) \\ &\quad -\sigma_1 \tilde{W}_1^{\mathrm{T}} \hat{W}_1 \\ &\leq -k_1 \underline{b}_{11} r_1^2 - \frac{r_1^2 \left(\hat{W}_1^{\mathrm{T}} S_1(Z_1)\right)^2}{|r_1 \hat{W}_1^{\mathrm{T}} S_1(Z_1)| + \delta_1} + |r_1 \hat{W}_1^{\mathrm{T}} S_1(Z_1)| + |r_1||\varepsilon_1(Z_1)| \\ &\quad -\sigma_1 \tilde{W}_1^{\mathrm{T}} \hat{W}_1 \end{aligned} \tag{5.18}$$

Noting that

$$-\frac{r_1^2 \left(\hat{W}_1^{\mathrm{T}} S_1(Z_1)\right)^2}{|r_1 \hat{W}_1^{\mathrm{T}} S_1(Z_1)| + \delta_1} + |r_1 \hat{W}_1^{\mathrm{T}} S_1(Z_1)| = \frac{|r_1 \hat{W}_1^{\mathrm{T}} S_1(Z_1)| \delta_1}{|r_1 \hat{W}_1^{\mathrm{T}} S_1(Z_1)| + \delta_1} \tag{5.19}$$

According to Lemma 5.4, we can obtain from (5.19) that

$$-\frac{r_1^2\left(\hat{W}_1^{\mathrm{T}} S_1(Z_1)\right)^2}{|r_1 \hat{W}_1^{\mathrm{T}} S_1(Z_1)| + \delta_1} + |r_1 \hat{W}_1^{\mathrm{T}} S_1(Z_1)| \le \delta_1 \tag{5.20}$$

By completion of squares and using Young's inequality, the following inequalities hold:

$$-\sigma_1 \tilde{W}_1^{\mathrm{T}} \hat{W}_1 \le -\frac{\sigma_1}{2}\|\tilde{W}_1\|^2 + \frac{\sigma_1}{2}\|W_1^*\|^2 \tag{5.21}$$

$$|r_1||\varepsilon_1(Z_1)| \le \frac{r_1^2}{2c_1} + \frac{c_1\varepsilon_1^2(Z_1)}{2} \le \frac{r_1^2}{2c_1} + \frac{c_1\bar{\varepsilon}_1^2}{2} \tag{5.22}$$

where $c_1$ is a positive constant. Substituting the above inequalities (5.20)–(5.22) into (5.18) leads to

$$\begin{aligned} \dot{V}_1 &\le -\left(k_1\underline{b}_{11} - \frac{1}{2c_1}\right) r_1^2 - \frac{\sigma_1}{2}\|\tilde{W}_1\|^2 + \delta_1 + \frac{c_1}{2}\bar{\varepsilon}_1^2 + \frac{\sigma_1}{2} w_{1m}^2 \\ &\le -\lambda_{10} V_1 + \mu_{10} \end{aligned} \tag{5.23}$$

where $\lambda_{10} = \min\left\{(2k_1\underline{b}_{11} - 1/c_1)/d_{11}, \sigma_1/\lambda_{\max}(\Gamma_1^{-1})\right\}$, $\mu_{10} = \delta_1 + \frac{c_1}{2}\bar{\varepsilon}_1^2 + \frac{\sigma_1}{2} w_{1m}^2$.

□ *$q_2$-subsystem*

Similar to Sect. 5.3.1, since $\dot{q}_2 = \dot{q}_{2r} + r_2$, $\ddot{q}_2 = \ddot{q}_{2r} + \dot{r}_2$, (5.5) becomes

$$\frac{d_{22}(q_3)d_{33} - d_{23}^2}{d_{33}} \dot{r}_2 + c_{22}(q_3, \dot{q}_3) r_2 = b_{22}(\dot{q}_3)\tau_2 - f_{S2,1} \tag{5.24}$$

where

$$\begin{aligned} f_{S2,1} = {} & \frac{d_{22}(q_3)d_{33} - d_{23}^2}{d_{33}} \ddot{q}_{2r} + c_{22}(q_3, \dot{q}_3)\dot{q}_{2r} + c_{23}(q_3, \dot{q}_2)\dot{q}_3 + \Delta_2(q, \dot{q}) \\ & + \frac{d_{23}}{d_{33}}(b_{31}(\dot{q}_3)\tau_1 - c_{32}(q_3, \dot{q}_2)\dot{q}_2 - f_3(\dot{q}_3) - g_3 - \Delta_3(q, \dot{q})) \end{aligned}$$

is an unknown function, which is approximated by RBFNN to arbitrarily any accuracy as

$$f_{S2,1} = W_2^{*\mathrm{T}} S_2(Z_2) + \varepsilon_2(Z_2) \tag{5.25}$$

where the input vector $Z_2 = [\tau_1, q_1, \dot{q}_1, q_2, \dot{q}_2, q_3, \dot{q}_3, q_{2d}, \dot{q}_{2d}, \ddot{q}_{2d}]^{\mathrm{T}} \in \Omega_{Z_2} \subset R^{10}$, $\varepsilon_2(Z_2)$ is the approximation error satisfying $|\varepsilon_2(Z_2)| \le \bar{\varepsilon}_2$, where $\bar{\varepsilon}_2$ is an unknown positive constant; $W_2^*$ are unknown ideal constant weights satisfying $\|W_2^*\| \le w_{2m}$, where $w_{2m}$ is an unknown positive constant; and $S_2(Z_2)$ are the basis functions. By using $\hat{W}_2$ to approximate $W_2^*$, the error between the actual and the ideal RBFNNs can be expressed as

$$\hat{W}_2^{\mathrm{T}} S_2(Z_2) - W_2^{*\mathrm{T}} S_2(Z_2) = \tilde{W}_2^{\mathrm{T}} S_2(Z_2) \tag{5.26}$$

where $\tilde{W}_2 = \hat{W}_2 - W_2^*$.

To analyze the closed loop stability for the $q_2$-subsystem, let

$$V_2 = \frac{1}{2}\frac{d_{22}(q_3)d_{33} - d_{23}^2}{d_{33}} r_2^2 + \frac{1}{2}\tilde{W}_2^{\mathrm{T}} \Gamma_2^{-1} \tilde{W}_2 \tag{5.27}$$

**Lemma 5.6.** *The function $V_2$ (5.27) is positive definite and decrescent, in the sense that there exist two time-invariant positive definite functions $\underline{V}_2(r_2, \tilde{W}_2)$ and $\bar{V}_2(r_2, \tilde{W}_2)$, such that*

$$\underline{V}_2(r_2, \tilde{W}_2) \le V_2 \le \bar{V}_2(r_2, \tilde{W}_2)$$

*Proof.* Noting that the particular choice of $V_2$ in (5.27), a function of $r_2$, $\tilde{W}_2$ and $d_{22}(q_3)$, is to establish the stability for $r_2$ and $\tilde{W}_2$ only, therefore, we regard $d_{22}(q_3)$ as a function of time. From Assumptions 5.1 and 5.4, we know that

$$0 < \frac{\left|\underline{d}_{22}|d_{33}| - d_{23}^2\right|}{|d_{33}|} < \left|\frac{d_{22}(q_3)d_{33} - d_{23}^2}{d_{33}}\right| \le \frac{\bar{d}_{22}|d_{33}| + d_{23}^2}{|d_{33}|} \tag{5.28}$$

Therefore, there also exist time-invariant positive definite functions $\underline{V}_2(r_2, \tilde{W}_2)$ and $\bar{V}_2(r_2, \tilde{W}_2)$, such that $\underline{V}_2(r_2, \tilde{W}_2) \le V_2 \le \bar{V}_2(r_2, \tilde{W}_2)$, which implies that $V_2$ is also positive definite and decrescent, according to [94]. This completes the proof. □

The time derivative of (5.27) is given as

$$\dot{V}_2 = \frac{1}{2}\dot{d}_{22}(q_3) r_2^2 + \frac{d_{22}(q_3)d_{33} - d_{23}^2}{d_{33}} r_2 \dot{r}_2 + \tilde{W}_2^{\mathrm{T}} \Gamma_2^{-1} \dot{\tilde{W}}_2 \tag{5.29}$$

According to Assumption 5.2, (5.29) becomes

$$\dot{V}_2 = r_2 \left[ \frac{d_{22}(q_3)d_{33} - d_{23}^2}{d_{33}} \dot{r}_2 + c_{22}(q_3, \dot{q}_3) r_2 \right] + \tilde{W}_2^{\mathrm{T}} \Gamma_2^{-1} \dot{\tilde{W}}_2 \tag{5.30}$$

As $W_2^*$ is a constant vector, it is easy to obtain that

$$\dot{\tilde{W}}_2 = \dot{\hat{W}}_2 \tag{5.31}$$

Substituting (5.24), (5.25) and (5.31) into (5.30), we have

$$\dot{V}_2 = r_2\left[b_{22}(\dot{q}_3)\tau_2 - W_2^{*\mathrm{T}} S_2(Z_2) - \varepsilon_2(Z_2)\right] + \tilde{W}_2^{\mathrm{T}} \Gamma_2^{-1} \dot{\hat{W}}_2 \tag{5.32}$$

Consider the following RBFNN-based control law and RBFNN weight adaption law:

$$\tau_2 = k_2 r_2 + \frac{r_2\left(\hat{W}_2^{\mathrm{T}} S_2(Z_2)\right)^2}{\underline{b}_{22}\left(|r_2 \hat{W}_2^{\mathrm{T}} S_2(Z_2)| + \delta_2\right)} \tag{5.33}$$

$$\dot{\hat{W}}_2 = -\Gamma_2\left[S_2(Z_2) r_2 + \sigma_2 \hat{W}_2\right] \tag{5.34}$$

where $k_2 > 0$, $\delta_2 > 0$, $\Gamma_2 = \Gamma_2^{\mathrm{T}} > 0$ and $\sigma_2 > 0$. Substituting (5.33) and (5.34) into (5.32), we have

$$\begin{aligned}
\dot{V}_2 &= k_2 b_{22}(\dot{q}_3) r_1^2 + \frac{b_{22}(\dot{q}_3)}{\underline{b}_{22}} \frac{r_2^2\left(\hat{W}_2^{\mathrm{T}} S_2(Z_2)\right)^2}{|r_2 \hat{W}_2^{\mathrm{T}} S_2(Z_2)| + \delta_2} - r_2 W_2^{*\mathrm{T}} S_2(Z_2) - r_2 \varepsilon_2(Z_2) \\
&\quad - r_2 \tilde{W}_2^{\mathrm{T}} S_2(Z_2) - \sigma_2 \tilde{W}_2^{\mathrm{T}} \hat{W}_2
\end{aligned} \tag{5.35}$$

According to Assumption 5.3 and (5.26), we can rewrite (5.35) as

$$\begin{aligned}
\dot{V}_2 &\le -k_2 \underline{b}_{22} r_2^2 - \frac{r_2^2\left(\hat{W}_2^{\mathrm{T}} S_2(Z_2)\right)^2}{|r_2 \hat{W}_2^{\mathrm{T}} S_2(Z_2)| + \delta_2} - r_2 \hat{W}_2^{\mathrm{T}} S_2(Z_2) - r_2 \varepsilon_2(Z_2) \\
&\quad - \sigma_2 \tilde{W}_2^{\mathrm{T}} \hat{W}_2 \\
&\le -k_2 \underline{b}_{22} r_2^2 - \frac{r_2^2\left(\hat{W}_2^{\mathrm{T}} S_2(Z_2)\right)^2}{|r_2 \hat{W}_2^{\mathrm{T}} S_2(Z_2)| + \delta_2} + |r_2 \hat{W}_2^{\mathrm{T}} S_2(Z_2)| + |r_2||\varepsilon_2(Z_2)| \\
&\quad - \sigma_2 \tilde{W}_2^{\mathrm{T}} \hat{W}_2
\end{aligned} \tag{5.36}$$

Similar to (5.20), we have

$$-\frac{r_2^2\left(\hat{W}_2^{\mathrm{T}} S_2(Z_2)\right)^2}{|r_2 \hat{W}_2^{\mathrm{T}} S_2(Z_2)| + \delta_2} + |r_2 \hat{W}_2^{\mathrm{T}} S_2(Z_2)| \le \delta_2 \tag{5.37}$$

By completion of squares and using Young's inequality, the following inequalities hold:

$$-\sigma_2 \tilde{W}_2^{\mathrm{T}} \hat{W}_2 \leq -\frac{\sigma_2}{2} \|\tilde{W}_2\|^2 + \frac{\sigma_2}{2} \|W_2^*\|^2 \tag{5.38}$$

$$|r_2||\varepsilon_2(Z_2)| \leq \frac{r_2^2}{2c_2} + \frac{c_2 \varepsilon_2^2(Z_2)}{2} \leq \frac{r_2^2}{2c_2} + \frac{c_2 \bar{\varepsilon}_2^2}{2} \tag{5.39}$$

where $c_2$ is a positive constant. Substituting the above inequalities (5.37)–(5.39) into (5.36) leads to

$$\begin{aligned} \dot{V}_2 &\leq -\left(k_2 \underline{b}_{22} - \frac{1}{2c_2}\right) r_2^2 - \frac{\sigma_2}{2} \|\tilde{W}_2\|^2 + \delta_2 + \frac{c_2}{2} \bar{\varepsilon}_2^2 + \frac{\sigma_2}{2} w_{2m}^2 \\ &\leq -\lambda_{20} V_2 + \mu_{20} \end{aligned} \tag{5.40}$$

where $\lambda_{20} = \min\left\{(2k_2\underline{b}_{22} - 1/c_2)|d_{33}|/(\bar{d}_{22}|d_{33}| + d_{23}^2), \sigma_2/\lambda_{\max}(\Gamma_2^{-1})\right\}$, $\mu_{20} = \delta_2 + \frac{c_2}{2}\bar{\varepsilon}_2^2 + \frac{\sigma_2}{2} w_{2m}^2$.

□ *$q_3$-subsystem*

Finally, using the designed control laws (5.15) and (5.33), the $q_3$-subsystem (5.6) can be rewritten as

$$\dot{\eta} = \psi(\xi, \eta, u) \tag{5.41}$$

where $\eta = [q_3, \dot{q}_3]^{\mathrm{T}}$, $\xi = [q_1, q_2, \dot{q}_1, \dot{q}_2]^{\mathrm{T}}$, $u = [\tau_1, \tau_2]^{\mathrm{T}}$.

Then, the zero dynamics can be addressed as [35]

$$\dot{\eta} = \psi(0, \eta, u^*(0, \eta)) \tag{5.42}$$

where $u^* = [\tau_1^*, \tau_2^*]^{\mathrm{T}}$.

**Assumption 5.6.** [35] System (5.4)–(5.6) is hyperbolically minimum-phase, i.e., zero dynamics (5.42) is exponentially stable. In addition, assume that the control input $u$ is designed as a function of the states ($\xi$, $\eta$) and the reference signal satisfying Assumption 5.5, and the function $f(\xi, \eta, u)$ is Lipschitz in $\xi$, i.e., there exist constants $L_\xi$ and $L_f$ for $f(\xi, \eta, u)$ such that

$$\|f(\xi, \eta, u) - f(0, \eta, u_\eta)\| \leq L_\xi \|\xi\| + L_f \tag{5.43}$$

where $u_\eta = u^*(0, \eta)$.

Under Assumption 5.6, by the Converse Theorem of Lyapunov [52], there exists a Lyapunov function $V_0(\eta)$ which satisfies

$$\gamma_a \|\eta\|^2 \leq V_0(\eta) \leq \gamma_b \|\eta\|^2 \tag{5.44}$$

$$\frac{\partial V_0}{\partial \eta} f(0, \eta, u_\eta) \leq -\lambda_a \|\eta\|^2 \tag{5.45}$$

$$\|\frac{\partial V_0}{\partial \eta}\| \leq \lambda_b \|\eta\| \tag{5.46}$$

where $\gamma_a$, $\gamma_b$, $\lambda_a$ and $\lambda_b$ are positive constants.

**Lemma 5.7.** *[35] For the internal dynamics $\dot{\eta} = f(\xi, \eta, u)$ of the system, if Assumption 5.6 is satisfied, and the states $\xi$ are bounded by a positive constant $\|\xi\|_{\max}$, i.e., $\|\xi\| \leq \|\xi\|_{\max}$, then there exist positive constants $L_\eta$ and $T_0$, such that*

$$\|\eta(t)\| \leq L_\eta, \quad \forall t > T_0 \tag{5.47}$$

*Proof.* According to Assumption 5.6, there exists a Lyapunov function $V_0(\eta)$. Differentiating $V_0(\eta)$ along (5.4)–(5.6) yields

$$\begin{aligned}
\dot{V}_0(\eta) &= \frac{\partial V_0}{\partial \eta} f(\xi, \eta, u) \\
&= \frac{\partial V_0}{\partial \eta} f(0, \eta, u_\eta) + \frac{\partial V_0}{\partial \eta} \left[ f(\xi, \eta, u) - f(0, \eta, u_\eta) \right]
\end{aligned} \tag{5.48}$$

Noting (5.43)–(5.46), (5.48) can be written as

$$\begin{aligned}
\dot{V}_0(\eta) &\leq -\lambda_a \|\eta\|^2 + \lambda_b \|\eta\| (L_\xi \|\xi\| + L_f) \\
&\leq -\lambda_a \|\eta\|^2 + \lambda_b \|\eta\| (L_\xi \|\xi\|_{\max} + L_f)
\end{aligned}$$

Therefore, $\dot{V}_0(\eta) \leq 0$, whenever

$$\|\eta\| \geq \frac{\lambda_b}{\lambda_a} (L_\xi \|\xi\|_{\max} + L_f)$$

By letting $L_\eta = \frac{\lambda_b}{\lambda_a}(L_\xi \|\xi\|_{\max} + L_f)$, we conclude that there exists a positive constant $T_0$, such that (5.47) holds. □

The following Theorem shows the stability and control performance of the closed loop system.

**Theorem 5.8.** *Consider the closed-loop system consisting of the subsystems (5.4)–(5.6), the control laws (5.15), (5.33) and adaptation laws (5.16), (5.34). Under Assumptions 5.1–5.6, the overall closed-loop neural control system is Semi-Globally Uniformly Ultimately Bounded (SGUUB) in the sense that all of the signals in the closed-loop system are bounded, and the tracking errors and neural weights converge to the following regions,*

$$
\begin{aligned}
|e_1| &\le |e_1(0)| + \frac{1}{\lambda_1}\sqrt{\frac{2\mu_1}{d_{11}}} \\
\|\hat{W}_1\| &\le \sqrt{\frac{2\mu_1}{\lambda_{\min}(\Gamma_1^{-1})}} + w_{1m} \\
|e_2| &\le |e_2(0)| + \frac{1}{\lambda_2}\sqrt{\frac{2|d_{33}|\mu_2}{\left|\underline{d}_{22}|d_{33}| - d_{23}^2\right|}} \\
\|\hat{W}_2\| &\le \sqrt{\frac{2\mu_2}{\lambda_{\min}(\Gamma_2^{-1})}} + w_{2m}
\end{aligned}
\tag{5.49}
$$

*with*

$$
\begin{aligned}
\mu_i &= \frac{\mu_{i0}}{\lambda_{i0}} + V_i(0), \quad \mu_{i0} = \delta_i + \frac{1}{2}\bar{\varepsilon}_i^2 + \frac{\sigma_i}{2}w_{im}^2, \quad i = 1,2 \\
\lambda_{10} &= \min\left\{(2k_1\underline{b}_{11} - 1/c_1)/d_{11}, \sigma_1/\lambda_{\max}(\Gamma_1^{-1})\right\} \\
\lambda_{20} &= \min\left\{(2k_2 - 1/c_2)|d_{33}|/(\bar{d}_{22}|d_{33}| + d_{23}^2), \sigma_2/\lambda_{\max}(\Gamma_2^{-1})\right\}
\end{aligned}
$$

*where $e_i(0)$ and $V_i(0)$ are initial values of $e_i(t)$ and $V_i(t)$, respectively.*

*Proof.* Based on the previous analysis, the proof proceeds by studying each subsystem in order. First, the closed loop stability analysis of the $q_1$-subsystem (5.4) with control $\tau_1$ (5.15) and adaptation law (5.16) is made by use of Lyapunov synthesis. Second, the similar closed loop stability will be achieved on the $q_2$-subsystem (5.5) with $\tau_2$ (5.33) and adaptation law (5.34). Finally, the stability analysis of internal dynamics of the $q_3$-subsystem (5.6) is made based on the stability of the previous two subsystems.

*$q_1$- subsystem:*

Solving the inequality (5.23), we have $0 \le V_1(t) \le \mu_1$ with $\mu_1 = \frac{\mu_{10}}{\lambda_{10}} + V_1(0)$. Then, from the definition of $V_1(t)$ (5.12), we can obtain

$$
|r_1| \le \sqrt{\frac{2\mu_1}{d_{11}}}, \quad \|\tilde{W}_1\| \le \sqrt{\frac{2\mu_1}{\lambda_{\min}(\Gamma_1^{-1})}} \tag{5.50}
$$

Since $\dot{e}_1 = -\lambda_1 e_1 + r_1$, solving this equation results in

$$
e_1 = \mathrm{e}^{-\lambda_1 t} e_1(0) + \int_0^t \mathrm{e}^{-\lambda_1(t-\tau)} r_1 \mathrm{d}\tau \tag{5.51}
$$

According to (5.50) and (5.51), we have

$$|e_1| \le |e_1(0)| + \frac{1}{\lambda_1}\sqrt{\frac{2\mu_1}{d_{11}}} \tag{5.52}$$

Noting $q_1 = e_1 + q_{1d}$, $\hat{W}_1 = \tilde{W}_1 + W_1^*$, $\|W_1^*\| \le w_{1m}$ and Assumption 5.5, we obtain

$$|q_1| \le |e_1| + |q_{1d}| \le |e_1(0)| + \frac{1}{\lambda_1}\sqrt{\frac{2\mu_1}{d_{11}}} + |q_{1d}| \in L_\infty$$

$$\|\hat{W}_1\| \le \|\tilde{W}_1\| + \|W_1^*\| \le \sqrt{\frac{2\mu_1}{\lambda_{\min}(\Gamma_1^{-1})}} + w_{1m} \in L_\infty$$

Since the control $\tau_1$ is a function of $r_1$ and $\hat{W}_1$, its boundedness is also assured.

*$q_2$- subsystem:*

Similar to the analysis of $q_1$- subsystem, we have

$$|r_2| \le \sqrt{\frac{2|d_{33}|\mu_2}{\left|\underline{d}_{22}|d_{33}| - d_{23}^2\right|}}, \quad \|\tilde{W}_2\| \le \sqrt{\frac{2\mu_2}{\lambda_{\min}(\Gamma_2^{-1})}} \tag{5.53}$$

Furthermore, we obtain

$$|e_2| \le |e_2(0)| + \frac{1}{\lambda_2}\sqrt{\frac{2|d_{33}|\mu_2}{\left|\underline{d}_{22}|d_{33}| - d_{23}^2\right|}}$$

$$|q_2| \le |e_2| + |q_{2d}| \le |e_2(0)| + \frac{1}{\lambda_2}\sqrt{\frac{2|d_{33}|\mu_2}{\left|\underline{d}_{22}|d_{33}| - d_{23}^2\right|}} + |q_{2d}| \in L_\infty$$

$$\|\hat{W}_2\| \le \|\tilde{W}_2\| + \|W_2^*\| \le \sqrt{\frac{2\mu_2}{\lambda_{\min}(\Gamma_2^{-1})}} + w_{2m} \in L_\infty \tag{5.54}$$

and thus the boundedness of control $\tau_2$.

*$q_3$- subsystem:*

From the previous stability analysis about the $q_1$-subsystem and the $q_2$-subsystem, we know that $q_1$, $q_2$, $\dot{q}_1$, $\dot{q}_2$ are bounded. Accordingly, $\xi$ are bounded. According to Lemma 5, we know that the internal dynamics are stable, i.e., $\eta$ ($q_3$ and $\dot{q}_3$) are bounded. All the signals in the closed-loop system are bounded. This completes the proof. □

### 5.3.2 MNN-Based Control

Nonlinearly parameterized approximators, such as the MNN, can be linearized by Taylor series expansions, with the higher order terms being taken as part of the modeling error. Due to the nonlinear parameterizations, the control design and stability analysis involving the MNN is more complex than the previous one based on the linearly parameterized network, i.e., the RBFNN, but still can follow the similar procedures as the afore-mentioned RBFNN-based one.

□ $q_1$*-subsystem*

Similar to the RBFNN case in Sect. 5.3.1, (5.4) is written as

$$d_{11}\dot{r}_1 = b_{11}(\dot{q}_3)\tau_1 - f_{S1,1} \tag{5.55}$$

where the unknown continuous function

$$f_{S1,1} = d_{11}\ddot{q}_{1r} + f_1(\dot{q}_3) + g_1 + \Delta_1(q,\dot{q}) \tag{5.56}$$

is approximated by MNN to arbitrarily any accuracy as

$$f_{S1,1} = W_1^{*\mathrm{T}} S_1(V_1^{*\mathrm{T}} Z_1) + \varepsilon_1(Z_1) \tag{5.57}$$

where the input vector $Z_1 = [q_1, \dot{q}_1, q_2, \dot{q}_2, q_3, \dot{q}_3, \dot{q}_{1d}, \ddot{q}_{1d}, 1]^{\mathrm{T}} \in \Omega_{Z_1} \subset R^9$; $\varepsilon_1(Z_1)$ is the approximation error satisfying $|\varepsilon_1(Z_1)| \le \bar{\varepsilon}_1$, where $\bar{\varepsilon}_1$ is a positive constant; $W_1^*$ and $V_1^*$ are unknown ideal constant weights satisfying $\|W_1^*\| \le w_{1m}, \|V_1^*\|_F \le v_{1m}$, which are positive constants. By using $\hat{W}_1^{\mathrm{T}} S_1(\hat{V}_1^{\mathrm{T}} Z_1)$ to approximate $W_1^{*\mathrm{T}} S_1(V_1^{*\mathrm{T}} Z_1)$, the error between the actual and the ideal MNN can be expressed as

$$\hat{W}_1^{\mathrm{T}} S(\hat{V}_1^{\mathrm{T}} Z_1) - W_1^{*\mathrm{T}} S(V_1^{*\mathrm{T}} Z_1) = \tilde{W}_1^{\mathrm{T}}(\hat{S}_1 - \hat{S}_1'\hat{V}_1^{\mathrm{T}} Z_1) + \hat{W}_1^{\mathrm{T}}\hat{S}_1'\tilde{V}_1^{\mathrm{T}} Z_1 + d_{u1} \tag{5.58}$$

where $\hat{S}_1 = S(\hat{V}_1^{\mathrm{T}} Z_1)$, $\hat{S}_1' = \mathrm{diag}\,\{\hat{s}_1', \hat{s}_2', \ldots, \hat{s}_l'\}$ with

$$\hat{s}_i' = s'(\hat{v}_i^{\mathrm{T}} Z) = \frac{\mathrm{d}[s(z_a)]}{\mathrm{d}z_a}\Big|_{z_a = \hat{v}_i^{\mathrm{T}} Z}$$

the residual term $d_{u1}$ is bounded by

$$|d_{u1}| \le \|V_1^*\|_F \|Z_1 \hat{W}_1^{\mathrm{T}} \hat{S}_1'\|_F + \|W_1^*\| \|\hat{S}_1'\hat{V}_1^{\mathrm{T}} Z_1\| + |W_1^*|_1 \tag{5.59}$$

and the weight estimation errors $\tilde{W}_1 = \hat{W}_1 - W_1^*$, $\tilde{V}_1 = \hat{V}_1 - V_1^*$.

Consider the following Lyapunov function candidate

$$V_1(r_1, \tilde{W}_1, \tilde{V}_1) = \frac{1}{2} d_{11} r_1^2 + \frac{1}{2} \tilde{W}_1^{\mathrm{T}} \Gamma_{W1}^{-1} \tilde{W}_1 + \frac{1}{2} \mathrm{tr}\left\{ \tilde{V}_1^{\mathrm{T}} \Gamma_{V1}^{-1} \tilde{V}_1 \right\} \tag{5.60}$$

The time derivative of (5.60) along (5.55) and (5.57) is given by

$$\begin{aligned} \dot{V}_1 &= r_1 \left[ b_{11}(\dot{q}_3)\tau_1 - W_1^{*\mathrm{T}} S_1(V_1^{*\mathrm{T}} Z_1) - \varepsilon_1(Z_1) \right] + \tilde{W}_1^{\mathrm{T}} \Gamma_{W1}^{-1} \dot{\tilde{W}}_1 \\ &\quad + \mathrm{tr}\left\{ \tilde{V}_1^{\mathrm{T}} \Gamma_{V1}^{-1} \dot{\tilde{V}}_1 \right\} \end{aligned} \tag{5.61}$$

As $W_1^*$, $V_1^*$ are constant vectors, it is easy to obtain that

$$\dot{\tilde{W}}_1 = \dot{\hat{W}}_1, \quad \dot{\tilde{V}}_1 = \dot{\hat{V}}_1 \tag{5.62}$$

Substituting (5.62) into (5.61), we have

$$\begin{aligned} \dot{V}_1 &= r_1 \left[ b_{11}(\dot{q}_3)\tau_1 - W_1^{*\mathrm{T}} S_1(V_1^{*\mathrm{T}} Z_1) - \varepsilon_1(Z_1) \right] + \tilde{W}_1^{\mathrm{T}} \Gamma_{W1}^{-1} \dot{\hat{W}}_1 \\ &\quad + \mathrm{tr}\{ \tilde{V}_1^{\mathrm{T}} \Gamma_{V1}^{-1} \dot{\hat{V}}_1 \} \end{aligned} \tag{5.63}$$

Consider the following MNN-based control law and MNN weight adaption laws:

$$\begin{aligned} \tau_1 &= -k_1 r_1 - \frac{r_1 \left( \hat{W}_1^{\mathrm{T}} S(\hat{V}_1^{\mathrm{T}} Z_1) \right)^2}{\underline{b}_{11} \left( |r_1 \hat{W}_1^{\mathrm{T}} S(\hat{V}_1^{\mathrm{T}} Z_1)| + \delta_1 \right)} - \frac{k_1 r_1}{\underline{b}_{11}} (\| Z_1 \hat{W}_1^{\mathrm{T}} \hat{S}_1' \|_F^2 \\ &\quad + \| \hat{S}_1' \hat{V}_1^{\mathrm{T}} Z_1 \|^2) \end{aligned} \tag{5.64}$$

$$\dot{\hat{W}}_1 = -\Gamma_{W1} [(\hat{S}_1 - \hat{S}_1' \hat{V}_1^{\mathrm{T}} Z_1) r_1 + \sigma_{W1} \hat{W}_1] \tag{5.65}$$

$$\dot{\hat{V}}_1 = -\Gamma_{V1} [Z_1 \hat{W}_1^{\mathrm{T}} \hat{S}_1' r_1 + \sigma_{V1} \hat{V}_1] \tag{5.66}$$

where $k_1 > 0$, $\delta_1 > 0$, $\Gamma_{W1} = \Gamma_{W1}^{\mathrm{T}} > 0$, $\Gamma_{V1} = \Gamma_{V1}^{\mathrm{T}} > 0$, $\sigma_{W1} > 0$, $\sigma_{V1} > 0$.

Substituting (5.64)–(5.66) in (5.63), we have

$$\begin{aligned} \dot{V}_1 &= -k_1 b_{11}(\dot{q}_3) r_1^2 - \frac{b_{11}(\dot{q}_3)}{\underline{b}_{11}} \frac{r_1^2 \left( \hat{W}_1^{\mathrm{T}} S(\hat{V}_1^{\mathrm{T}} Z_1) \right)^2}{\left( |r_1 \hat{W}_1^{\mathrm{T}} S(\hat{V}_1^{\mathrm{T}} Z_1)| + \delta_1 \right)} \\ &\quad - \frac{b_{11}(\dot{q}_3)}{\underline{b}_{11}} k_1 r_1^2 \left( \| Z_1 \hat{W}_1^{\mathrm{T}} \hat{S}_1' \|_F^2 + \| \hat{S}_1' \hat{V}_1^{\mathrm{T}} Z_1 \|^2 \right) - r_1 W_1^{*\mathrm{T}} S_1(V_1^{*\mathrm{T}} Z_1) \\ &\quad - r_1 \varepsilon_1(Z_1) - r_1 \tilde{W}_1^{\mathrm{T}} (\hat{S}_1 - \hat{S}_1' \hat{V}_1^{\mathrm{T}} Z_1) - \sigma_{W1} \tilde{W}_1^{\mathrm{T}} \hat{W}_1 \\ &\quad - \mathrm{tr}\{ \tilde{V}_1^{\mathrm{T}} Z_1 \hat{W}_1^{\mathrm{T}} \hat{S}_1' r_1 \} - \sigma_{V1} \mathrm{tr}\{ \tilde{V}_1^{\mathrm{T}} \hat{V}_1 \} \end{aligned} \tag{5.67}$$

Noting Assumption 5.3 and the fact that $\mathrm{tr}\{\tilde{V}_1{}^{\mathrm{T}} Z_1 \hat{W}_1^{\mathrm{T}} \hat{S}_1' r_1\} = r_1 \hat{W}_1^{\mathrm{T}} \hat{S}_1' \tilde{V}_1{}^{\mathrm{T}} Z_1$, (5.67) becomes

$$
\begin{aligned}
\dot{V}_1 \leq & -k_1 \underline{b}_{11} r_1^2 - \frac{r_1^2 \left( \hat{W}_1^{\mathrm{T}} S(\hat{V}_1^{\mathrm{T}} Z_1) \right)^2}{\left( |r_1 \hat{W}_1^{\mathrm{T}} S(\hat{V}_1^{\mathrm{T}} Z_1)| + \delta_1 \right)} \\
& -k_1 r_1^2 \left( \| Z_1 \hat{W}_1^{\mathrm{T}} \hat{S}_1' \|_F^2 + \| \hat{S}_1' \hat{V}_1^{\mathrm{T}} Z_1 \|^2 \right) \\
& + |r_1| |\varepsilon_1(Z_1)| - r_1 W_1^{*\mathrm{T}} S_1(V_1^{*\mathrm{T}} Z_1) - r_1 \tilde{W}_1^{\mathrm{T}} (\hat{S}_1 - \hat{S}_1' \hat{V}_1^{\mathrm{T}} Z_1) \\
& - r_1 \hat{W}_1^{\mathrm{T}} \hat{S}_1' \tilde{V}_1{}^{\mathrm{T}} Z_1 - \sigma_{W1} \tilde{W}_1^{\mathrm{T}} \hat{W}_1 - \sigma_{V1} \mathrm{tr}\{\tilde{V}_1{}^{\mathrm{T}} \hat{V}_1\}
\end{aligned} \tag{5.68}
$$

From (5.58) and (5.59), we know

$$
\begin{aligned}
& -r_1 W_1^{*\mathrm{T}} S_1(V_1^{*\mathrm{T}} Z_1) - r_1 \tilde{W}_1^{\mathrm{T}} (\hat{S}_1 - \hat{S}_1' \hat{V}_1^{\mathrm{T}} Z_1) - r_1 \hat{W}_1^{\mathrm{T}} \hat{S}_1' \tilde{V}_1{}^{\mathrm{T}} Z_1 \\
& = -r_1 \hat{W}_1^{\mathrm{T}} S(\hat{V}_1^{\mathrm{T}} Z_1) - r_1 d_{u1} \\
& \leq |r_1 \hat{W}_1^{\mathrm{T}} S(\hat{V}_1^{\mathrm{T}} Z_1)| + |r_1| \| V_1^* \|_F \| Z_1 \hat{W}_1^{\mathrm{T}} \hat{S}_1' \|_F + |r_1| \| W_1^* \| \| \hat{S}_1' \hat{V}_1^{\mathrm{T}} Z_1 \| \\
& \quad + |r_1| |W_1^*|_1
\end{aligned} \tag{5.69}
$$

Substituting (5.69) in (5.68) leads to

$$
\begin{aligned}
\dot{V}_1 \leq & -k_1 \underline{b}_{11} r_1^2 - \frac{r_1^2 \left( \hat{W}_1^{\mathrm{T}} S(\hat{V}_1^{\mathrm{T}} Z_1) \right)^2}{\left( |r_1 \hat{W}_1^{\mathrm{T}} S(\hat{V}_1^{\mathrm{T}} Z_1)| + \delta_1 \right)} + |r_1 \hat{W}_1^{\mathrm{T}} S(\hat{V}_1^{\mathrm{T}} Z_1)| \\
& -k_1 r_1^2 \left( \| Z_1 \hat{W}_1^{\mathrm{T}} \hat{S}_1' \|_F^2 + \| \hat{S}_1' \hat{V}_1^{\mathrm{T}} Z_1 \|^2 \right) + |r_1| |\varepsilon_1(Z_1)| \\
& + |r_1| \| V_1^* \|_F \| Z_1 \hat{W}_1^{\mathrm{T}} \hat{S}_1' \|_F + |r_1| \| W_1^* \| \| \hat{S}_1' \hat{V}_1^{\mathrm{T}} Z_1 \| \\
& + |r_1| |W_1^*|_1 - \sigma_{W1} \tilde{W}_1^{\mathrm{T}} \hat{W}_1 - \sigma_{V1} \mathrm{tr}\{\tilde{V}_1{}^{\mathrm{T}} \hat{V}_1\}
\end{aligned} \tag{5.70}
$$

According to Lemma 5.4,

$$
-\frac{r_1^2 \left( \hat{W}_1^{\mathrm{T}} S(\hat{V}_1^{\mathrm{T}} Z_1) \right)^2}{|r_1 \hat{W}_1^{\mathrm{T}} S(\hat{V}_1^{\mathrm{T}} Z_1)| + \delta_1} + |r_1 \hat{W}_1^{\mathrm{T}} S(\hat{V}_1^{\mathrm{T}} Z_1)| = \frac{|r_1 \hat{W}_1^{\mathrm{T}} S(\hat{V}_1^{\mathrm{T}} Z_1)| \delta_1}{|r_1 \hat{W}_1^{\mathrm{T}} S(\hat{V}_1^{\mathrm{T}} Z_1)| + \delta_1} \leq \delta_1 \tag{5.71}
$$

By completion of squares and using Young's inequality, the following inequalities hold:

$$|r_1||\varepsilon_1(Z_1)| \le \frac{r_1^2}{2c_{11}} + \frac{c_{11}\bar{\varepsilon}_1^2}{2} \tag{5.72}$$

$$|r_1|\|V_1^*\|_F\|Z_1\hat{W}_1^{\mathrm{T}}\hat{S}_1'\|_F \le k_1 r_1^2\|Z_1\hat{W}_1^{\mathrm{T}}\hat{S}_1'\|_F^2 + \frac{1}{4k_1}\|V_1^*\|_F^2 \tag{5.73}$$

$$|r_1|\|W_1^*\|\|\hat{S}_1'\hat{V}_1^{\mathrm{T}}Z_1\| \le k_1 r_1^2\|\hat{S}_1'\hat{V}_1^{\mathrm{T}}Z_1\|^2 + \frac{1}{4k_1}\|W_1^*\|^2 \tag{5.74}$$

$$|r_1||W_1^*|_1 \le \frac{r_1^2}{2c_{12}} + \frac{c_{12}|W_1^*|_1^2}{2} \tag{5.75}$$

$$-\sigma_{W1}\tilde{W}_1^{\mathrm{T}}\hat{W}_1 \le -\frac{\sigma_{W1}}{2}\|\tilde{W}_1\|^2 + \frac{\sigma_{W1}}{2}\|W_1^*\|^2 \tag{5.76}$$

$$-\sigma_{V1}\mathrm{tr}\{\tilde{V}_1{}^{\mathrm{T}}\hat{V}_1\} \le -\frac{\sigma_{V1}}{2}\|\tilde{V}_1\|_F^2 + \frac{\sigma_{V1}}{2}\|V_1^*\|_F^2 \tag{5.77}$$

Substituting (5.71)–(5.77) into (5.70), we have

$$\begin{aligned}
\dot{V}_1 \le & -\left(k_1\underline{b}_{11} - \frac{1}{2c_{11}} - \frac{1}{2c_{12}}\right)r_1^2 - \frac{\sigma_{W1}}{2}\|\tilde{W}_1\|^2 - \frac{\sigma_{V1}}{2}\|\tilde{V}_1\|_F^2 + \delta_1 \\
& + \left(\frac{\sigma_{W1}}{2} + \frac{1}{4k_1}\right)\|W_1^*\|^2 + \left(\frac{\sigma_{V1}}{2} + \frac{1}{4k_1}\right)\|V_1^*\|_F^2 + \frac{c_{11}}{2}\bar{\varepsilon}_1^2 + \frac{c_{12}|W_1^*|_1^2}{2} \\
\le & -\lambda_{10}V_1 + \mu_{10}
\end{aligned} \tag{5.78}$$

where $\lambda_{10} = \min\left\{(2k_1\underline{b}_{11} - 1/c_{11} - 1/c_{12})/d_{11}, \sigma_{W1}/\lambda_{\max}(\Gamma_{W1}^{-1}), \sigma_{V1}/\lambda_{\max}(\Gamma_{V1}^{-1})\right\}$, $\mu_{10} = \delta_1 + (\frac{\sigma_{W1}}{2} + \frac{1}{4k_1})\|W_1^*\|^2 + (\frac{\sigma_{V1}}{2} + \frac{1}{4k_1})\|V_1^*\|_F^2 + \frac{c_{11}}{2}\bar{\varepsilon}_1^2 + \frac{c_{12}|W_1^*|_1^2}{2}$.

□ *$q_2$-subsystem*

Similar to Sect. 5.3.1, (5.5) becomes

$$\frac{d_{22}(q_3)d_{33} - d_{23}^2}{d_{33}}\dot{r}_2 + c_{22}(q_3,\dot{q}_3)r_2 = b_{22}(\dot{q}_3)\tau_2 - f_{S2,1} \tag{5.79}$$

where the unknown function

$$\begin{aligned}
f_{S2,1} = & \frac{d_{22}(q_3)d_{33} - d_{23}^2}{d_{33}}\ddot{q}_{2r} + c_{22}(q_3,\dot{q}_3)\dot{q}_{2r} + c_{23}(q_3,\dot{q}_2)\dot{q}_3 + \Delta_2(q,\dot{q}) \\
& + \frac{d_{23}}{d_{33}}(b_{31}(\dot{q}_3)\tau_1 - c_{32}(q_3,\dot{q}_2)\dot{q}_2 - f_3(\dot{q}_3) - g_3 - \Delta_3(q,\dot{q}))
\end{aligned}$$

is approximated by MNN to arbitrarily any accuracy as

$$f_{S2,1} = W_2^{*\mathrm{T}}S_2(V_2^{*\mathrm{T}}Z_2) + \varepsilon_2(Z_2)$$

where the input vector $Z_2 = [\tau_1, q_1, \dot{q}_1, q_2, \dot{q}_2, q_3, \dot{q}_3, q_{2d}, \dot{q}_{2d}, \ddot{q}_{2d}, 1]^{\mathrm{T}} \in \Omega_{Z_2} \subset R^{11}$, $\varepsilon_2(Z_2)$ is the approximation error satisfying $|\varepsilon_2(Z_2)| \le \bar{\varepsilon}_2$, where $\bar{\varepsilon}_2$ is a positive constant; $W_2^*$ and $V_2^*$ are ideal constant weights satisfying $\|W_2^*\| \le w_{2m}, \|V_2^*\|_F \le v_{2m}$, which are positive constants. By using $\hat{W}_2^{\mathrm{T}} S_2(\hat{V}_2^{\mathrm{T}} Z_2)$ to approximate $W_2^{*\mathrm{T}} S_2(V_2^{*\mathrm{T}} Z_2)$, the error between the actual and the ideal MNN can be expressed as

$$\hat{W}_2^{\mathrm{T}} S(\hat{V}_2^{\mathrm{T}} Z_2) - W_2^{*\mathrm{T}} S(V_2^{*\mathrm{T}} Z_2) = \tilde{W}_2^{\mathrm{T}}(\hat{S}_2 - \hat{S}_2' \hat{V}_2^{\mathrm{T}} Z_2) + \hat{W}_2^{\mathrm{T}} \hat{S}_2' \tilde{V}_2^{\mathrm{T}} Z_2 + d_{u2} \tag{5.80}$$

where $\hat{S}_2 = S(\hat{V}_2^{\mathrm{T}} Z_2)$, $\hat{S}_2' = \mathrm{diag}\,\{\hat{s}_1', \hat{s}_2', \ldots, \hat{s}_l'\}$ with

$$\hat{s}_i' = s'(\hat{v}_i^{\mathrm{T}} Z_2) = \frac{\mathrm{d}[s(z_a)]}{\mathrm{d}z_a}\Big|_{z_a = \hat{v}_i^{\mathrm{T}} Z_2}$$

and the residual term $d_{u2}$ is bounded by

$$|d_{u2}| \le \|V_2^*\|_F \|Z_2 \hat{W}_2^{\mathrm{T}} \hat{S}_2'\|_F + \|W_2^*\| \|\hat{S}_2' \hat{V}_2^{\mathrm{T}} Z_2\| + |W_2^*|_1 \tag{5.81}$$

and the weight estimation errors $\tilde{W}_2 = \hat{W}_2 - W_2^*$, $\tilde{V}_2 = \hat{V}_2 - V_2^*$.

To analyze the closed loop stability for the $q_2$-subsystem, consider the following Lyapunov function candidate

$$V_2(r_2, \tilde{W}_2, \tilde{V}_2) = \frac{1}{2}\frac{d_{22}(q_3)d_{33} - d_{23}^2}{d_{33}} r_2^2 + \frac{1}{2}\tilde{W}_2^{\mathrm{T}} \Gamma_{W2}^{-1} \tilde{W}_2 + \frac{1}{2}\mathrm{tr}\{\tilde{V}_2^{\mathrm{T}} \Gamma_{V2}^{-1} \tilde{V}_2\} \tag{5.82}$$

**Lemma 5.9.** *The function $V_2$ (5.82) is positive definite and decrescent, in the sense that there exist two time-invariant positive definite functions $\underline{V}_2(r_2, \tilde{W}_2, \tilde{V}_2)$ and $\bar{V}_2(r_2, \tilde{W}_2, \tilde{V}_2)$, such that*

$$\underline{V}_2(r_2, \tilde{W}_2, \tilde{V}_2) \le V_2 \le \bar{V}_2(r_2, \tilde{W}_2, \tilde{V}_2)$$

*Proof.* The proof follows the same approach as Lemma 5.6 and is omitted here for conciseness. □

The time derivative of (5.82) is given as

$$\dot{V}_2 = \frac{1}{2}\dot{d}_{22}(q_3) r_2^2 + \frac{d_{22}(q_3)d_{33} - d_{23}^2}{d_{33}} r_2 \dot{r}_2 + \tilde{W}_2^{\mathrm{T}} \Gamma_2^{-1} \dot{\hat{W}}_2 + \mathrm{tr}\{\tilde{V}_2^{\mathrm{T}} \Gamma_{V2}^{-1} \dot{\hat{V}}_2\} \tag{5.83}$$

According to Assumption 5.2, (5.83) becomes

$$\dot{V}_2 = r_2\left[\frac{d_{22}(q_3)d_{33} - d_{23}^2}{d_{33}}\dot{r}_2 + c_{22}(q_3, \dot{q}_3)r_2\right] + \tilde{W}_2^{\mathrm{T}}\Gamma_2^{-1}\dot{\tilde{W}}_2 + \mathrm{tr}\{\tilde{V}_2^{\mathrm{T}}\Gamma_{V2}^{-1}\dot{\tilde{V}}_2\} \tag{5.84}$$

As $W_2^*$, $V_2^*$ are constant vectors, it is easy to obtain that

$$\dot{\tilde{W}}_2 = \dot{\hat{W}}_2, \quad \dot{\tilde{V}}_2 = \dot{\hat{V}}_2 \tag{5.85}$$

Substituting (5.79),(5.80), and (5.85) into (5.89), we have

$$\begin{aligned}\dot{V}_2 = {} & r_2\left[b_{22}(\dot{q}_3)\tau_2 - W_2^{*\mathrm{T}}S_2(V_2^{*\mathrm{T}}Z_2) - \varepsilon_2(Z_2)\right] \\ & +\tilde{W}_2^{\mathrm{T}}\Gamma_{W2}^{-1}\dot{\hat{W}}_2 + \mathrm{tr}\{\tilde{V}_2^{\mathrm{T}}\Gamma_{V2}^{-1}\dot{\hat{V}}_2\}\end{aligned} \tag{5.86}$$

Consider the following MNN-based control law and MNN weight adaption laws:

$$\begin{aligned}\tau_2 = {} & k_2r_2 + \frac{r_2\left(\hat{W}_2^{\mathrm{T}}S(\hat{V}_2^{\mathrm{T}}Z_2)\right)^2}{\underline{b}_{22}\left(|r_2\hat{W}_2^{\mathrm{T}}S(\hat{V}_2^{\mathrm{T}}Z_2)| + \delta_2\right)} \\ & +\frac{k_2r_2}{\underline{b}_{22}}(\|Z_2\hat{W}_2^{\mathrm{T}}\hat{S}_2'\|_F^2 + \|\hat{S}_2'\hat{V}_2^{\mathrm{T}}Z_2\|^2)\end{aligned} \tag{5.87}$$

$$\dot{\hat{W}}_2 = -\Gamma_{W2}[(\hat{S}_2 - \hat{S}_2'\hat{V}_2^{\mathrm{T}}Z_2)r_2 + \sigma_{W2}\hat{W}_2] \tag{5.88}$$

$$\dot{\hat{V}}_2 = -\Gamma_{V2}[Z_2\hat{W}_2^{\mathrm{T}}\hat{S}_2'r_2 + \sigma_{V2}\hat{V}_2] \tag{5.89}$$

where $k_2 > 0$, $\delta_2 > 0$, $\Gamma_{W2} = \Gamma_{W2}^{\mathrm{T}} > 0$, $\Gamma_{V2} = \Gamma_{V2}^{\mathrm{T}} > 0$, $\sigma_{W2} > 0$, $\sigma_{V2} > 0$.

Substituting (5.87)–(5.89) into (5.86), we have

$$\begin{aligned}\dot{V}_2 = {} & k_2b_{22}(\dot{q}_3)r_2^2 + \frac{b_{22}(\dot{q}_3)}{\underline{b}_{22}}\frac{r_2^2\left(\hat{W}_2^{\mathrm{T}}S(\hat{V}_2^{\mathrm{T}}Z_2)\right)^2}{\left(|r_2\hat{W}_2^{\mathrm{T}}S(\hat{V}_2^{\mathrm{T}}Z_2)| + \delta_2\right)} \\ & +\frac{b_{22}(\dot{q}_3)}{\underline{b}_{22}}k_2r_2^2\left(\|Z_2\hat{W}_2^{\mathrm{T}}\hat{S}_2'\|_F^2 + \|\hat{S}_2'\hat{V}_2^{\mathrm{T}}Z_2\|^2\right) - r_2W_2^{*\mathrm{T}}S_2(V_2^{*\mathrm{T}}Z_2) \\ & -r_2\varepsilon_2(Z_2) - r_2\tilde{W}_2^{\mathrm{T}}(\hat{S}_2 - \hat{S}_2'\hat{V}_2^{\mathrm{T}}Z_2) - \sigma_{W2}\tilde{W}_2^{\mathrm{T}}\hat{W}_2 \\ & -\mathrm{tr}\{\tilde{V}_2^{\mathrm{T}}Z_2\hat{W}_2^{\mathrm{T}}\hat{S}_2'r_2\} - \sigma_{V2}\mathrm{tr}\{\tilde{V}_2^{\mathrm{T}}\hat{V}_2\}\end{aligned} \tag{5.90}$$

Noting Assumption 5.3 and the fact that $\mathrm{tr}\{\tilde{V}_2^{\mathrm{T}} Z_2 \hat{W}_2^{\mathrm{T}} \hat{S}_2' r_2\} = r_2 \hat{W}_2^{\mathrm{T}} \hat{S}_2' \tilde{V}_2^{\mathrm{T}} Z_2$, (5.90) becomes

$$\begin{aligned}
\dot{V}_2 \leq & -k_2 \underline{b}_{22} r_1^2 - \frac{r_2^2\left(\hat{W}_2^{\mathrm{T}} S(\hat{V}_2^{\mathrm{T}} Z_2)\right)^2}{\left(|r_2 \hat{W}_2^{\mathrm{T}} S(\hat{V}_2^{\mathrm{T}} Z_2)| + \delta_2\right)} - k_2 r_2^2\left(\|Z_2 \hat{W}_2^{\mathrm{T}} \hat{S}_2'\|_F^2 + \|\hat{S}_2' \hat{V}_2^{\mathrm{T}} Z_2\|^2\right) \\
& + |r_2||\varepsilon_2(Z_2)| - r_2 W_2^{*\mathrm{T}} S_2(V_2^{*\mathrm{T}} Z_2) - r_2 \tilde{W}_2^{\mathrm{T}}(\hat{S}_2 - \hat{S}_2' \hat{V}_2^{\mathrm{T}} Z_2) \\
& - r_2 \hat{W}_2^{\mathrm{T}} \hat{S}_2' \tilde{V}_2^{\mathrm{T}} Z_2 - \sigma_{W2} \tilde{W}_2^{\mathrm{T}} \hat{W}_2 - \sigma_{V2} \mathrm{tr}\{\tilde{V}_2^{\mathrm{T}} \hat{V}_2\}
\end{aligned} \tag{5.91}$$

From (5.80) and (5.81), we know that

$$\begin{aligned}
& -r_2 W_2^{*\mathrm{T}} S_2(V_2^{*\mathrm{T}} Z_2) - r_2 \tilde{W}_2^{\mathrm{T}}(\hat{S}_2 - \hat{S}_2' \hat{V}_2^{\mathrm{T}} Z_2) - r_2 \hat{W}_2^{\mathrm{T}} \hat{S}_2' \tilde{V}_2^{\mathrm{T}} Z_2 \\
& = -r_2 \hat{W}_2^{\mathrm{T}} S(\hat{V}_2^{\mathrm{T}} Z_2) - r_2 d_{u2} \\
& \leq |r_2 \hat{W}_2^{\mathrm{T}} S(\hat{V}_2^{\mathrm{T}} Z_2)| + |r_2|\|V_2^*\|_F \|Z_2 \hat{W}_2^{\mathrm{T}} \hat{S}_2'\|_F + |r_2|\|W_2^*\|\|\hat{S}_2' \hat{V}_2^{\mathrm{T}} Z_2\| \\
& \quad + |r_2||W_2^*|_1
\end{aligned} \tag{5.92}$$

Substituting (5.92) into (5.91) leads to

$$\begin{aligned}
\dot{V}_2 \leq & -k_2 \underline{b}_{22} r_2^2 - \frac{r_2^2\left(\hat{W}_2^{\mathrm{T}} S(\hat{V}_2^{\mathrm{T}} Z_2)\right)^2}{\left(|r_2 \hat{W}_2^{\mathrm{T}} S(\hat{V}_2^{\mathrm{T}} Z_2)| + \delta_2\right)} + |r_2 \hat{W}_2^{\mathrm{T}} S(\hat{V}_2^{\mathrm{T}} Z_2)| \\
& - k_2 r_2^2\left(\|Z_2 \hat{W}_2^{\mathrm{T}} \hat{S}_2'\|_F^2 + \|\hat{S}_2' \hat{V}_2^{\mathrm{T}} Z_2\|^2\right) + |r_2||\varepsilon_2(Z_2)| \\
& + |r_2|\|V_2^*\|_F \|Z_2 \hat{W}_2^{\mathrm{T}} \hat{S}_2'\|_F + |r_2|\|W_2^*\|\|\hat{S}_2' \hat{V}_2^{\mathrm{T}} Z_2\| + |r_2||W_2^*|_1 \\
& - \sigma_{W2} \tilde{W}_2^{\mathrm{T}} \hat{W}_2 - \sigma_{V2} \mathrm{tr}\{\tilde{V}_2^{\mathrm{T}} \hat{V}_2\}
\end{aligned} \tag{5.93}$$

According to Lemma 5.4,

$$-\frac{r_2^2\left(\hat{W}_2^{\mathrm{T}} S(\hat{V}_2^{\mathrm{T}} Z_2)\right)^2}{\left(|r_2 \hat{W}_2^{\mathrm{T}} S(\hat{V}_2^{\mathrm{T}} Z_2)| + \delta_2\right)} + |r_2 \hat{W}_2^{\mathrm{T}} S(\hat{V}_2^{\mathrm{T}} Z_2)| = \frac{|r_2 \hat{W}_2^{\mathrm{T}} S(\hat{V}_2^{\mathrm{T}} Z_2)|\delta_2}{|r_2 \hat{W}_2^{\mathrm{T}} S(\hat{V}_2^{\mathrm{T}} Z_2)| + \delta_2} \leq \delta_2 \tag{5.94}$$

By completion of squares and using Young's inequality, the following inequalities hold:

$$|r_2||\varepsilon_2(Z_2)| \leq \frac{r_2^2}{2c_{21}} + \frac{c_{21}\bar{\varepsilon}_2^2}{2} \tag{5.95}$$

$$|r_2|\|V_2^*\|_F \|Z_2 \hat{W}_2^{\mathrm{T}} \hat{S}_2'\|_F \leq k_2 r_2^2 \|Z_2 \hat{W}_2^{\mathrm{T}} \hat{S}_2'\|_F^2 + \frac{1}{4k_2}\|V_2^*\|_F^2 \tag{5.96}$$

$$|r_2|\|W_2^*\|\|\hat{S}_2'\hat{V}_2^{\mathrm{T}}Z_2\| \le k_2 r_2^2\|\hat{S}_2'\hat{V}_2^{\mathrm{T}}Z_2\|^2 + \frac{1}{4k_2}\|W_2^*\|^2 \tag{5.97}$$

$$|r_2||W_2^*|_1 \le \frac{r_2^2}{2c_{22}} + \frac{c_{22}|W_2^*|_1^2}{2} \tag{5.98}$$

$$-\sigma_{W2}\tilde{W}_2^{\mathrm{T}}\hat{W}_2 \le -\frac{\sigma_{W2}}{2}\|\tilde{W}_2\|^2 + \frac{\sigma_{W2}}{2}\|W_1^*\|^2 \tag{5.99}$$

$$-\sigma_{V2}\mathrm{tr}\{\tilde{V_2}^{\mathrm{T}}\hat{V}_2\} \le -\frac{\sigma_{V2}}{2}\|\tilde{V}_2\|_F^2 + \frac{\sigma_{V2}}{2}\|V_2^*\|_F^2 \tag{5.100}$$

Substituting (5.94)–(5.100) into (5.93), we have

$$\begin{aligned}\dot{V}_2 \le &-\left(k_2\underline{b}_{22} - \frac{1}{2c_{21}} - \frac{1}{2c_{22}}\right) r_2^2 - \frac{\sigma_{W2}}{2}\|\tilde{W}_2\|^2 - \frac{\sigma_{V2}}{2}\|\tilde{V}_2\|_F^2 + \delta_2 \\ &+\left(\frac{\sigma_{W2}}{2} + \frac{1}{4k_2}\right)\|W_2^*\|^2 + \left(\frac{\sigma_{V2}}{2} + \frac{1}{4k_2}\right)\|V_2^*\|_F^2 + \frac{c_{21}}{2}\bar{\varepsilon}_2^2 + \frac{c_{22}|W_2^*|_1^2}{2} \\ \le &-\lambda_{20}V_2 + \mu_{20}\end{aligned} \tag{5.101}$$

where $\lambda_{20} = \min\left\{(2k_2\underline{b}_{22} - 1/c_{21} - 1/c_{22})|d_{33}|/(\bar{d}_{22}|d_{33}| + d_{23}^2), \sigma_{W2}/\lambda_{\max}(\Gamma_{W2}^{-1}), \sigma_{V2}/\lambda_{\max}(\Gamma_{V2}^{-1})\right\}$, $\mu_{20} = \delta_2 + (\frac{\sigma_{W2}}{2} + \frac{1}{4k_2})\|W_2^*\|^2 + (\frac{\sigma_{V2}}{2} + \frac{1}{4k_21})\|V_2^*\|_F^2 + \frac{c_{21}}{2}\bar{\varepsilon}_2^2 + \frac{c_{22}|W_2^*|_1^2}{2}$.

□ *$q_3$-subsystem*

Finally, for the system (5.4)–(5.6) under control laws (5.64) and (5.87), we can obtain similar internal dynamics to Sect. 5.3.1.

The main result in this section can be summarized as the following theorem:

**Theorem 5.10.** *Consider the closed-loop system consisting of the subsystems (5.4)–(5.6), the control laws (5.64), (5.87), and adaptation laws (5.65)–(5.66), (5.88)–(5.89). Under Assumptions 5.1–5.6, the overall closed-loop neural control system is SGUUB in the sense that all of the signals in the closed-loop system are bounded, and the tracking errors and weights converge to the following regions,*

$$|e_1| \le |e_1(0)| + \frac{1}{\lambda_1}\sqrt{\frac{2\mu_1}{d_{11}}}, \quad |e_2| \le |e_2(0)| + \frac{1}{\lambda_2}\sqrt{\frac{2|d_{33}|\mu_2}{\left|\underline{d}_{22}|d_{33}| - d_{23}^2\right|}},$$

$$\|\hat{V}_1\|_F \le \sqrt{\frac{2\mu_1}{\lambda_{\min}(\Gamma_1^{-1})}} + v_{1m}, \quad \|\hat{V}_2\|_F \le \sqrt{\frac{2\mu_2}{\lambda_{\min}(\Gamma_2^{-1})}} + v_{2m}$$

$$\|\hat{W}_1\| \le \sqrt{\frac{2\mu_1}{\lambda_{\min}(\Gamma_1^{-1})}} + w_{1m}, \quad \|\hat{W}_2\| \le \sqrt{\frac{2\mu_2}{\lambda_{\min}(\Gamma_2^{-1})}} + w_{2m},$$

*with*

$$\mu_i = \frac{\mu_{i0}}{\lambda_{i0}} + V_i(0)$$

$$\mu_{i0} = \delta_i + \left(\frac{\sigma_{Wi}}{2} + \frac{1}{4k_i}\right)\|W_i^*\|^2 + \left(\frac{\sigma_{Vi}}{2} + \frac{1}{4k_i 1}\right)\|V_i^*\|_F^2 + \frac{c_{i1}}{2}\bar{\varepsilon}_i^2 + \frac{c_{i2}|W_i^*|_1^2}{2},$$
$$i = 1,2$$

$$\lambda_{10} = \min\left\{(2k_1\underline{b}_{11} - 1/c_{11} - 1/c_{12})/d_{11}, \sigma_{W1}/\lambda_{\max}(\Gamma_{W1}^{-1}), \sigma_{V1}/\lambda_{\max}(\Gamma_{V1}^{-1})\right\}$$

$$\lambda_{20} = \min\left\{(2k_2\underline{b}_{22} - 1/c_{21} - 1/c_{22})|d_{33}|/(\bar{d}_{22}|d_{33}| + d_{23}^2), \sigma_{W2}/\lambda_{\max}(\Gamma_{W2}^{-1}),\right.$$
$$\left.\sigma_{V2}/\lambda_{\max}(\Gamma_{V2}^{-1})\right\}$$

*where $e_i(0)$ and $V_i(0)$ are initial values of $e_i(t)$ and $V_i(t)$, respectively.*

*Proof.* The proof of Theorem 5.10 follows the same approach as Theorem 5.8, and will be omitted here for conciseness. □

## 5.4 Simulation Study

To illustrate the proposed adaptive neural control, we consider the VARIO helicopter mounted on a platform [104], with the dynamic model as (5.1) and the following parameters $d_{11} = 7.5$, $d_{22}(q_3) = 0.4305 + 0.0003\cos^2(-4.143q_3)$, $d_{23} = 0.108$, $d_{33} = 0.4993$, $c_{22}(q_3, \dot{q}_3) = 0.0006214\sin(-8.286q_3)\dot{q}_3$, $c_{23}(q_3, \dot{q}_2) = c_{32}(q_3, \dot{q}_2) = 0.0006214\sin(-8.286q_3)\dot{q}_2$, $g_1 = -77.259$, $g_3 = -2.642$, $f_1(\dot{q}_3) = -0.6004\dot{q}_3$, $f_3(\dot{q}_3) = -0.0001206\dot{q}_3^2$, $b_{11}(\dot{q}_3) = 3.411\dot{q}_3^2$, $b_{22}(\dot{q}_3) = -0.1525\dot{q}_3^2$, $b_{31}(\dot{q}_3) = 12.01\dot{q}_3 + 10^5$, and all quantities are expressed in S.I. units. The control objective is to track the uniformly bounded desired trajectories given in [104] as follows:

$$q_{1d} = \begin{cases} -0.2 & 0 \le t \le 50\,\text{s} \\ 0.3[\mathrm{e}^{-(t-50)^2/350} - 1] - 0.2 & 50 < t \le 130\,\text{s} \\ 0.1\cos[(t-130)/10] - 0.6 & 130 < t \le 20\pi + 130 \\ -0.5 & t \ge 20\pi + 130 \end{cases}$$

$$q_{2d} = \begin{cases} 0 & t < 50\,\text{s} \\ 1 - \mathrm{e}^{-(t-50)^2/350} & 50 \le t < 120\,\text{s} \\ \mathrm{e}^{-(t-120)^2/350} & 120 \le t < 180 \\ -1 + \mathrm{e}^{-(t-180)^2/350} & t \ge 180 \end{cases}$$

### 5.4.1 Internal Dynamics Stability Analysis

In this section, we analyze the stability of the internal dynamics according to the related discussion in [104]. For conciseness, we consider the RBFNN-based control case only, which can be easily extended to the MNN-based control case without any difficulties. For the RBFNN-based control case, we substitute (5.10), (5.15), (5.25) and (5.33) into the $q_3$-subsystem (5.6). According to the definition of the zero dynamics [35], we set $r_1$, $r_2$, $\tilde{W}_1^{\mathrm{T}}$, $\tilde{W}_2^{\mathrm{T}}$, $\varepsilon_1(Z_1)$ and $\varepsilon_2(Z_2)$ to zero, and the desired trajectories and initial data can be chosen in such a way that terms including $\dot{q}_2^2$, $\ddot{q}_{1d}$, $\ddot{q}_{2d}$ can be neglected [104], so we have

$$\ddot{q}_3 = \frac{1}{d_{33}}\left[\frac{b_{31}(\dot{q}_3)}{b_{11}(\dot{q}_3)}(f_1(\dot{q}_3) + g_1) - f_3(\dot{q}_3) - g_3\right] \tag{5.102}$$

Substituting the term values given in the beginning of Sect. 5.4 into (5.102) and analyzing the values of the main rotor angular velocity from which the main rotor angular acceleration is zero, we have

$$4.1137 \times 10^{-4}\dot{q}_3^4 + 1.8011\dot{q}_3^2 - 60968\dot{q}_3 - 7725900 = 0$$

Its solutions are $\dot{q}_3^* = -124.63, -219.5 \pm 468.16i$ and $563.64$ rad/s. Only the first value $\dot{q}_3^* = -124.63$ has a physical meaning for the system. If we linearize (5.102) around the equilibrium point $\dot{q}_3^* = -124.63$, we can obtain an eigenvalue $-2.44$. Therefore, according to [52], all initials of $\dot{q}_3$ sufficiently near $\dot{q}_3^* = -124.63$ can converge to $-124.63$. It then follows that the internal dynamics of the helicopter system in (5.1) have a stable behavior.

The simulation result in Fig. 5.1 also shows that the internal dynamics using RBFNN-based control are indeed stable. From Fig. 5.1, we can observe that the main rotor angular velocity $\dot{q}_3$ converges to the nominal value $-124.63$ rad/s for different initial conditions ranging from $-40$ rad/s to $-150$ rad/s, which includes the typical operating values more than sufficiently. These results are expected from the previous stability analysis, and also consistent with the results in [104]. In particular, we also notice that the further the initial condition starts from the nominal value $-124.63$ rad/s, the longer the settling time takes, and the more serious the transient oscillations become. This is reasonable in practice. If some preliminary knowledge about the nominal value is known in advance, the initial condition can be set closer to achieve better performance.

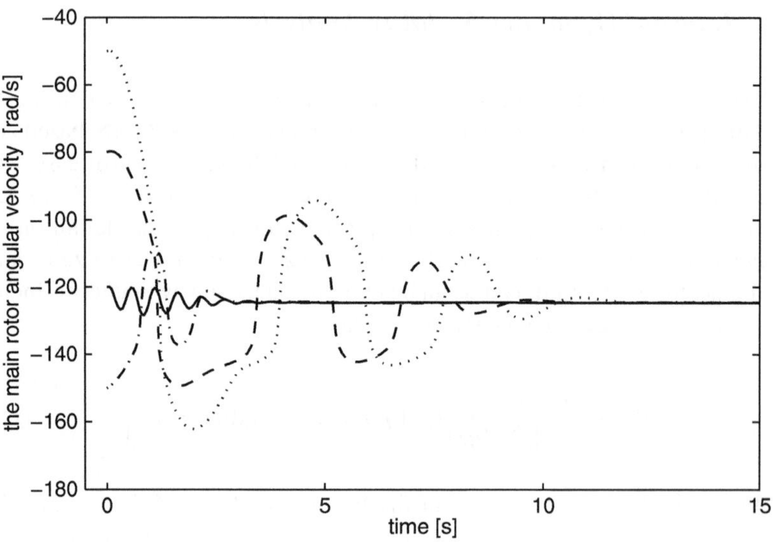


**Fig. 5.1** Main rotor angular velocity behavior for different initial conditions using RBFNN-based control

## *5.4.2 Performance Comparison Results Between Approximation-Based Control and Model-Based Control*

In this subsection, we will compare the altitude and yaw angle tracking performance using RBFNN-based control, MNN-based control and the model-based control adopted in [104]. If all the parameters and functions in (5.1) are known exactly, and the unmodeled uncertainties $\Delta(\cdot) = 0$, the perfect tracking performance can be achieved using model-based control, which has been shown in the work [104]. However, in practice, there always exist some model uncertainties, which may be caused by unmodeled dynamics or aerodynamical disturbances from the environment. To this end, we assume $\Delta(\cdot) \neq 0$, in particular, $\Delta(\cdot) = [2.0, \;\; 0, \;\; 0.0001206\dot{q}_3^2 + 0.142]^{\mathrm{T}}$.

The control parameters for the RBFNN control laws (5.15) (5.33) and adaptation laws (5.16) (5.34) are chosen as follows: $k_1 = 0.000085$, $\Lambda_1 = 0.2$, $k_2 = 0.0002$, $\Lambda_2 = 1.0$, $\Gamma_1 = 0.001I$, $\Gamma_2 = 0.0001I$, $\sigma_1 = 0.001$, $\sigma_2 = 0.001$. NNs $\hat{W}_1^{\mathrm{T}} S_1(Z_1)$ contains $3^8$ nodes (i.e., $l_1 = 2187$), with centers $\mu_l (l = 1, ..., l_1)$ evenly spaced in $[-1.0, 1.0] \times [-0.1, 0.1] \times [-10.0, -10.0] \times [-40000.0, 0.0] \times [-1.0, 1.0] \times [-150.0, -40.0] \times [-0.1, 0.1] \times [-0.01, 0.01]$, and widths $\eta_l = 1.0 (l = 1, ..., l_1)$. NNs $\hat{W}_2^{\mathrm{T}} S_2(Z_2)$ contains $3^{10}$ nodes (i.e., $l_2 = 59049$), with centers $\mu_l (l = 1, ..., l_2)$ evenly spaced in $[-0.005, 0.005] \times [-1.0, 1.0] \times [-0.1, 0.1] \times [-10.0, -10.0] \times [-40000.0, 0.0] \times [-1.0, 1.0] \times [-150.0, -40.0] \times [-10.0, 10.0] \times [-1.0, 1.0] \times [-0.01, 0.01]$, and widths $\eta_l = 1.0 (l = 1, ..., l_2)$. The initial conditions are:

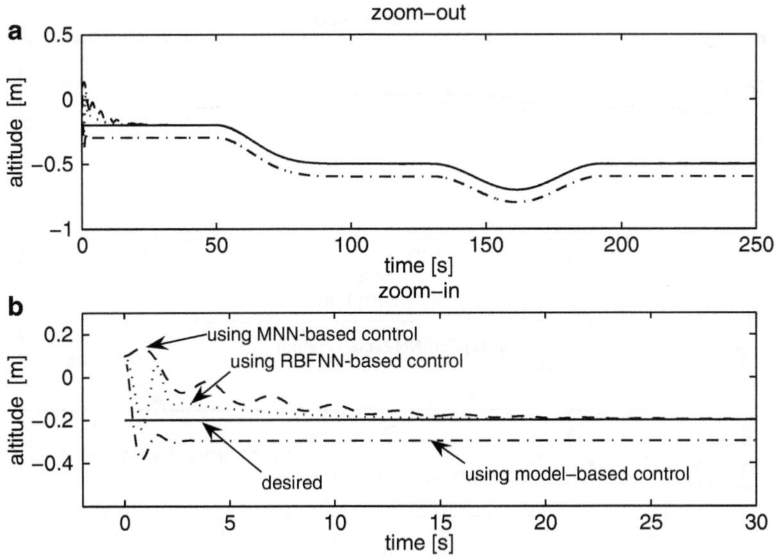


**Fig. 5.2** Altitude tracking performance in the presence of model uncertainties

$q_1(0) = 0.1$ m, $\dot{q}_1(0) = 0.0$ m/s, $q_2(0) = -\pi$ rad, $\dot{q}_2(0) = 0.0$ rad/s, $q_3(0) = -\pi$ rad, $\dot{q}_3(0) = -120.0$ rad/s, $\tau_1 = 0.0$ m, $\tau_2 = 0.0$ m, $\hat{W}_1(0) = 0.0$, $\hat{W}_2(0) = 0.0$.

For the MNN control laws (5.64) and (5.87) and adaptation laws (5.65), (5.66), (5.88) and (5.89), the design parameters are chosen as: $k_1 = 0.00016$, $\Lambda_1 = 1.2$, $k_2 = 0.0002$, $\Lambda_2 = 1.0$, $\Gamma_{W1} = 0.0002I$, $\Gamma_{V1} = 0.03I$, $\delta_{W1} = 0.0$, $\sigma_{V1} = 0.0$, $\Gamma_{W2} = 0.0001I$, $\Gamma_{V2} = 0.01I$, $\sigma_{W2} = 0.0$, $\sigma_{V2} = 0.0$. NNs $\hat{W}_1^{\mathrm{T}} S_1(\hat{V}_1^{\mathrm{T}} \bar{z}_1)$ contains five nodes and NNs $\hat{W}_2^{\mathrm{T}} S_2(\hat{V}_2^{\mathrm{T}} \bar{z}_2)$ contains 15 nodes. The initial conditions are: $q_1(0) = 0.1$ m, $\dot{q}_1(0) = 0$ m/s, $q_2(0) = -\pi$ rad, $\dot{q}_2(0) = 0.0$ rad/s, $q_3(0) = -\pi$ rad, $\dot{q}_3(0) = -120.0$ rad/s, $\tau_1 = 0.0$ m, $\tau_2 = 0.0$ m, $\hat{W}_1(0) = 0.0$, $\hat{V}_1(0) = 0.0$, $\hat{W}_2(0) = 0.0$, $\hat{V}_2(0) = 0.0$.

From Figs. 5.2 and 5.3, we can observe that due to the existence of model uncertainties, both the altitude tracking and yaw angle tracking using model-based control have some offsets to the desired trajectories for the whole period. This means that model-based control depends on the accuracy of the model heavily and cannot deal with the uncertainties well. For the tracking performance using the RBFNN-based control and MNN-based control, though there are also some oscillations at the initial period, the tracking errors can converge to a very small neighborhood of desired trajectories in a short time of about 20 s. This is because the model uncertainties can be learnt by RBFNN and MNN during the beginning 25 s. After that period, the uncertainties can be compensated for, and thus, the robustness of uncertainties is improved and good tracking performance is achieved. In addition, Figs. 5.4 and 5.5 indicates norms of neural weights for approximation-based control and control actions for three control methods.

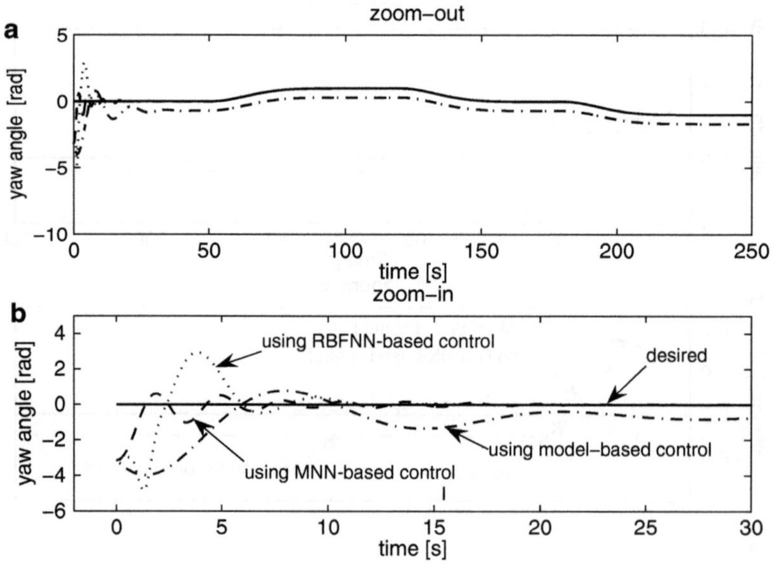


**Fig. 5.3** Yaw angle tracking performance in the presence of model uncertainties

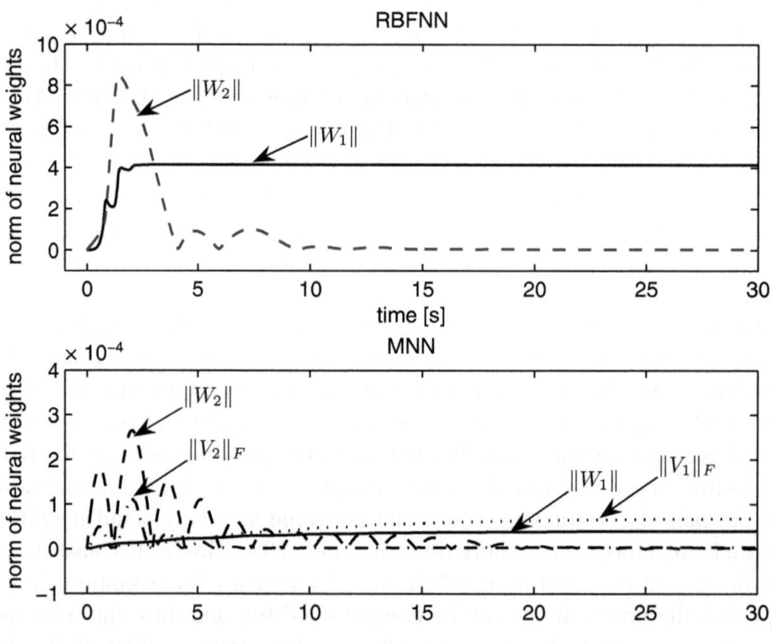


**Fig. 5.4** Norm of neural weights

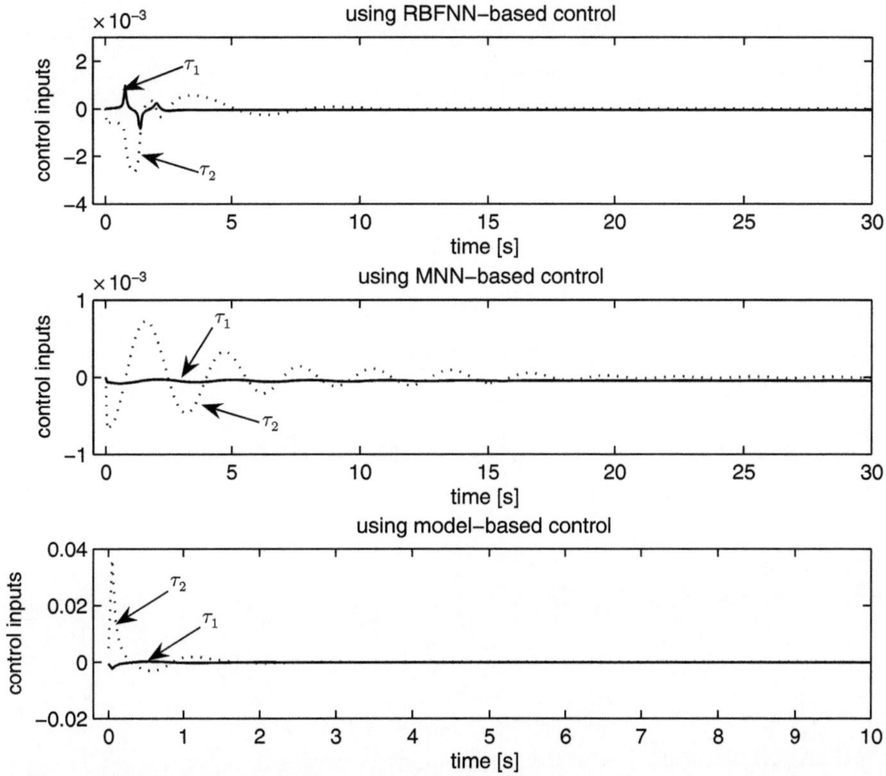


**Fig. 5.5** Control inputs for altitude and yaw angle tracking in the presence of model uncertainties

## 5.5 Conclusion

In this chapter, NN approximation-based control was investigated for the MIMO helicopter altitude and yaw angle tracking in the presence of model uncertainties. Compared with the model-based control, which is sensitive to the accuracy of the model representation, NN approximation-based control is tolerant of model uncertainties, and can be viewed as a key advantage over model-based control of helicopters, for which accurate modeling of helicopter dynamics is difficult, time-consuming and costly. Simulation results demonstrated that the helicopter is able to track altitude and yaw angle reference signals satisfactorily, with all other closed-loop signals bounded.

# Chapter 6
# Attitude Control of Uncertain Helicopters with Actuator Dynamics

## 6.1 Introduction

In a helicopter flight control system, control signals are produced by the control command via actuator servo dynamics. The flight control performance can be improved if the actuator servo dynamics are explicitly considered in the control design. To the best of our knowledge, there are few works in the literature that take into account actuator dynamics in the control design, which is not only practically relevant, but also more challenging. Notable examples are aggressive control of helicopters in presence of parametric and dynamical uncertainties [67], and adaptive output feedback control under known actuator characteristics including actuator dynamics and saturation [57].

While previous chapters in this monograph have considered actuators with instantaneous response, this chapter takes into account actuator dynamics in the first-order low-pass filter form. We tackle the attitude control problem for helicopter systems with uncertain multi-input multi-output (MIMO) dynamics. Attitude regulation is important for helicopter security monitoring and helicopter hovering flight, where the main objective is to maintain the orientation and altitude for the helicopter at specified values. Roll, pitch, and yaw angles are used to describe how the helicopter is orientated in space.

Attitude control for helicopters is a challenging problem due to the nonlinearity of the dynamics and strong interactions between variables, and calls for advanced techniques to achieve good performance. Examples include a fuzzy gain-scheduler [46], as well as the use of nonlinear feedback elements in the control system for the Bell 412 Helicopter [106]. In this chapter, we contribute an approximation-based attitude control method, based on neural networks, to handle the nonlinearities, model uncertainties (e.g., unknown moment coefficients and mass) and external disturbances. Rigorous stability analysis and extensive simulations results show the effectiveness and robustness of the proposed attitude control.

B. Ren et al., *Modeling, Control and Coordination of Helicopter Systems*,
DOI 10.1007/978-1-4614-1563-3_6, 

## 6.2 Problem Formulation

Assuming the flight positions and velocities along $x$ and $y$ axes are very small, i,e., $x = 0$, $y = 0$, $u = 0$ and $v = 0$, the attitude/altitude dynamics of the helicopter can be derived from the six degrees of freedom [64, 77] and represented in the nonlinear form of

$$\dot{z} = w \cos\phi \cos\theta \tag{6.1}$$

$$\dot{\phi} = p + q \sin\phi \tan\theta + r \cos\phi \tan\theta \tag{6.2}$$

$$\dot{\theta} = q \cos\phi - r \sin\phi \tag{6.3}$$

$$\dot{\psi} = q \sin\phi \sec\theta + r \cos\phi \sec\theta \tag{6.4}$$

$$\dot{w} = g \cos\phi \cos\theta + Z/m \tag{6.5}$$

$$\dot{p} = (c_1 r + c_2 p)q + c_3 L + c_4 N \tag{6.6}$$

$$\dot{q} = c_5 pr - c_6(p^2 - r^2) + c_7 M \tag{6.7}$$

$$\dot{r} = (c_8 p - c_2 r)q + c_4 L + c_9 N \tag{6.8}$$

where $z$ is the helicopter altitude and $w$ is the linear velocity along the $z$ axis; $m$ is the mass of helicopter; $p, q, r$ are the fuselage coordination system angular velocity components, respectively; $\phi, \theta, \psi$ are Euler angles, i.e., fuselage attitude angles, respectively; $Z$ is the aerodynamic force, and $L, M, N$ are aerodynamic moments about center of gravity, respectively. The coefficients of moment equations are given by [64]

$$c_1 = \frac{(I_{yy} - I_{zz})I_{zz} - I_{xz}^2}{\Gamma}$$

$$c_2 = \frac{(I_{xx} - I_{yy} + I_{zz})I_{xz}}{\Gamma}$$

$$c_3 = \frac{I_{zz}}{\Gamma}$$

$$c_4 = \frac{I_{xz}}{\Gamma}$$

$$c_5 = \frac{I_{zz} - I_{xx}}{I_{yy}}$$

$$c_6 = \frac{I_{xz}}{I_{yy}}$$

$$c_7 = \frac{1}{I_{yy}}$$

$$c_8 = \frac{I_{xx}(I_{xx} - I_{yy}) + I_{xz}^2}{\Gamma}$$

$$c_9 = \frac{I_{xx}}{\Gamma}$$

$$\Gamma = I_{xx} I_{zz} - I_{xz}^2$$

where $I_{xx}, I_{xz}, I_{yy}, I_{zz}$ are the inertia moments of the helicopter.

Moments $L, M, N$ including the contributions from aerodynamics and propulsion can be written as [66, 67]

$$\begin{aligned} L &= L_R + Y_R h_R + Z_R y_R + Y_T h_T \\ M &= M_R - X_R h_R + Z_R l_R \\ N &= N_R - Y_R l_R - Y_T l_T \\ Z &= Z_R \end{aligned} \tag{6.9}$$

where the subscripts denote the main rotor (R) and the tail rotor (T). $(l_R, y_R, h_R)$ and $(l_T, y_T, h_T)$ are the coordinates of the main rotor and tail rotor shafts relative to center of helicopter mass, respectively. The forces $X_R, Y_R, Z_R, Y_T$ and torques $L_R, M_R, N_R$ can be expressed as [66, 67]

$$\begin{aligned} X_R &= -T_R \sin a_{1s} \\ Y_R &= -T_R \sin b_{1s} \\ Z_R &= -T_R \cos a_{1s} \cos b_{1s} \\ Y_T &= -T_T \\ L_R &= C_b^R b_{1s} - Q_R \sin a_{1s} \\ M_R &= C_a^R a_{1s} + Q_R \sin b_{1s} \\ N_R &= -Q_R \cos a_{1s} \cos b_{1s} \end{aligned} \tag{6.10}$$

where $a_{1s}$ and $b_{1s}$ are the longitudinal and lateral inclination of the tip path plane of the main rotor; $C_a^R$ and $C_b^R$ are physical parameters modeling the flapping dynamic of the main rotor; $Q_R$ is the total main rotor torque; $T_R$ and $T_T$ are the thrusts generated by the main and the tail rotors which can be computed as shown in [66,67]

$$\begin{aligned} T_R &= K_{T_M} \theta_M w_e^2 \\ T_T &= K_{T_T} \theta_T w_e^2 \end{aligned} \tag{6.11}$$

where $\theta_M$ and $\theta_T$ are the collective pitches of the main and tail rotors, respectively; $w_e$ denotes the angular velocity of the main rotor; and $K_{T_M}$ and $K_{T_T}$ are the aerodynamics constants of the rotor's blades, respectively.

Based on (6.1)–(6.11), the attitude dynamics of a helicopter with model uncertainty and external disturbance can be written as

$$\begin{aligned}\dot{x}_1 &= J(x_1)x_2\\ \dot{x}_2 &= F(x_1,x_2) + H(x_3) + \Delta F_1(x_1,x_2) + D(x_1,x_2,t)\\ y &= x_1\end{aligned} \tag{6.12}$$

where

$$x_1 = \begin{bmatrix} z\\ \phi\\ \theta\\ \psi\end{bmatrix},\ x_2 = \begin{bmatrix} w\\ p\\ q\\ r\end{bmatrix},\ x_3 = \begin{bmatrix} b_{1s}\\ a_{1s}\\ \theta_{\mathrm{T}}\\ \theta_{\mathrm{M}}\end{bmatrix}$$

$$J(x_1) = \begin{bmatrix} \cos\phi\cos\theta & 0 & 0 & 0\\ 0 & 1 & \sin\phi\tan\theta & \cos\phi\tan\theta\\ 0 & 0 & \cos\phi & -\sin\phi\\ 0 & 0 & \sin\phi\sec\theta & \cos\phi\sec\theta\end{bmatrix}$$

$$F(x_1,x_2) = \begin{bmatrix} g\cos\phi\cos\theta\\ (c_1 r + c_2 p)q\\ c_5 pr - c_6(p^2 - r^2)\\ (c_8 p - c_2 r)q\end{bmatrix}$$

$H(x_3) = [H_1(x_3), H_2(x_3), H_3(x_3), H_4(x_3)]^{\mathrm{T}}$, $\Delta F_1(x_1,x_2) = [\Delta f_{11}(x_1,x_2), \Delta f_{12}(x_1,x_2), \Delta f_{13}(x_1,x_2), \Delta f_{14}(x_1,x_2)]^{\mathrm{T}}$, $\Delta f_{1i}(x), i = 1,2,3,4$, are the system modeling uncertainties, $D(x_1,x_2,t) = [D_1(x_1,x_2,t), D_2(x_1,x_2,t), D_3(x_1,x_2,t), D_4(x_1,x_2,t)]^{\mathrm{T}}$, $D_i(x_1,x_2,t), i = 1,2,3,4$, are the system external disturbances such as wind disturbance, and $y$ is the system output. $H_1(x_3)$, $H_2(x_3)$, $H_3(x_3)$ and $H_4(x_3)$ are given by

$$\begin{aligned}
H_1(x_3) &= -K_{T_{\mathrm{M}}}\theta_{\mathrm{M}} w_e^2 \cos a_{1s}\cos b_{1s}/m\\
H_2(x_3) &= c_3 C_b^{\mathrm{R}} b_{1s} - c_3 Q_{\mathrm{R}}\sin a_{1s} - c_3 K_{T_{\mathrm{M}}}\theta_{\mathrm{M}} w_e^2 \sin b_{1s} h_{\mathrm{R}}\\
&\quad - c_3 K_{T_{\mathrm{M}}}\theta_{\mathrm{M}} w_e^2\cos a_{1s}\cos b_{1s} y_{\mathrm{R}} - c_3 K_{T_{\mathrm{T}}}\theta_{\mathrm{T}} w_e^2 h_{\mathrm{T}}\\
&\quad - c_7 Q_{\mathrm{R}}\cos a_{1s}\cos b_{1s} + c_7 K_{T_{\mathrm{M}}}\theta_{\mathrm{M}} w_e^2\sin b_{1s} l_{\mathrm{R}} + c_7 K_{T_{\mathrm{T}}}\theta_{\mathrm{T}} w_e^2 l_{\mathrm{T}}\\
H_3(x_3) &= c_7 C_a^{\mathrm{R}} a_{1s} + c_7 Q_{\mathrm{R}}\sin b_{1s} + c_7 K_{T_{\mathrm{M}}}\theta_{\mathrm{M}} w_e^2 h_{\mathrm{R}}\\
&\quad - c_7 K_{T_{\mathrm{M}}}\theta_{\mathrm{M}} w_e^2\cos a_{1s}\cos b_{1s} l_{\mathrm{R}}\\
H_4(x_3) &= c_4 C_b^{\mathrm{R}} b_{1s} - c_4 Q_{\mathrm{R}}\sin a_{1s} - c_4 K_{T_{\mathrm{M}}}\theta_{\mathrm{M}} w_e^2\sin b_{1s} h_{\mathrm{R}}\\
&\quad - c_4 K_{T_{\mathrm{M}}}\theta_{\mathrm{M}} w_e^2\cos a_{1s}\cos b_{1s} y_{\mathrm{R}} - c_4 K_{T_{\mathrm{T}}}\theta_{\mathrm{T}} w_e^2 h_{\mathrm{T}}\\
&\quad - c_9 Q_{\mathrm{R}}\cos a_{1s}\cos b_{1s} + c_9 K_{T_{\mathrm{M}}}\theta_{\mathrm{M}} w_e^2\sin b_{1s} l_{\mathrm{R}} + c_9 K_{T_{\mathrm{T}}}\theta_{\mathrm{T}} w_e^2 l_{\mathrm{T}}
\end{aligned}$$

From (6.12), it is difficult to directly design the robust attitude control for helicopters due to the input implicit function $H(x_3)$. To expediently design model-based attitude control, $H(x_3)$ is separated into two parts including the linear part $Gx_3$ and the nonlinear part $F_2(x_3)$. On the other hand, to improve the closed-loop system control performance, we include the influence of actuator dynamics in the control design, where the actuator dynamics are assumed to be in the first-order low-pass filter form. Considering (6.12), the attitude dynamics of a helicopter including the actuator dynamics, the model uncertainties and the external disturbances can be rewritten as

$$\begin{aligned} \dot{x}_1 &= J(x_1)x_2 \\ \dot{x}_2 &= F(x_1, x_2) + Gx_3 + \Delta F_1(x_1, x_2) + F_2(x_3) + D(x_1, x_2, t) \\ \dot{x}_3 &= -\lambda(x_3 - u) \\ y &= x_1 \end{aligned} \tag{6.13}$$

where

$$G = \begin{bmatrix} \frac{\tau_1}{m} & \frac{\tau_2}{m} & 0 & \frac{\tau_3}{m} \\ c_3\tau_4 & c_3\tau_5 & c_3\tau_6 + c_4\tau_8 & c_3\tau_7 + c_4\tau_9 \\ 0 & 0 & c_7\tau_8 & c_7\tau_9 \\ c_4\tau_4 & c_4\tau_5 & c_4\tau_6 + c_9\tau_8 & c_4\tau_7 + c_9\tau_9 \end{bmatrix}, \ \tau_j > 0$$

are control gain parameters and $u$ is the system input. The 3rd sub-equation of (6.13) is the actuator dynamics and $\lambda$ is the actuator gain. $F_2(x_3) = [f_{21}(x_3), f_{22}(x_3), f_{23}(x_3), f_{24}(x_3)]^{\mathrm{T}}$, $f_{2i}(x_3)$, $i = 1, 2, 3, 4$ are given by

$$F_2(x_3) = \begin{bmatrix} 1 & 0 & 0 & 0 \\ 0 & c_3 & 0 & c_4 \\ 0 & 0 & 0 & c_7 \\ 0 & c_4 & 0 & c_9 \end{bmatrix} \begin{bmatrix} f_{31}(x_3) \\ f_{32}(x_3) \\ f_{33}(x_3) \\ f_{34}(x_3) \end{bmatrix}$$

$$\begin{aligned} f_{31}(x_3) &= -\tau_1 b_{1s} - \tau_2 a_{1s} - \tau_3 \theta_{\mathrm{M}} - K_{T_{\mathrm{M}}} \theta_{\mathrm{M}} w_e^2 \\ f_{32}(x_3) &= (C_b^{\mathrm{R}} - \tau_4) b_{1s} - (Q_{\mathrm{R}} + \tau_5) a_{1s} - K_{T_{\mathrm{M}}} \theta_{\mathrm{M}} w_e^2 b_{1s} h_{\mathrm{R}} \\ &\quad -(K_{T_{\mathrm{M}}} w_e^2 y_{\mathrm{R}} + \tau_7)\theta_{\mathrm{M}} - (K_{T_{\mathrm{T}}} w_e^2 h_{\mathrm{T}} + \tau_6)\theta_{\mathrm{T}} \\ f_{33}(x_3) &= C_a^{\mathrm{R}} a_{1s} + Q_{\mathrm{R}} b_{1s} - K_{T_{\mathrm{M}}} \theta_{\mathrm{M}} w_e^2 a_{1s} h_{\mathrm{R}} - K_{T_{\mathrm{M}}} \theta_{\mathrm{M}} w_e^2 l_{\mathrm{R}} \\ f_{34}(x_3) &= -Q_{\mathrm{R}} + (K_{T_{\mathrm{M}}} w_e^2 b_{1s} l_{\mathrm{R}} - \tau_9)\theta_{\mathrm{M}} + (K_{T_{\mathrm{T}}} w_e^2 l_{\mathrm{T}} - \tau_8)\theta_{\mathrm{T}} \end{aligned} \tag{6.14}$$

It is easy to know that $J(x_1)$ is invertible for all $\theta \in (-\pi/2, \pi/2)$ and $\phi \in (-\pi/2, \pi/2)$. To facilitate control system design, we assume that all states of the helicopter attitude dynamics (6.13) are measurable. Moreover, the following assumptions are needed for the subsequent developments.

**Assumption 6.1.** [101] For the continuous functions $D_i(x_1, x_2, t) : \mathcal{R}^4 \times \mathcal{R}^4 \times \mathcal{R} \rightarrow \mathcal{R}, i = 1, 2, 3, 4$, there exist positive, smooth, nondecreasing functions $h_i(x_1, x_2) : \mathcal{R}^4 \times \mathcal{R}^4 \rightarrow \mathcal{R}^+$ and time-dependent functions $n_i(t) : \mathcal{R}^+ \rightarrow \mathcal{R}^+, i = 1, 2, 3$ such that

$$|D_i(x_1, x_2, t)| \leq h_i(x_1, x_2) + n_i(t) \tag{6.15}$$

where

$$n_i(t) \leq \bar{n}_i$$

with unknown constants $\bar{n}_i \in \mathcal{R}^+, \ \forall t > t_0$.

**Lemma 6.1.** *[34, 101] For bounded initial conditions, if there exists a $C^1$ continuous and positive definite Lyapunov function $V(x)$ satisfying $\gamma_1(\|x\|) \leq V(x) \leq \gamma_2(\|x\|)$, such that $\dot{V}(x) \leq -\kappa V(x) + c$, where $\gamma_1, \gamma_2 : \mathcal{R}^n \rightarrow \mathcal{R}$ are class $K$ functions and $c$ is a positive constant, then the solution $x(t)$ is uniformly bounded.*

The control objective is to keep the desired altitude/attitude of helicopter in the presence of model uncertainty and environmental disturbance. Thus, the proposed control must render the helicopter track a desired attitude $x_{1d}$ such that the tracking errors converge to a very small neighborhood of the origin, i.e., $\lim_{t\rightarrow\infty} \|y - x_{1d}\| < \epsilon$ with $\epsilon > 0$.

**Assumption 6.2.** [101] For all $t > 0$, there exist $\delta_{11} > 0$, $\delta_{21} > 0$ and $\delta_{31} > 0$ such that $\|\dot{x}_{1d}(t)\| \leq \delta_{11}$, $\|\ddot{x}_{1d}(t)\| \leq \delta_{21}$ and $\|x_{1d}^{(3)}(t)\| \leq \delta_{31}$.

In the following sections, we start from the model-based attitude control for a nominal plant in Sect. 6.3 with the free of the uncertainty $\Delta F_1(x_1, x_2)$ and external disturbance $D(x_1, x_2, t)$ of system (6.13). However, the model-based control is sensitive to the disturbance and system uncertainty. When the disturbance and system uncertainty exist in helicopter dynamics, the closed-loop system control performance will be degraded, and will even lead to closed-loop system instability. To improve the robustness of the attitude control, the robust attitude control of helicopters using neural networks is designed in Sect. 6.4. Furthermore, in Sect. 6.5, we consider the case where the nominal plant is unknown also which makes the attitude control design more complicated, and approximation-based control in combination with backstepping technique is employed to keep the desired attitude of the helicopter system (6.13).

## 6.3 Model-Based Attitude Control for Nominal Plant

In this section, we assume that the moment coefficients and mass of the helicopter are known and neglect the uncertainty $\Delta F_1(x_1, x_2)$ and external disturbance $D(x_1, x_2, t)$ of system (6.13). Then, model-based backstepping attitude control is developed for the nominal dynamics of the helicopter. Rigorous analysis through

Lyapunov analysis is given to show the stability of the closed-loop system. To develop the model-based attitude control, we define error variables $z_1 = x_1 - x_{1d}$, $z_2 = x_2 - \alpha_1$ and $z_3 = x_3 - \alpha_2$, where $\alpha_1 \in \mathcal{R}^4$ and $\alpha_2 \in \mathcal{R}^4$ are virtual control laws.

*Step 1:* Considering (6.13) and differentiating $z_1$ with respect to time yields

$$\dot{z}_1 = \dot{x}_1 - \dot{x}_{1d} = J(x_1)(z_2 + \alpha_1) - \dot{x}_{1d} \tag{6.16}$$

Due to the non-singularity of $J(x_1)$, the virtual control law $\alpha_1$ is chosen as

$$\alpha_1 = J^{-1}(x_1)(-K_1 z_1 + \dot{x}_{1d}) \tag{6.17}$$

where $K_1 = K_1^{\mathrm{T}} > 0$.

Substituting (6.17) into (6.16), we obtain

$$\dot{z}_1 = J(x_1)z_2 - K_1 z_1 \tag{6.18}$$

Consider the Lyapunov function candidate $V_1 = \frac{1}{2}z_1^{\mathrm{T}} z_1$. The time derivative of $V_1$ is

$$\dot{V}_1 = -z_1^{\mathrm{T}} K_1 z_1 + z_1^{\mathrm{T}} J(x_1) z_2 \tag{6.19}$$

The first term on the right-hand side is negative, and the second term will be canceled in the next step.

*Step 2:* Differentiating $z_2$ with respect to time yields

$$\dot{z}_2 = \dot{x}_2 - \dot{\alpha}_1 = F(x_1, x_2) + Gx_3 + F_2(x_3) - \dot{\alpha}_1 \tag{6.20}$$

where $\dot{\alpha}_1 = \dot{J}(x_1)^{-1}(-K_1 z_1 + \dot{x}_{1d}) + J(x_1)^{-1}(-K_1 \dot{z}_1 + \ddot{x}_{1d})$.

Consider the Lyapunov function candidate

$$V_2 = V_1 + \frac{1}{2} z_2^{\mathrm{T}} z_2 \tag{6.21}$$

Invoking (6.20), the time derivative of $V_2$ is

$$\begin{aligned}\dot{V}_2 = {} & -z_1^{\mathrm{T}} K_1 z_1 + z_1^{\mathrm{T}} J(x_1) z_2 + z_2^{\mathrm{T}} F(x_1, x_2) + z_2^{\mathrm{T}} G(z_3 + \alpha_2) + z_2^{\mathrm{T}} F_2(x_3) \\ & -z_2^{\mathrm{T}} \dot{\alpha}_1\end{aligned} \tag{6.22}$$

The virtual control law $\alpha_2$ is proposed as follows:

$$\alpha_2 = Q^{+}(Z_1)[\dot{\alpha}_1 - K_2 z_2 - F(x_1, x_2) - J^{\mathrm{T}}(x_1) z_1] \tag{6.23}$$

where

$$Q^{+} = \mathrm{QG}^{\mathrm{T}}(\mathrm{GQG}^{\mathrm{T}})^{-1} \tag{6.24}$$

with $Q$ being chosen such that $GQG^T$ is nonsingular. When the matrix $G$ is determined, we always can find an appropriate matrix $Q$ that renders $GQG^T$ nonsingular.

Substituting (6.23) and (6.24) into (6.22), we have

$$\dot{V}_2 \leq -z_1^T K_1 z_1 - z_2^T K_2 z_2 + z_2^T G z_3 + z_2^T F_2(x_3) \tag{6.25}$$

*Step 3:* Differentiating $z_3$ with respect to time yields

$$\dot{z}_3 = \dot{x}_3 - \dot{\alpha}_2 = -\lambda(x_3 - u) - \dot{\alpha}_2 \tag{6.26}$$

Consider the Lyapunov function candidate

$$V_3 = V_2 + \frac{1}{2} z_3^T z_3 \tag{6.27}$$

Considering (6.25) and (6.26), the time derivative of $V_3$ is given by

$$\dot{V}_3 \leq -z_1^T K_1 z_1 - z_2^T K_2 z_2 + z_2^T G z_3 + z_2^T F_2(x_3) - \lambda z_3^T x_3 + \lambda z_3^T u - z_3^T \dot{\alpha}_2 \tag{6.28}$$

The input control $u$ is proposed as follows:

$$u = \begin{cases} x_3 - \lambda^{-1}\left(K_3 z_3 - \dot{\alpha}_2 + G^T z_2 + \frac{z_3 z_2^T F_2(x_3)}{\|z_3\|^2}\right), & \|z_3\| \geq \varepsilon_3 \\ 0, & \|z_3\| < \varepsilon_3 \end{cases} \tag{6.29}$$

where $K_3 = K_3^T > 0$ and $\varepsilon_3 > 0$ are the design parameters.

The above design procedure can be summarized in the following theorem.

**Theorem 6.2.** *Considering the nominal attitude dynamics of the helicopter system (6.13), the model-based control law is designed according to (6.29). Under the proposed model-based control and for any bounded initial condition, the closed-loop signals $z_1$, $z_2$ and $z_3$ are bounded. Namely, there exist design parameters $K_1 = K_1^T > 0$, $K_2 = K_2^T > 0$ and $K_3 = K_3^T > 0$ such that the overall closed-loop control system is semi-globally stable. Furthermore, the tracking error $z_1$ converges to a compact set and the control objective is obtained.*

*Proof.* When $\|z_3\| \geq \varepsilon_3$, substituting the first equation of (6.29) into (6.28), we obtain

$$\dot{V}_3 \leq -z_1^T K_1 z_1 - z_2^T K_2 z_2 - z_3^T K_3 z_3 \leq -\lambda_{\min}(K_i) V_3 \tag{6.30}$$

where $i = 1, 2, 3$ and $\lambda_{\min}(.)$ denotes the smallest eigenvalues of a matrix. Therefore, we know that the closed-loop system is stable when $\|z_3\| \geq \varepsilon_3$ according to (6.30). If $\|z_3\| < \varepsilon_3$, $z_3$ approximates to zero. This means that $x_3 = \alpha_2$. From

(6.14), we know that $F_2(x_3)$ is bounded due to the bounded available deflexion angles of the main rotor and the tail rotor. Based on the boundary of $F_2(x_3)$, we can conclude all signals of the closed-loop system are bounded according to Lemma 6.1 if only appropriate design parameters $K_1$ and $K_2$ are chosen. This concludes the proof. □

*Remark 6.3.* To handle the nonlinear term $F_2(x_3)$ in (6.13), the model-based control is proposed as a discontinuous form which can excite the chattering phenomenon. However, we can adjust design parameter $\varepsilon_3$ to decrease the chattering phenomenon and improve the control performance. Furthermore, the $\dot{\alpha}_2$ is used in the model-based control law (6.29). From (6.23), we can see the $\alpha_2$ is continuous and differentiable in which the design matrix $Q$ is introduced to avoid the potential singularity of $G$ . Since $G$ is known which is independent of system states and can be designed according to the control demand, we can always choose an appropriate design matrix $Q$ and parameters $\tau_i$ to make $\mathrm{GQG}^{\mathrm{T}}$ nonsingular.

## 6.4 Robust Attitude Control of Helicopters with Uncertainties and Disturbances

In this section, we present a robust attitude control in combination with a radial basis function neural network (RBFNN) to keep the desired attitude of a helicopter system (6.13) in the presence of model uncertainty and external disturbance. The RBFNN can be considered as a two-layer network in which the hidden layer performs a fixed nonlinear transformation with no adjustable parameters, i.e., the input space is mapped into a new space. The output layer then combines the outputs in the latter space linearly. Therefore, they belong to a class of linearly parameterized networks.

Define the error variables $z_1 = x_1 - x_{1d}$, $z_2 = x_2 - \alpha_1$ and $z_3 = x_3 - \alpha_2$ which are the same as the related definitions of variables in Sect. 6.3. Since there are no model uncertainties and disturbances in the altitude and attitude angle equations, the design of Step 1 is similar to the case of the model-based attitude control for nominal helicopter dynamics in Sect. 3. Here, we only give the design processes of Steps 2 and 3 for the robust attitude control.

*Step 2:* Considering (6.13) and differentiating $z_2$ with respect to time yields

$$\begin{aligned}\dot{z}_2 &= \dot{x}_2 - \dot{\alpha}_1 \\ &= F(x_1, x_2) + Gx_3 + \Delta F_1(x_1, x_2) + F_2(x_3) + D(x_1, x_2, t) - \dot{\alpha}_1 \end{aligned} \tag{6.31}$$

where $\alpha_1$ is defined in (6.17).

Consider the Lyapunov function candidate

$$V_2^* = V_1 + \frac{1}{2} z_2^{\mathrm{T}} z_2 \tag{6.32}$$

Due to (6.19), (6.31) and Assumption 6.1, the time derivative of $V_2^*$ is given by

$$\begin{aligned}\dot{V}_2^* \leq& -z_1^{\mathrm{T}} K_1 z_1 + z_1^{\mathrm{T}} J(x_1) z_2 + z_2^{\mathrm{T}} F(x_1, x_2) + z_2^{\mathrm{T}} G(z_3 + \alpha_2) + z_2^{\mathrm{T}} F_2(x_3) \\ & - z_2^{\mathrm{T}} \rho(Z) - z_2^{\mathrm{T}} \dot{\alpha}_1 \end{aligned} \tag{6.33}$$

where $\rho(Z) = -\Delta F_1(x_1, x_2) - \mathrm{Sgn}(z_2)(h(x_1, x_2) + \bar{n})$, $Z = [x_1^{\mathrm{T}}, x_2^{\mathrm{T}}, \alpha_1^{\mathrm{T}}]$, $\mathrm{Sgn}(z_2) := \mathrm{diag}\{\mathrm{sgn}(z_{2j})\}$, $h(x_1, x_2) := [h_1(x_1, x_2), h_2(x_1, x_2), h_3(x_1, x_2), h_4(x_1, x_2)]^{\mathrm{T}}$, and $\bar{n} = [\bar{n}_1, \bar{n}_2, \bar{n}_3, \bar{n}_4]^{\mathrm{T}}$, $j = 1, 2, 3, 4$. Since $\Delta F_1(x_1, x_2)$, $h(x_1, x_2)$ and $\bar{n}$ are all unknown, the model-based control cannot be directly designed. To overcome this problem, we utilize RBFNNs in [90] to approximate the unknown term $\rho(Z)$ which is expressed as

$$\hat{\rho}(Z) = \hat{\Theta}^{\mathrm{T}} S(Z) \tag{6.34}$$

where $\hat{\Theta} \in \mathcal{R}^{l \times 4}$ is the approximation parameter, $S(Z) = [s_1(Z), s_2(Z), ..., s_l(Z)]^{\mathrm{T}} \in \mathcal{R}^{l \times 1}$ represents the vector of the smooth basis function, with the NN node number $l > 1$ and $s_i(Z)$ being chosen as the commonly used Gaussian functions $s_i(Z) = \exp\left[\frac{-(Z-\mu_i)^{\mathrm{T}}(Z-\mu_i)}{\eta_i^2}\right]$, $i = 1, 2, ..., l$, where $\mu_i$ is the center of the receptive field and $\eta_i$ is the width of the Gaussian function. $\hat{\Theta}^{\mathrm{T}} S(Z)$ approximates $\Theta^{*\mathrm{T}} S(Z)$ given by

$$\Theta^{*\mathrm{T}} S(Z) + \varepsilon = \rho(Z) \tag{6.35}$$

where $\Theta^*$ is the optimal weight value of RBFNN.

Substituting (6.35) into (6.33), we obtain

$$\begin{aligned}\dot{V}_2^* \leq& -z_1^{\mathrm{T}} K_1 z_1 + z_1^{\mathrm{T}} J(x_1) z_2 + z_2^{\mathrm{T}} F(x_1, x_2) + z_2^{\mathrm{T}} G z_3 - z_2^{\mathrm{T}} \dot{\alpha}_1 \\ & + z_2^{\mathrm{T}} G \alpha_2 + z_2^{\mathrm{T}} F_2(x_3) + z_2^{\mathrm{T}}(-\Theta^{*\mathrm{T}} S(Z) - \varepsilon)\end{aligned} \tag{6.36}$$

The virtual control law $\alpha_2$ is proposed based on the RBFNNs as follows.

$$\alpha_2 = Q^+(Z_1)[\dot{\alpha}_1 - K_2 z_2 + \hat{\rho}_1(Z_1) - F(x_1, x_2) - J^{\mathrm{T}}(x_1) z_1] \tag{6.37}$$

where $Q^+$ is defined in (6.24).

Substituting (6.37) into (6.36), we have

$$\dot{V}_2^* \leq -z_1^{\mathrm{T}} K_1 z_1 - z_2^{\mathrm{T}} K_2 z_2 + z_2^{\mathrm{T}} G z_3 + z_2^{\mathrm{T}} F_2(x_3) + z_2^{\mathrm{T}} \tilde{\Theta}^{\mathrm{T}} S(Z) - z_2^{\mathrm{T}} \varepsilon \tag{6.38}$$

where $\tilde{\Theta} = \hat{\Theta} - \Theta^*$.

Considering the stability of error signals $\tilde{\Theta}$, the augmented Lyapunov function candidate can be written as

$$V_2 = V_2^* + \frac{1}{2}\mathrm{tr}(\tilde{\Theta}^{\mathrm{T}} \Lambda^{-1} \tilde{\Theta}) \tag{6.39}$$

where $\Lambda = \Lambda^{\mathrm{T}} > 0$.

The time derivative of $V_2$ along (6.38) is

$$\dot{V}_2 \le -z_1^{\mathrm{T}} K_1 z_1 - z_2^{\mathrm{T}} K_2 z_2 + z_2^{\mathrm{T}} G z_3 + z_2^{\mathrm{T}} F_2(x_3) + z_2^{\mathrm{T}} \tilde{\Theta}^{\mathrm{T}} S(Z) - z_2^{\mathrm{T}} \varepsilon + \mathrm{tr}(\tilde{\Theta}^{\mathrm{T}} \Lambda^{-1} \dot{\tilde{\Theta}}) \tag{6.40}$$

Consider the adaptive laws for $\hat{\Theta}$ as

$$\dot{\hat{\Theta}} = -\Lambda(S(Z_1) z_2^{\mathrm{T}} + \sigma \hat{\Theta}) \tag{6.41}$$

where $\sigma > 0$.

Noting the following facts

$$-z_2^{\mathrm{T}} \varepsilon \le \frac{1}{2}\|\varepsilon\|^2 + \frac{1}{2}\|z_2\|^2 \tag{6.42}$$

$$2\tilde{\Theta}^{\mathrm{T}} \hat{\Theta} = \|\tilde{\Theta}\|^2 + \|\hat{\Theta}\|^2 - \|\Theta^*\|^2 \ge \|\tilde{\Theta}\|^2 - \|\Theta^*\|^2 \tag{6.43}$$

and considering (6.41), we have

$$\dot{V}_2 \le -z_1^{\mathrm{T}} K_1 z_1 - z_2^{\mathrm{T}} K_2 z_2 + z_2^{\mathrm{T}} G z_3 + z_2^{\mathrm{T}} F_2(x_3) + \frac{1}{2}\|\varepsilon\|^2 + \frac{1}{2}\|z_2\|^2 - \frac{\sigma}{2}\|\tilde{\Theta}\|^2 + \frac{\sigma}{2}\|\Theta^*\|^2 \tag{6.44}$$

*Step 3:* Differentiating $z_3$ with respect to time yields

$$\dot{z}_3 = \dot{x}_3 - \dot{\alpha}_2 = -\lambda(x_3 - u) - \dot{\alpha}_2 \tag{6.45}$$

Consider the Lyapunov function candidate

$$V_3 = V_2 + \frac{1}{2} z_3^{\mathrm{T}} z_3 \tag{6.46}$$

Considering (6.45), the time derivative of $V_3$ is given by

$$\dot{V}_3 \le -z_1^{\mathrm{T}} K_1 z_1 - z_2^{\mathrm{T}} K_2 z_2 + z_2^{\mathrm{T}} G z_3 + z_2^{\mathrm{T}} F_2(x_3) + \frac{1}{2}\|\varepsilon\|^2 + \frac{1}{2}\|z_2\|^2 - \frac{\sigma}{2}\|\tilde{\Theta}\|^2 + \frac{\sigma}{2}\|\Theta^*\|^2 - \lambda z_3^{\mathrm{T}} x_3 + \lambda z_3^{\mathrm{T}} u - z_3^{\mathrm{T}} \dot{\alpha}_2 \tag{6.47}$$

The control law $u$ is proposed as follows:

$$u = \begin{cases} x_3 - \lambda^{-1}\left(K_3 z_3 - \dot{\alpha}_2 + G^{\mathrm{T}} z_2 + \frac{z_3 z_2^{\mathrm{T}} F_2(x_3)}{\|z_3\|^2}\right), & \|z_3\| \ge \varepsilon_3 \\ 0, & \|z_3\| < \varepsilon_3 \end{cases} \tag{6.48}$$

where $K_3 = K_3^{\mathrm{T}} > 0$ and $\varepsilon_3$ are the design parameters.

The above design procedure can be summarized in the following theorem.

**Theorem 6.4.** *Consider the helicopter attitude dynamics (6.13) that satisfy the Assumptions 6.1–6.2. The robust attitude control is designed according to (6.48) using neural networks and parameter updated law is chosen as (6.41). For bounded initial conditions, there exist design parameters* $\sigma > 0$, $\Lambda = \Lambda^T > 0$, $K_1 = K_1^T > 0$, $K_2 = K_2^T > 0$ *and* $K_3 = K_3^T > 0$ *such that the overall closed-loop control system is semi-globally stable in the sense that all of the closed-loop signals* $z_1$, $z_2$, $z_3$ *and* $\tilde{\Theta}$ *are bounded. Furthermore, the tracking error* $z_1$ *converges to the compact set* $\Omega_{z_1} := \left\{ z_1 \in R^4 | \; \|z_1\| \leq \sqrt{D} \right\}$ *where* $D = 2(V_3(0) + \frac{C}{\kappa})$.

*Proof.* When $\|z_3\| \geq \varepsilon_3$, substituting (6.48) into (6.47), we obtain

$$\begin{aligned}\dot{V}_3 &\leq -z_1^T K_1 z_1 - z_2^T \left( K_2 - \frac{1}{2} I_{3\times 3} \right) z_2 - z_3^T K_3 z_3 + \frac{1}{2}\|\varepsilon\|^2 - \frac{\sigma}{2}\|\tilde{\Theta}\|^2 + \frac{\sigma}{2}\|\Theta^*\|^2 \\ &\leq -\kappa V_3 + C \end{aligned} \quad (6.49)$$

where

$$\begin{aligned}\kappa &:= \min\left( 2\lambda_{\min}(K_1), 2\lambda_{\min}\left( K_2 - \frac{1}{2} I_{3\times 3} \right), 2\lambda_{\min}(K_3), \frac{2\sigma}{\lambda_{\max}(\Lambda^{-1})} \right) \\ C &:= \frac{1}{2}\|\bar{\varepsilon}\|^2 + \frac{\sigma}{2}\|\Theta^*\|^2 \end{aligned} \quad (6.50)$$

To ensure that $\kappa > 0$, the design parameter $K_2$ must make $K_2 - \frac{1}{2} I_{3\times 3} > 0$.
Multiplying (6.49) by $e^{\kappa t}$ yields

$$\frac{d}{dt}\left( V_3(t) e^{\kappa t} \right) \leq e^{\kappa t} C \quad (6.51)$$

Integrating (6.51) over $[0, \; t]$, we obtain

$$0 \leq V_3(t) \leq \frac{C}{\kappa} + \left[ V_3(0) - \frac{C}{\kappa} \right] e^{-\kappa t} \quad (6.52)$$

According to (6.51) and (6.52), we can prove the bounded stability of the closed-loop system when $\|z_3\| \geq \varepsilon_3$. When $\|z_3\| < \varepsilon_3$, we can also conclude all signals of the closed-loop system are bounded based on Lemma 6.1 if only appropriate design parameters $K_1$ and $K_2$ are chosen according to the bounded $F_2(x_3)$. Therefore, all signals of the closed-loop system, i.e., $z_1$, $z_2$, $z_3$ and $\tilde{\Theta}$ are uniformly ultimately bounded. From (6.46), we know that for any given design parameters $\sigma$, $\Lambda$, $K_1$, $K_2$ and $K_3$ can be used to adjust the closed-loop system performance. This concludes the proof. □

## 6.5 Approximation-Based Attitude Control of Helicopters

In this section, we consider the case where all moment coefficients and the helicopter mass are unknown which make the attitude control design of (6.13) more complicated. Approximation-based control in combination with backstepping technique is employed to keep the desired attitude of helicopter system (6.13). Define an auxiliary design variable $\xi = J(x_1)x_2$ and the error variables $z_1 = x_1 - x_{1d}$, $z_2 = \xi - \alpha_1$ and $z_3 = x_3 - \alpha_2$, where $\alpha_1 \in \mathcal{R}^4$ and $\alpha_2 \in \mathcal{R}^4$ are virtual control laws. It is apparent that $x_2 \to 0$ if $\xi \to 0$ due to the non-singularity of $J(x_1)$.

*Step 1:* Differentiating $z_1$ in (6.13) with respect to time yields

$$\dot{z}_1 = \dot{x}_1 - \dot{x}_{1d} = \xi - \dot{x}_{1d} = z_2 + \alpha_1 - \dot{x}_{1d} \tag{6.53}$$

The virtual control law $\alpha_1$ is chosen as

$$\alpha_1 = -K_1 z_1 + \dot{x}_{1d} \tag{6.54}$$

where $K_1 = K_1^{\mathrm{T}} > 0$.

Substituting (6.54) into (6.53), we obtain

$$\dot{z}_1 = -K_1 z_1 + z_2 \tag{6.55}$$

Consider the Lyapunov function candidate $V_1 = \frac{1}{2} z_1^{\mathrm{T}} z_1$. The time derivative of $V_1$ is

$$\dot{V}_1 = -z_1^{\mathrm{T}} K_1 z_1 + z_1^{\mathrm{T}} z_2 \tag{6.56}$$

*Step 2:* Differentiating $z_2$ with respect to time yields

$$\begin{aligned}\dot{z}_2 = \dot{\xi} - \dot{\alpha}_1 &= \dot{J}(x_1)x_2 + J(x_1)\dot{x}_2 - \dot{\alpha}_1 \\ &= \dot{J}(x_1)x_2 + J(x_1)F(x_1, x_2) + J(x_1)Gx_3 + J(x_1)\Delta F_1(x_1, x_2) \\ &\quad + J(x_1)F_2(x_3) + J(x_1)D(x_1, x_2, t) - \dot{\alpha}_1\end{aligned} \tag{6.57}$$

where $\dot{\alpha}_1 = \ddot{x}_{1d} - K_1 \dot{z}_1$.

Consider the Lyapunov function candidate

$$V_2^* = V_1 + \frac{1}{2} z_2^{\mathrm{T}} z_2 \tag{6.58}$$

Due to (6.57) and Assumption 6.1, the time derivative of $V_2^*$ is given by

$$\begin{aligned}\dot{V}_2^* \le &-z_1^{\mathrm{T}} K_1 z_1 + z_1^{\mathrm{T}} z_2 + z_2^{\mathrm{T}} J(x_1) F_2(x_3) - z_2^{\mathrm{T}} \rho_1(Z_1) - z_2^{\mathrm{T}} \rho_2(Z_2)\alpha_2 \\ &- z_2^{\mathrm{T}} \rho_2(Z_2) z_3 - z_2^{\mathrm{T}} \dot{\alpha}_1\end{aligned} \tag{6.59}$$

where $\rho_1(Z_1) = -\dot{J}(x_1)x_2 - J(x_1)F(x_1,x_2) - J(x_1)\Delta F_1(x_1,x_2) - J(x_1)\mathrm{Sgn}(\zeta_2)(h(x_1,x_2)+\bar{n})$, $\rho_2(Z_2) = -J(x_1)G$, $Z_1 = [x_1^{\mathrm{T}}, x_2^{\mathrm{T}}, \alpha_1^{\mathrm{T}}]$, $Z_2 = x_1$, $\mathrm{Sgn}(\zeta_2) := \mathrm{diag}\{\mathrm{sgn}(\xi_{2j})\}$, $\zeta_2 = z_2^{\mathrm{T}} J(x_1)$, $h(x_1,x_2) := [h_1(x_1,x_2), h_2(x_1,x_2), h_3(x_1,x_2), h_4(x_1,x_2)]^{\mathrm{T}}$, and $\bar{n} = [\bar{n}_1, \bar{n}_2, \bar{n}_3, \bar{n}_4]^{\mathrm{T}}$, $j = 1,2,3,4$. Since $F(x_1,x_2)$ and $G$ are all unknown, the previous proposed robust attitude control cannot be implemented. To overcome this problem, we utilize the RBFNNs in [90] to approximate the unknown terms $\rho_1(Z_1)$ and $\rho_2(Z_2)$ as

$$\hat{\rho}_1(Z_1) = \hat{\Theta}_1^{\mathrm{T}} S_1(Z_1) \tag{6.60}$$

$$\hat{\rho}_2(Z_2) = \hat{\Theta}_2^{\mathrm{T}} S_2(Z_2) \tag{6.61}$$

where $\hat{\Theta}_i$ are the approximation parameters, and $S_i(Z_i)$ represents the basis functions, $i = 1,2$. The optimal approximation $\Theta_i^{*\mathrm{T}} S_i(Z_i)$ is given by

$$\Theta_1^{*\mathrm{T}} S_1(Z_1) + \varepsilon_1 = \rho_1(Z_1) \tag{6.62}$$

$$\Theta_2^{*\mathrm{T}} S_2(Z_2) + \varepsilon_2 = \rho_2(Z_2) \tag{6.63}$$

where $\Theta_i^*$ are optimal weight values of RBFNNs and $\varepsilon_i$ is the approximation error satisfying $|\varepsilon_i| \le \bar{\varepsilon}_i$ with constant $\bar{\varepsilon}_i > 0$, $i = 1,2$.

Substituting (6.62) and (6.63) into (6.59), we obtain

$$\begin{aligned}\dot{V}_2^* \le & -z_1^{\mathrm{T}} K_1 z_1 + z_1^{\mathrm{T}} z_2 + z_2^{\mathrm{T}} J(x_1) F_2(x_3) + z_2^{\mathrm{T}} \tilde{\Theta}_1^{\mathrm{T}} S_1(Z_1) - z_2^{\mathrm{T}} \varepsilon_1 \\ & + z_2^{\mathrm{T}} \tilde{\Theta}_2^{\mathrm{T}} S_2(Z_2)\alpha_2 - z_2^{\mathrm{T}} \varepsilon_2 \alpha_2 - z_2^{\mathrm{T}} \hat{\Theta}_1^{\mathrm{T}} S_1(Z_1) - z_2^{\mathrm{T}} \hat{\Theta}_2^{\mathrm{T}} S_2(Z_1)\alpha_2 \\ & - z_2^{\mathrm{T}} \rho_2(Z_1) z_3 - z_2^{\mathrm{T}} \dot{\alpha}_1\end{aligned} \tag{6.64}$$

where $\tilde{\Theta}_1 = \hat{\Theta}_1 - \Theta_1^*$ and $\tilde{\Theta}_2 = \hat{\Theta}_2 - \Theta_2^*$.

The virtual control law $\alpha_2$ is proposed based on the NNs as follows.

$$\alpha_2 = \hat{\rho}_2^{+}(Z_2)[K_2 z_2 - \dot{\alpha}_1 - \hat{\rho}_1(Z_1) + z_1] \tag{6.65}$$

where

$$\alpha_{20} = K_2 z_2 - \dot{\alpha}_1 - \hat{\rho}_1(Z_1) + z_1 \tag{6.66}$$

$$\hat{\rho}_2^{+}(Z_2) = \hat{\rho}_2^{\mathrm{T}}(Z_2)[\delta I_{4\times 4} + \hat{\rho}_2(Z_2)\hat{\rho}_2^{\mathrm{T}}(Z_2)]^{-1} \tag{6.67}$$

herein $K_2 = K_2^{\mathrm{T}} > 0$ and $\delta > 0$ are design parameters.

It is clear that we have

$$\hat{\rho}_2(Z_2)\hat{\rho}_2^{\mathrm{T}}(Z_2)[\delta I_{3\times 3} + \hat{\rho}_2(Z_2)\hat{\rho}_2^{\mathrm{T}}(Z_2)]^{-1} = I - \delta[\delta I_{4\times 4} + \hat{\rho}_2(Z_2)\hat{\rho}_2^{\mathrm{T}}(Z_2)]^{-1}$$

Substituting (6.65) and (6.68) into (6.64), we have

$$\dot{V}_2^* \leq -z_1^{\mathrm{T}} K_1 z_1 - z_2^{\mathrm{T}} K_2 z_2 + z_2^{\mathrm{T}} \tilde{\Theta}_1^{\mathrm{T}} S_1(Z_1) - z_2^{\mathrm{T}} \varepsilon_1 + z_2^{\mathrm{T}} \tilde{\Theta}_2^{\mathrm{T}} S_2(Z_2) \alpha_2$$
$$- z_2^{\mathrm{T}} \varepsilon_2 \alpha_2 + \delta z_2^{\mathrm{T}} [\delta I_{4\times4} + \hat{\rho}_2(Z_2) \hat{\rho}_2^{\mathrm{T}}(Z_2)]^{-1} \alpha_{20} - z_2^{\mathrm{T}} \rho_2(Z_2) z_3$$
$$+ z_2^{\mathrm{T}} J(x_1) F_2(x_3) \tag{6.68}$$

Considering the stability of error signals $\tilde{\Theta}_1$ and $\tilde{\Theta}_2$, the augmented Lyapunov function candidate can be written as

$$V_2 = V_2^* + \frac{1}{2}\mathrm{tr}(\tilde{\Theta}_1^{\mathrm{T}} \Lambda_1^{-1} \tilde{\Theta}_1) + \frac{1}{2}\mathrm{tr}(\tilde{\Theta}_2^{\mathrm{T}} \Lambda_2^{-1} \tilde{\Theta}_2) \tag{6.69}$$

where $\Lambda_1 = \Lambda_1^{\mathrm{T}} > 0$ and $\Lambda_2 = \Lambda_2^{\mathrm{T}} > 0$.

The time derivative of $V_2$ along (6.68) is

$$\dot{V}_2 \leq -z_1^{\mathrm{T}} K_1 z_1 - z_2^{\mathrm{T}} K_2 z_2 + z_2^{\mathrm{T}} J(x_1) F_2(x_3) + z_2^{\mathrm{T}} \tilde{\Theta}_1^{\mathrm{T}} S_1(Z_1) - z_2^{\mathrm{T}} \varepsilon_1$$
$$+ z_2^{\mathrm{T}} \tilde{\Theta}_2^{\mathrm{T}} S_2(Z_2) \alpha_2 - z_2^{\mathrm{T}} \varepsilon_2 \alpha_2 + \delta z_2^{\mathrm{T}} [\delta I_{4\times4} + \hat{\rho}_2(Z_2) \hat{\rho}_2^{\mathrm{T}}(Z_2)]^{-1} \alpha_{20}$$
$$- z_2^{\mathrm{T}} \rho_2(Z_2) z_3 + \mathrm{tr}(\tilde{\Theta}_1^{\mathrm{T}} \Lambda_1^{-1} \dot{\hat{\Theta}}_1) + \mathrm{tr}(\tilde{\Theta}_2^{\mathrm{T}} \Lambda_2^{-1} \dot{\hat{\Theta}}_2) \tag{6.70}$$

Consider the adaptive laws for $\hat{\Theta}_1$ and $\hat{\Theta}_2$ as

$$\dot{\hat{\Theta}}_1 = -\Lambda_1 (S_1(Z_1) z_2^{\mathrm{T}} + \sigma_1 \hat{\Theta}_1) \tag{6.71}$$

$$\dot{\hat{\Theta}}_2 = -\Lambda_2 (S_2(Z_1) \alpha_2 z_2^{\mathrm{T}} + \sigma_2 \hat{\Theta}_2) \tag{6.72}$$

where $\sigma_1 > 0$ and $\sigma_2 > 0$.

Noting the following facts

$$-z_2^{\mathrm{T}} \varepsilon_1 \leq \frac{1}{2}\|\varepsilon_1\|^2 + \frac{1}{2}\|z_2\|^2 \tag{6.73}$$

$$-z_2^{\mathrm{T}} \varepsilon_2 \alpha_2 \leq \frac{1}{2}\|\varepsilon_2\|^2 + \frac{1}{2}\|z_2\|^2 \|\alpha_2\|^2 \tag{6.74}$$

$$2\tilde{\Theta}_1^{\mathrm{T}} \hat{\Theta}_1 = \|\tilde{\Theta}_1\|^2 + \|\hat{\Theta}_1\|^2 - \|\Theta_1^*\|^2 \geq \|\tilde{\Theta}_1\|^2 - \|\Theta_1^*\|^2 \tag{6.75}$$

$$2\tilde{\Theta}_2^{\mathrm{T}} \hat{\Theta}_2 = \|\tilde{\Theta}_2\|^2 + \|\hat{\Theta}_2\|^2 - \|\Theta_2^*\|^2 \geq \|\tilde{\Theta}_2\|^2 - \|\Theta_2^*\|^2 \tag{6.76}$$

and considering (6.71) and (6.72), we have

$$\dot{V}_2 \leq -z_1^{\mathrm{T}} K_1 z_1 - z_2^{\mathrm{T}} K_2 z_2 + z_2^{\mathrm{T}} J(x_1) F_2(x_3) + \frac{1}{2}\|\varepsilon_1\|^2 + \frac{1}{2}\|z_2\|^2 + \frac{1}{2}\|\varepsilon_2\|^2$$
$$+ \frac{1}{2}\|z_2\|^2 \|\alpha_2\|^2 + \delta z_2^{\mathrm{T}} [\delta I_{4\times4} + \hat{\rho}_2(Z_2) \hat{\rho}_2^{\mathrm{T}}(Z_2)]^{-1} \alpha_{20} - z_2^{\mathrm{T}} \rho_2(Z_2) z_3$$
$$- \frac{\sigma_1}{2}\|\tilde{\Theta}_1\|^2 + \frac{\sigma_1}{2}\|\Theta_1^*\|^2 - \frac{\sigma_2}{2}\|\tilde{\Theta}_2\|^2 + \frac{\sigma_2}{2}\|\Theta_2^*\|^2 \tag{6.77}$$

*Step 3:* Differentiating $z_3$ with respect to time yields

$$\dot{z}_3 = \dot{x}_3 - \dot{\alpha}_2 = -\lambda(x_3 - u) - \dot{\alpha}_2 \tag{6.78}$$

Consider the Lyapunov function candidate

$$V_3^* = V_2 + \frac{1}{2} z_3^{\mathrm{T}} z_3 \tag{6.79}$$

Considering (6.78), the time derivative of $V_3^*$ is

$$\begin{aligned}\dot{V}_3^* \le& -z_1^{\mathrm{T}} K_1 z_1 - z_2^{\mathrm{T}} K_2 z_2 + z_2^{\mathrm{T}} J(x_1) F_2(x_3) + \frac{1}{2}\|\varepsilon_1\|^2 + \frac{1}{2}\|z_2\|^2 + \frac{1}{2}\|\varepsilon_2\|^2 \\ &+ \frac{1}{2}\|z_2\|^2\|\alpha_2\|^2 + \delta z_2^{\mathrm{T}}[\delta I_{4\times 4} + \hat{\rho}_2(Z_1)\hat{\rho}_2^{\mathrm{T}}(Z_1)]^{-1}\alpha_{20} - z_2^{\mathrm{T}} \rho_2(Z_1) z_3 \\ &- \frac{\sigma_1}{2}\|\tilde{\Theta}_1\|^2 + \frac{\sigma_1}{2}\|\Theta_1^*\|^2 - \frac{\sigma_2}{2}\|\tilde{\Theta}_2\|^2 + \frac{\sigma_2}{2}\|\Theta_2^*\|^2 - \lambda z_3^{\mathrm{T}} x_3 + \lambda z_3^{\mathrm{T}} u \\ &- z_3^{\mathrm{T}} \dot{\alpha}_2 \end{aligned} \tag{6.80}$$

From (6.14), we know that $F_2(x_3)$ is unknown due to the unknown moment coefficients. Thus, it cannot be used to design the attitude control. To conveniently develop the approximation-based attitude control, the following variables are given:

$$\bar{\zeta}_2 = [\zeta_{22}, \zeta_{23}, \zeta_{24}]^{\mathrm{T}} \tag{6.81}$$

$$\bar{F}_2(x_3) = [f_{22}(x_3), f_{23}(x_3), f_{24}(x_3)]^{\mathrm{T}} \tag{6.82}$$

where $\zeta_{2i}$ is the $i$th row of vector $z_2^{\mathrm{T}} R(x_1)$.

Considering (6.81) and (6.82), (6.80) can be written as

$$\begin{aligned}\dot{V}_3^* \le& -z_1^{\mathrm{T}} K_1 z_1 - z_2^{\mathrm{T}} K_2 z_2 + \zeta_{21} f_{31}(x_3) + \varsigma_2^{\mathrm{T}} \varphi + \frac{1}{2}\|\varepsilon_1\|^2 + \frac{1}{2}\|z_2\|^2 + \frac{1}{2}\|\varepsilon_2\|^2 \\ &+ \frac{1}{2}\|z_2\|^2\|\alpha_2\|^2 + \delta z_2^{\mathrm{T}}[\delta I_{4\times 4} + \hat{\rho}_2(Z_1)\hat{\rho}_2^{\mathrm{T}}(Z_1)]^{-1}\alpha_{20} - z_2^{\mathrm{T}} \rho_2(Z_1) z_3 \\ &- \frac{\sigma_1}{2}\|\tilde{\Theta}_1\|^2 + \frac{\sigma_1}{2}\|\Theta_1^*\|^2 - \frac{\sigma_2}{2}\|\tilde{\Theta}_2\|^2 + \frac{\sigma_2}{2}\|\Theta_2^*\|^2 - \lambda z_3^{\mathrm{T}} x_3 + \lambda z_3^{\mathrm{T}} u \\ &- z_3^{\mathrm{T}} \dot{\alpha}_2 \end{aligned} \tag{6.83}$$

where $\varsigma = [\zeta_{22} f_{32}, \zeta_{22} f_{34} + \zeta_{24} f_{32}, \zeta_{23} f_{34}, \zeta_{24} f_{34}]^{\mathrm{T}}$ and $\varphi = [c_3, c_4, c_7, c_9]^{\mathrm{T}}$.

The approximation-based attitude control $u$ is proposed as follows based on the NNs.

$$u = \begin{cases} u_0, & \|z_3\| \ge \varepsilon_3 \\ 0, & \|z_3\| < \varepsilon_3 \end{cases} \tag{6.84}$$

where $u_0 = x_3 - \frac{\lambda^{-1}(\delta z_2^{\mathrm{T}}[\delta I_{4\times 4}+\hat{\rho}_2(Z_1)\hat{\rho}_2^{\mathrm{T}}(Z_1)]^{-1}\alpha_{20}+\zeta_{21}f_{31}(x_3)+\varsigma_2^{\mathrm{T}}\hat{\varphi}+\|z_2\|^2+\frac{1}{2}\|z_2\|^2\|\alpha_2\|^2)}{\|z_3\|^2}z_3 - \lambda^{-1}(K_3z_3 + 2\gamma\|z_2\|^2z_3 + \frac{1}{2}\|z_2\|^2z_3 + \dot{\alpha}_2 + \hat{\rho}_2^{\mathrm{T}}(Z_1)z_2)$, $K_3 = K_3^{\mathrm{T}} > 0$, $\varepsilon_3 > 0$ and $\gamma > 0$ are design parameters.

Substituting (6.84) into (6.83), we obtain

$$\begin{aligned}
\dot{V}_3^* \le & -z_1^{\mathrm{T}}K_1z_1 - z_2^{\mathrm{T}}K_2z_2 - z_3^{\mathrm{T}}K_3z_3 - \varsigma_2^{\mathrm{T}}\tilde{\varphi} + \frac{1}{2}\|\varepsilon_1\|^2 + \frac{1}{2}\|\varepsilon_2\|^2 \\
& -2\gamma\|z_2\|^2\|z_3\|^2 - \frac{1}{2}\|z_2\|^2\|z_3\|^2 - z_2^{\mathrm{T}}\tilde{\Theta}_2^{\mathrm{T}}S_2(Z_1)z_3 - z_2^{\mathrm{T}}\varepsilon_2z_3 - \frac{1}{2}\|z_2\|^2 \\
& -\frac{\sigma_1}{2}\|\tilde{\Theta}_1\|^2 + \frac{\sigma_1}{2}\|\Theta_1^*\|^2 - \frac{\sigma_2}{2}\|\tilde{\Theta}_2\|^2 + \frac{\sigma_2}{2}\|\Theta_2^*\|^2
\end{aligned} \tag{6.85}$$

where $\tilde{\varphi} = \hat{\varphi} - \varphi$.

To achieve stability for error signals $\tilde{\varphi}$, the augmented Lyapunov function candidate can be chosen as

$$V_3 = V_3^* + \frac{1}{2}\tilde{\varphi}^{\mathrm{T}}\Lambda_3^{-1}\tilde{\varphi} \tag{6.86}$$

where $\Lambda_3 = \Lambda_3^{\mathrm{T}} > 0$.

Consider the adaptive law for $\hat{\varphi}$ as

$$\dot{\hat{\varphi}} = \Lambda_3(\varsigma_2 - \sigma_3\hat{\varphi}) \tag{6.87}$$

where $\sigma_3 > 0$.

Considering (6.87), the time derivative of $V_3$ along (6.85) is

$$\begin{aligned}
\dot{V}_3^* \le & -z_1^{\mathrm{T}}K_1z_1 - z_2^{\mathrm{T}}K_2z_2 - z_3^{\mathrm{T}}K_3z_3 + \frac{1}{2}\|\varepsilon_1\|^2 + \frac{1}{2}\|\varepsilon_2\|^2 \\
& -2\gamma\|z_2\|^2\|z_3\|^2 - \frac{1}{2}\|z_2\|^2\|z_3\|^2 - z_2^{\mathrm{T}}\tilde{\Theta}_2^{\mathrm{T}}S_2(Z_1)z_3 - z_2^{\mathrm{T}}\varepsilon_2z_3 - \frac{1}{2}\|z_2\|^2 \\
& -\frac{\sigma_1}{2}\|\tilde{\Theta}_1\|^2 + \frac{\sigma_1}{2}\|\Theta_1^*\|^2 - \frac{\sigma_2}{2}\|\tilde{\Theta}_2\|^2 + \frac{\sigma_2}{2}\|\Theta_2^*\|^2 - \frac{\sigma_3}{2}\|\tilde{\varphi}\|^2 + \frac{\sigma_3}{2}\|\varphi\|^2
\end{aligned} \tag{6.88}$$

It is apparent that the following facts

$$-z_2^{\mathrm{T}}\tilde{\Theta}_2^{\mathrm{T}}S_2(Z_1)z_3 \le \gamma\|z_2\|\|\tilde{\Theta}_2\|\|z_3\| \le 2\gamma\|z_2\|^2\|z_3\|^2 + \frac{\gamma\|\tilde{\Theta}_2\|^2}{2} \tag{6.89}$$

$$-z_2^{\mathrm{T}}\varepsilon_2z_3 \le \|z_2\|\|\varepsilon_2\|\|z_3\| \le \frac{1}{2}\|\varepsilon_2\|^2 + \frac{1}{2}\|z_2\|^2\|z_3\|^2 \tag{6.90}$$

hold, where $\gamma > 0$.

Substituting (6.89) and (6.90) into (6.88) yields

$$
\begin{aligned}
\dot{V}_3 \leq& -z_1^{\mathrm{T}} K_1 z_1 - z_2^{\mathrm{T}} K_2 z_2 - z_3^{\mathrm{T}} K_3 z_3 + \frac{1}{2}\|\varepsilon_1\|^2 + \|\varepsilon_2\|^2 - \frac{\sigma_1}{2}\|\tilde{\Theta}_1\|^2 + \frac{\sigma_1}{2}\|\Theta_1^*\|^2 \\
&- \frac{\sigma_2 - \gamma}{2}\|\tilde{\Theta}_2\|^2 + \frac{\sigma_2}{2}\|\Theta_2^*\|^2 - \frac{\sigma_3}{2}\|\tilde{\varphi}\|^2 + \frac{\sigma_3}{2}\|\varphi\|^2 \\
\leq& -\kappa V_3 + C
\end{aligned}
\tag{6.91}
$$

where

$$
\begin{aligned}
\kappa :=& \min\Bigg(2\lambda_{\min}(K_1), 2\lambda_{\min}(K_2), 2\lambda_{\min}(K_3), \frac{2\sigma_1}{\lambda_{\max}(\Lambda_1^{-1})}, \frac{\sigma_2 - \gamma}{\lambda_{\max}(\Lambda_2^{-1})}, \\
&\frac{2\sigma_3}{\lambda_{\max}(\Lambda_3^{-1})}\Bigg)
\end{aligned}
\tag{6.92}
$$

$$
C := \frac{1}{2}\|\bar{\varepsilon}_1\|^2 + \|\bar{\varepsilon}_2\|^2 + \frac{\sigma_1}{2}\|\Theta_1^*\|^2 + \frac{\sigma_2}{2}\|\Theta_2^*\|^2 + \frac{\sigma_3}{2}\|\varphi\|^2 \tag{6.93}
$$

To ensure that $\kappa > 0$, the design parameter $\sigma_2$ and $\gamma$ must make $\sigma_2 - \gamma > 0$. The above design procedure can be summarized in the following theorem.

**Theorem 6.5.** *Consider the helicopter attitude dynamics (6.2) with unknown moment coefficients and mass, model uncertainty and disturbance satisfy assumptions 6.1–6.2. The approximation-based flight control is proposed according to (6.84) using neural networks and parameter updated laws as chosen in (6.71), (6.72) and (6.87). For bounded initial conditions, there exist design parameters* $\sigma_i > 0$, $i = 1, 2, 3$, $\gamma$, $\Lambda_1 = \Lambda_1^{\mathrm{T}} > 0$, $\Lambda_2 = \Lambda_2^{\mathrm{T}} > 0$, $\Lambda_3 = \Lambda_3^{\mathrm{T}} > 0$, $K_1 = K_1^{\mathrm{T}} > 0$, $K_2 = K_2^{\mathrm{T}} > 0$ *and* $K_3 = K_3^{\mathrm{T}} > 0$ *such that the overall closed-loop control system is semi-globally stable in the sense that all of the closed-loop signals* $z_1$, $z_2$, $z_3$, $\tilde{\Theta}_1$, $\tilde{\Theta}_2$ *and* $\tilde{\varphi}$ *are bounded.*

*Proof.* The proof is similar to that of Theorem 6.4 and is omitted here. □

*Remark 6.6.* In the proposed approximation-based flight control of the helicopter, from (6.91) and (6.92), we can see that satisfactory closed-loop stability with suitable transient performance can be achieved by properly adjusting design parameters $K_i$, $\sigma_i$, $\Lambda_i$, and $\gamma$, $i = 1, 2, 3$. For example, the tracking error could be decreased by increasing the value of $K_i$, but that increase would also increase the control signal as in (6.54), (6.65), (6.84), and could excite unmodeled dynamics. Therefore, caution must be exercised in the choice of these parameters, due to the fact that there are some trade-offs between the control performance and other issues.

## 6.6 Simulation Results

In this section, extensive simulations are given to demonstrate the effectiveness of the proposed helicopter attitude control techniques. The APID MK-III helicopter model is used in our simulation, which is described by [6, 46]

$$\begin{aligned}
\ddot{z} &= \frac{1}{m}(Z_g - K_M\Omega^2\theta_M \cos\phi\cos\theta) \\
\ddot{\phi} &= -a\dot{\phi} + dK_M\Omega^2 b_{1s}\theta_M \\
\ddot{\theta} &= -b\dot{\theta} - eK_M\Omega^2 a_{1s}\theta_M \\
\ddot{\psi} &= -c\dot{\psi} + f(\theta_T + \psi_T) \\
\dot{b}_{1s} &= -\lambda(b_{1s} - u_{b_{1s}}) \\
\dot{a}_{1s} &= -\lambda(a_{1s} - u_{a_{1s}}) \\
\dot{\theta}_M &= -\lambda(\theta_M - u_{\theta_M}) \\
\dot{\theta}_T &= -\lambda(\theta_T - u_{\theta_T})
\end{aligned} \tag{6.94}$$

It is apparent that the dynamics of the APID MK-III helicopter shown in (6.94) have the same form of the model (6.1) if we neglect the flapping dynamics and engine dynamics. The helicopter's nominal parameters are shown in Table 6.1.
In this simulation, the control objective is to keep a certain desired altitude/attitude of the helicopter. In Sects. 6.6.1 and 6.6.2, we test the proposed flight control on the desired maneuver requiring aggressive attitude configurations and suppose that the desired altitude/attitude is $x_{1d} = [20, 0.2\sin(1.5t) + 0.5\cos(0.5t), 0.4(\sin(t) + 0.5\sin(0.5t)), 0.1\sin(1.5t) + 0.4\cos(0.5t)]^T$. Initial states $z = 15.0$, $\phi = 0.2$, $\theta = -0.1$, $\psi = 0.1$, $b_{1s} = 0.0$, $a_{1s} = 0.0$, $\theta_M = 0.0$ and $\theta_T = 0.0$. In Sect. 6.6.3, the hovering flight is used to illustrate the effectiveness of the proposed approximation-based attitude control. In all cases, the saturation values of the command control signals are chosen as $|u_{b_{1s}}| = |u_{a_{1s}}| = 1.8$ and $|u_{\theta_M}| = |u_{\theta_T}| = 1.0$. The control gain matrix $G$ and design matrix $W$ are chosen as follows:

$$G = \begin{bmatrix} 100 & 100 & -1703.4/m & 0 \\ 223.5824 & 0 & 223.5824 & 0 \\ 0 & -58.3258 & -58.3258 & 0 \\ 0 & 0 & 0 & 31.9065 \end{bmatrix}, W = \begin{bmatrix} 1 & 0 & 0 & 0 \\ 0 & 1 & 0 & 0 \\ 0 & 0 & 1 & 0 \\ 0 & 0 & 0 & 1 \end{bmatrix}$$

**Table 6.1** Parameters of the helicopter

| |
|---|
| $m = 50\,\text{kg}$, $a = 8.7072$, $b = 10.1815$, $c = 0.434$, $K_M\Omega^2 = 1703.4$, |
| $dK_M\Omega^2 = 223.5824$, $eK_M\Omega^2 = 58.3528$, $f = 31.9065$, $\lambda = 300$ |

**Table 6.2** Design parameters of the model-based attitude control

| |
|---|
| $K_1 = \text{diag}\{2, 2, 2, 2\}$, $K_2 = \text{diag}\{100, 100, 100, 100\}$, $K_3 = \text{diag}\{300, 300, 300, 300\}$ |

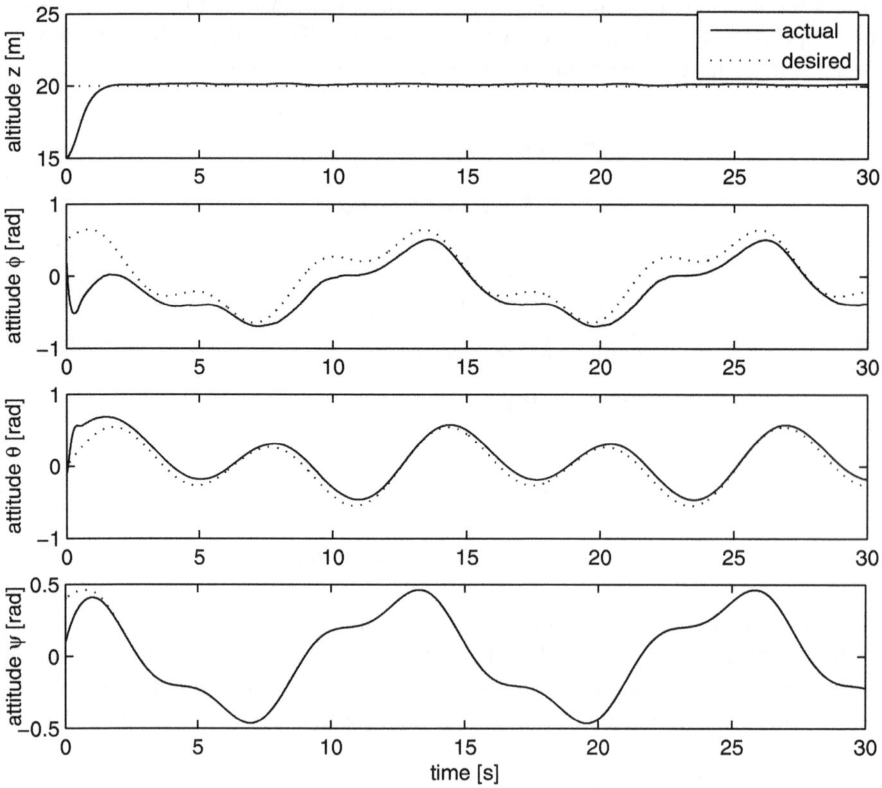


**Fig. 6.1** Altitude/attitude tracking performance using the model-based control

## 6.6.1 Model-Based Attitude Control

In this subsection, the model-based attitude control is designed according to (6.29). The control design parameters are shown in Table 6.2.

Under the proposed model-based attitude control, it can be observed from Fig. 6.1 that the maneuver altitude/attitude of the helicopter can be maintained within a small envelop of the desired altitude/attitude. From Fig. 6.2, we know that the velocity along $z$ axis and angular velocity under the model-based control converge to a small neighborhood of the origin. Figure 6.3 shows that the command control signals are saturated within the limits of the actuators. There exists chattering phenomenon in Figs. 6.2 and 6.3 which are caused by the discontinuous control input and the maneuver flight.

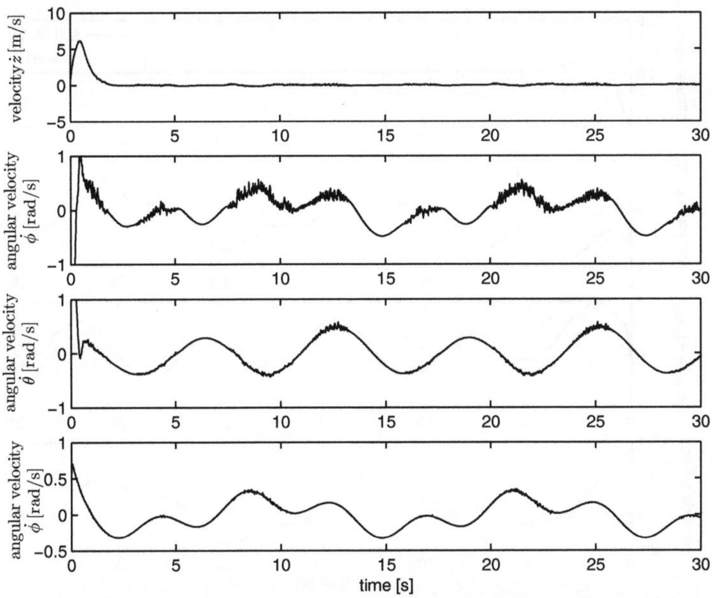


**Fig. 6.2** Velocity along $z$ axis and angular velocity signals using the model-based control

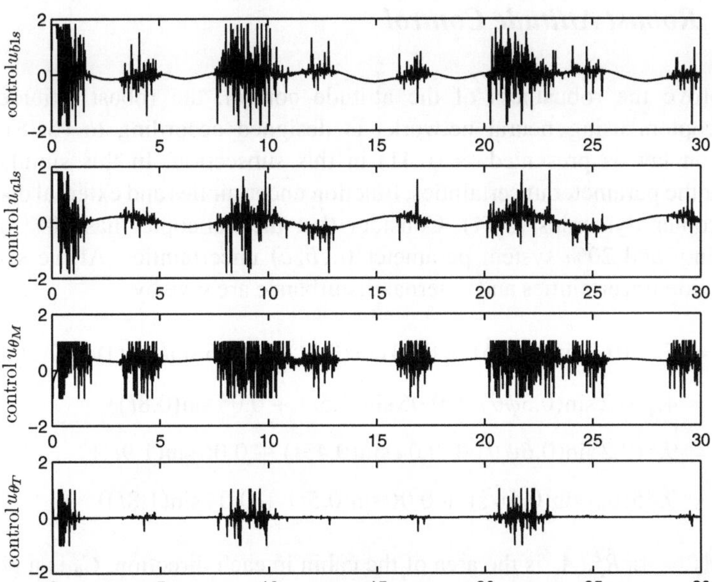


**Fig. 6.3** The command control signals using the model-based control

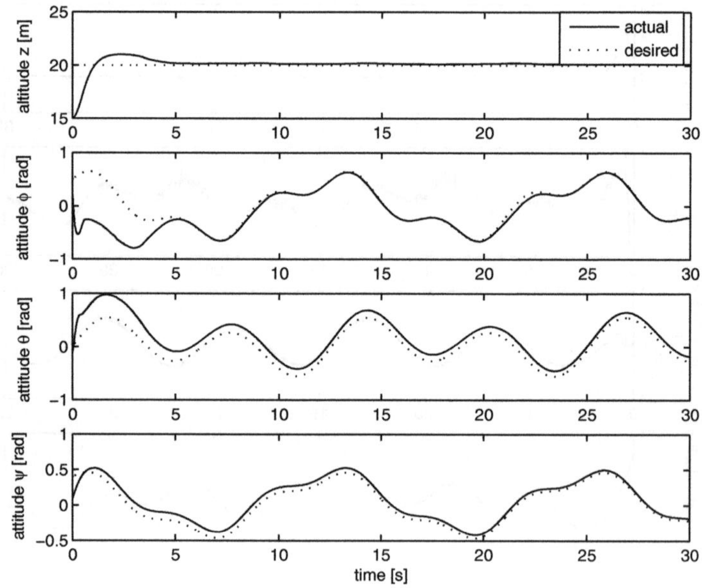


**Fig. 6.4** Altitude/attitude signals using the robust control with disturbance and uncertainty

## 6.6.2 *Robust Attitude Control*

To improve the robustness of the attitude control, the robust attitude control of helicopters using neural networks is designed according to (6.48) and the adaptation law is presented as (6.41) in this subsection. In this simulation, we consider the parameter uncertainties, function uncertainties and external disturbance in helicopter dynamics (6.94). Consider that the helicopter has 10% mass (m) uncertainty and 20% system parameter $(a, b, c)$ uncertainties. At the same time, the function uncertainties and external disturbance are give by

$$\begin{aligned}
\Delta F_1 &= 5.25(0.3\sin(0.6\dot{z}\dot{\phi}) + 0.04\cos(0.3t) + 0.06\sin(1.5t)) + 0.5C_{\rm d}A_{\rm c}V_{\rm W}^2 \\
\Delta F_2 &= 4.5(0.2\sin(0.5\dot{\phi}\dot{\theta}) + 0.05\sin(1.5t) + 0.05\sin(0.8t)) \\
\Delta F_3 &= 9.5(0.2\sin(0.6\dot{\psi}\dot{\theta}) + 0.04\sin(0.45t) + 0.06\sin(1.9t)) \\
\Delta F_4 &= 9.45(0.3\sin(0.5\dot{\psi}\dot{z}) + 0.06\sin(0.5t) + 0.04\sin(1.8t))
\end{aligned} \tag{6.95}$$

where $A_{\rm c} = 4\pi R_{\rm c}^2$. $A_{\rm c}$ is the area of the cabin in each direction, $C_{\rm d}$ is a given drag coefficient and $V_{\rm W}$ is the wind speed. Here, we choose wind speed $V_{\rm W} = 10\,{\rm m/s}$. The robust control design parameters are chosen as in Table 6.2.

The simulation results under the robust attitude control (6.41) are given in Figs. 6.4–6.6. It can be observed from Fig. 6.4 that the altitude/attitude of the helicopter can also be maintained within a small envelop of the desired maneuver

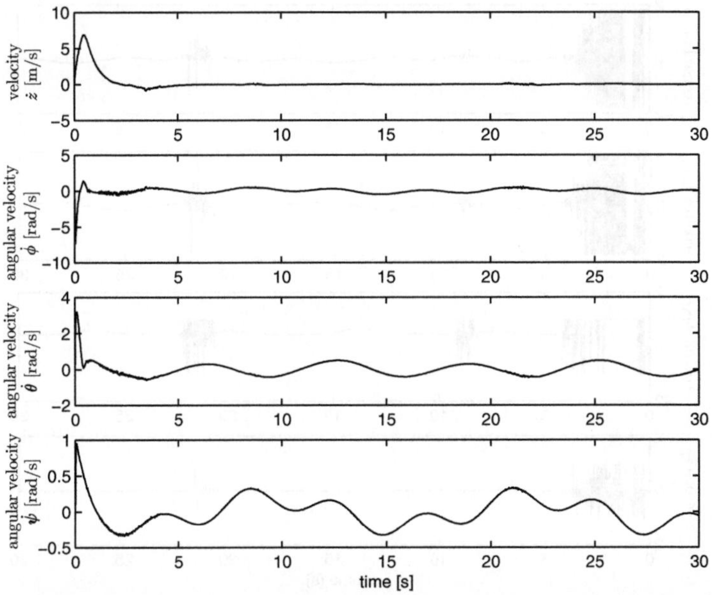


**Fig. 6.5** Velocity along $z$ axis and angular velocity signals using the robust attitude control with disturbance and uncertainty

altitude/attitude with the uncertainties and disturbances. From Fig. 6.5, the velocity along $z$ axis and angular velocity with the uncertainties and disturbances can converge to a compact set. Figure 6.6 shows that the command control signals are saturated within the limits of the actuators.

### 6.6.3 Approximation-Based Attitude Control

In this subsection, the approximation-based attitude control is designed according to (6.84). In the approximation-based attitude control, we would like to highlight that parametric uncertainties may exist in the helicopter model. All helicopter moment coefficients and helicopter mass are completely unknown. At the same time, the function uncertainties and external disturbance described in (6.95) are included. Here, we assume that the helicopter is in hovering flight. Thus, the desired altitude/attitude is given by $x_{1d} = [20, 0, 0, 0]^{\mathrm{T}}$. Initial states $z = 15.0$, $\phi = 0.01$, $\theta = 0$, $\psi = 0.04$, $b_{1s} = 0.0$, $a_{1s} = 0.0$, $\theta_{\mathrm{M}} = 0.0$ and $\theta_{\mathrm{T}} = 0.0$. The saturation values of the command control signals are the same as aforementioned.

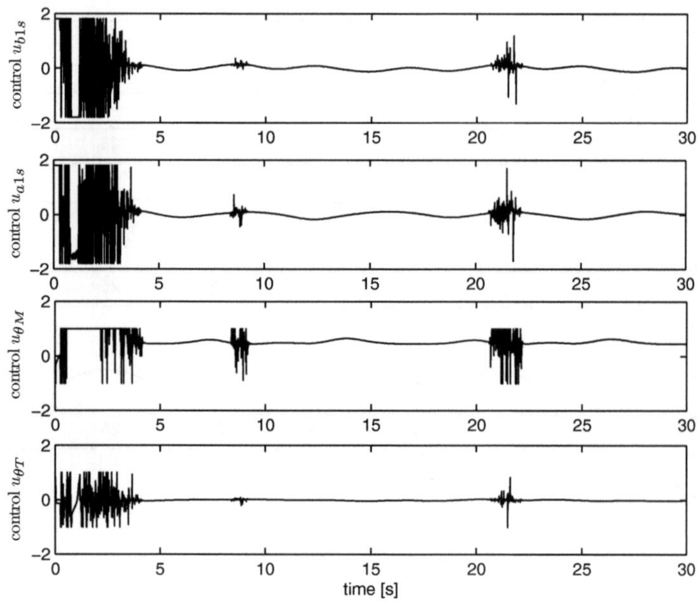


**Fig. 6.6** The command control signals using the robust attitude control

In practice, the selection of the centers and widths of RBF has a great influence on the performance of the designed controller. According to [90], Gaussian RBFNNs arranged on a regular lattice on $R^n$ can uniformly approximate sufficiently smooth functions on closed, bounded subsets. Accordingly, in the following simulation studies, the centers and widths are chosen on a regular lattice in the respective compact sets. Specifically, we employ 8 nodes for each input dimension of $\hat{W}_1^{\mathrm{T}} S(Z_1)$ and four nodes for each input dimension of $\hat{W}_2^{\mathrm{T}} S(Z_2)$, thus, we end up with 512 nodes (i.e., $l_1 = 512$) with centers $\mu_i$ $(i = 1, 2, ...l_1)$ evenly spaced in $[-1.0, +1.0]$ and widths $\eta_i = 2,000.0$ for neural network $\hat{W}_1^{\mathrm{T}} S(Z_1)$; and 64 nodes (i.e., $l_2 = 64$) with centers $\mu_j$ $(j = 1, 2, ...l_2)$ evenly spaced in $[-1.0, +1.0]$ and widths $\eta_j = 1{,}000.0$ for neural network $\hat{W}_2^{\mathrm{T}} S(Z_2)$.

Under the proposed approximation-based attitude control (6.84), we observe that the hovering flight is stable, i.e., the altitude/attitude of the helicopter can be maintained within a small envelop of the desired altitude/attitude in Fig. 6.7. From Fig. 6.8, it can be observed that the line velocity along the $z$ axis and the angular velocity of the hovering flight converge to a small neighborhood of the origin. The command control signals are shown in Fig. 6.9.

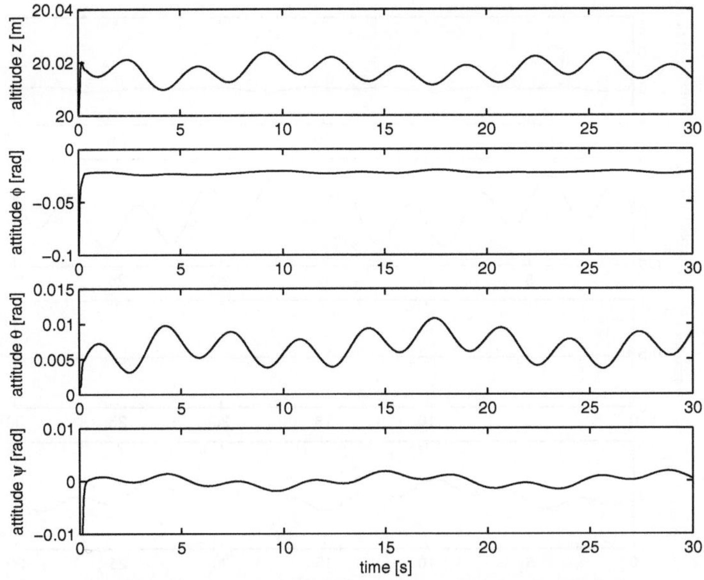


**Fig. 6.7** Altitude/attitude signals using approximation-based attitude control

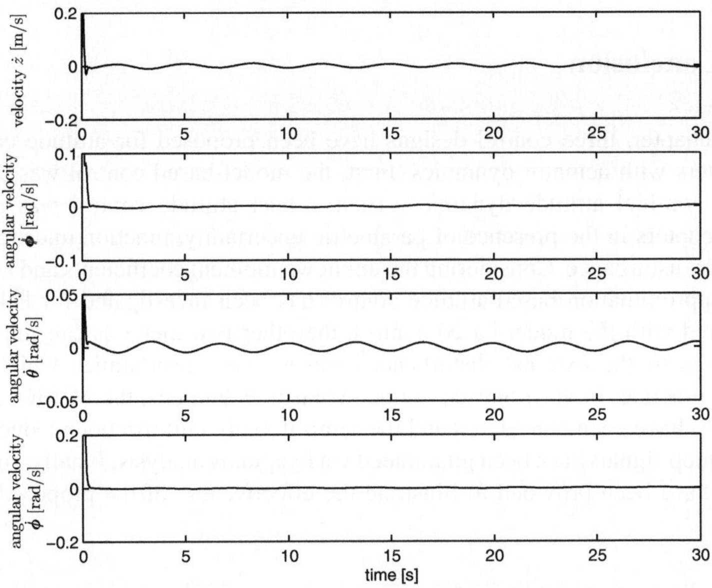


**Fig. 6.8** Velocity along $z$ axis and angular velocity signals using approximation-based attitude control

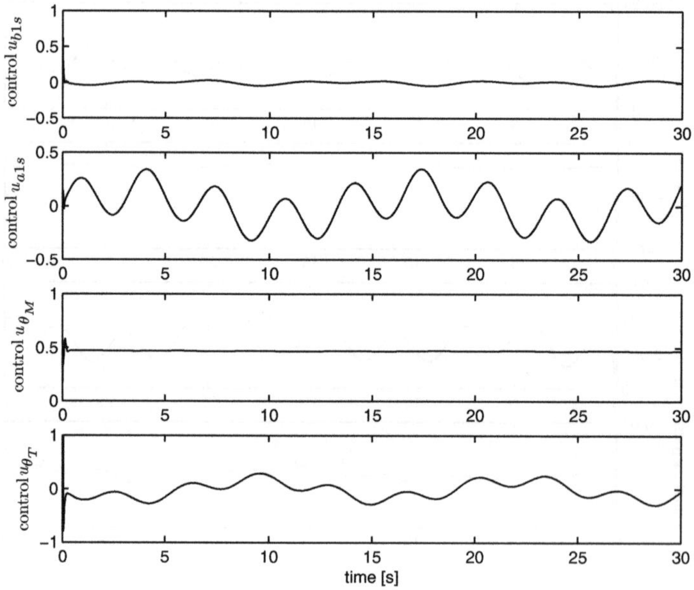


**Fig. 6.9** The command control signals using approximation-based attitude control

## 6.7 Conclusion

In this chapter, three control designs have been proposed for attitude control of helicopters with actuator dynamics. First, the model-based control was presented for the nominal attitude dynamics. Then, robust attitude control was proposed for helicopters in the presence of parametric uncertainty, function uncertainty and unknown disturbance. Considering the unknown moment coefficients and helicopter mass, approximation-based attitude control has been investigated for helicopters. Compared with the model-based control, the other two methods can improve the robustness to the external disturbances and system uncertainties which always exist in practice. In all proposed attitude control techniques, the MIMO nonlinear dynamics have been considered and the semi-globally uniform boundedness of the closed-loop signals have been guaranteed via Lyapunov analysis. Finally, simulation studies have been provided to illustrate the effectiveness of the proposed attitude control.

# Chapter 7
# Kinematic Formation Control Using Q-structures

## 7.1 Introduction

Research in multi-agent cooperative systems has been active in recent years, covering topics in consensus, high level decision making, and low level control mechanisms [24, 26, 29]. Multiple helicopter cooperation can accomplish complex tasks such as automated transportation, surveillance, and large area search and rescue. A fundamental problem in multi-helicopters cooperation is formation control, the structure in which the helicopters keep a desired formation configuration and at the same time complete the assigned tasks.

There are four approaches to formation control, namely behavioral [3], virtual structure [63], queues and artificial potential trenches [26, 29], and leader-following [107], all of which can be applied to multi-helicopter control. Firstly, in the behavioral approach [3], the control action for each helicopter is derived by a weighted average of each desired behavior, such as formation keeping, goal seeking and obstacle avoidance. Secondly, the virtual structure approach [63] treats the entire formation as a single rigid body, and derives the motion of each agent from the trajectory of a corresponding point on the structure.

Thirdly, for the queues and artificial potential trenches approach [29], the formation structures were presented by queues and artificial potential trenches, of which the explicit representation of every single node is not required and the scalability of the formation is improved when the team size changes. The original scheme [29] was extended to improve the performance of the scheme when only local communication is present, and resulted in a weakly connected network [26].

Last but not the least, in the leader–follower approach [89, 107], the leader tracks a predefined path and the follower maintains a desired geometric configuration with the leader. The follower can in turn be designated as a leader for another helicopter resulting in scalability of the formation. The advantage of this approach is that specifying a single quantity (the leader's motion) directs the group behavior. Therefore, it is simple since a reference trajectory is clearly defined by the leader

B. Ren et al., *Modeling, Control and Coordination of Helicopter Systems*,
DOI 10.1007/978-1-4614-1563-3_7, © Springer Science+Business Media, LLC 2012

and the internal formation stability is induced by the control laws of individual helicopters.

Beyond the methods of single-helicopter control detailed in Chaps. 4–6, this chapter touches on the kinematic formation control of multiple helicopters. This kinematic control can be viewed as a higher level decision making process that coordinates individual helicopter actions in order to produce an overall desired formation. The main purpose of the kinematic controller is to generate a desired reference plan and/or trajectory for each helicopter in the team, which will then be fed to the lower level helicopter controllers that take into account the dynamics of each helicopter. The control laws that govern the dynamics of a specific helicopter in the multi-agent system will be considered to be based on control strategies similar to those presented in earlier chapters of this book.

The kinematic control is based on the concept of a Q-structure, a novel and flexible methodology to define and support a large variety of formations. The Q-structure allows automatic scaling of formations according to changes in the overall size of the helicopter team. The chapter begins by exploring the use of the Q-structure for formation control where perfect communication is present between all members of the team. The second part of the chapter focuses on how the Q-structure can be adapted and used for teams where communication is imperfect.

## 7.2 Q-Structures and Formations

### 7.2.1 Assumptions

The following assumptions are made:

**Assumption 1.** Each helicopter is equipped with wireless broadcasting abilities. This enables the helicopter to broadcast information about itself to others within the broadcast range. The reaction of the team to failure of communication links to some of its members is described in Sects. 7.3.1 and 7.4.

**Assumption 2.** The team follows a single target (either virtual or real), and the position $q_t$, velocity $v_t$ and topside orientation vector $\kappa_t$ of the target are known, with $\|v_t\| < v_{\max}$, where $v_{\max}$ is the maximum velocity of each helicopter.[1] The unit vector $\kappa_t$ is normal to the top surface of the target, and together with $v_t$ specifies the overall orientation of the target in the 3-Dimensional space. This is shown in Fig. 7.1.

**Assumption 3.** A helicopter $r_i$ is able to localize itself, hence obtaining its position $q_i$. It can also estimate the values of $v_i$ and $\kappa_i$ in the world coordinate system. The various vectors are shown in Fig. 7.1.

[1]For this chapter, we shall be working in the $\mathbb{R}^3$-space.

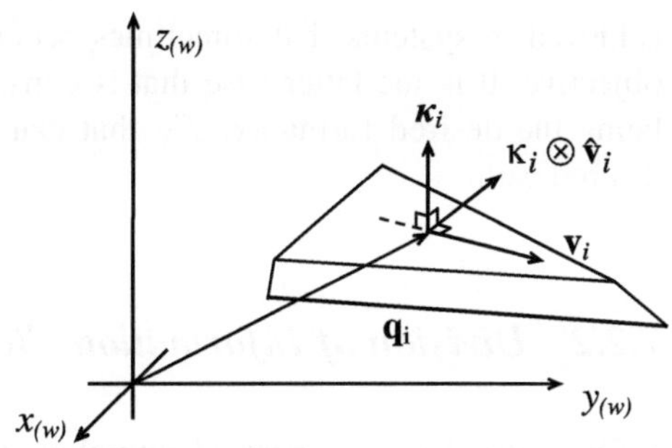


**Fig. 7.1** Vectors $\mathbf{q}_i$, $\mathbf{v}_i$ and $\kappa_i$ of $r_i$ in the world coordinate system

A formation $\mathcal{F}_N$ can be viewed as a set of constraints on the positions of each helicopter in relation to others in the team, such that each formation has a *specific appearance*. In conventional graphical notation, the constraints come in the form of nodes and edges in a connectivity graph (for instance in [69, 70]), where nodes represent the location of each helicopter and the edges represent the presence of communication links between helicopters. Consider an undirected graph $G_F = (V_g, E_g) \in G_N$, where $G_N$ is the space containing all possible graphs that can be formed from the set of $N$ vertices, given by $V_g$. Let the set of locations of all of the $N$ helicopters at any time $t$ be represented by

$$\mathcal{L}_T(t) = \{q_1(t), q_2(t), \ldots, q_N(t)\} \in C_N \tag{7.1}$$

where $C_N$ is the configuration space of the team as a whole. Define a function $\Phi_c : C_N \to G_N$ to give

$$G_c(t) = \Phi_c(\mathcal{L}_T(t)) = \big(\{q_1(t), \ldots, q_N(t)\}, \tag{7.2}$$

$$\{(q_i, q_j) | i \neq j \text{ and } \|q_i - q_j\| \leq d_{ij}\}\big) \tag{7.3}$$

with $d_{ij}$ being the cut-off distance beyond which there will not be communications between helicopters $r_i$ and $r_j$.

In a graphical representation, with converging formation controls/protocols

$$G_c(t) \to \mathcal{F}_N = G_{c,d}, \text{ as } t \to \infty \tag{7.4}$$

where $G_{c,d}$ is the connectivity graph of the desired formation. A general collection of helicopters in different (non-overlapping) positions does not constitute a formation.[2] It is also assumed that $G_{c,d}$ is weakly connected, although $G_c(t)$ can be disconnected. In [99], it was shown that the stability of agent flocks with a switching communication network can be maintained as long as the flock remains connected. It has also been shown (e.g., in [20]) that stability and convergence can still be

[2]This distinguishes formations from flocking/swarming.

achieved in systems that sometimes become disconnected, if there is a common objective. It is the latter case that is considered here, with the common objective being the desired formation $\mathcal{F}_N$ that can either be stationary or moving along a desired path.

### 7.2.2 Division of Information Flow

Information flow is separated into slow and fast time scales. The control of the formation takes place on these two levels based on the information available on each time scale. This reduces the amount of information that must be available to each helicopter for reactive decision making.

1. *Fast-time scale:* This facilitates time critical and reactive decision making, such as inter-helicopter collision avoidance and getting into formation. It only involves local communications between helicopters with limited communication range. Explicit controls governing the actual movements and paths of the helicopters occur at this level. Such decisions take place at a higher frequency when information is available.
2. *Slow-time scale:* This refers to the gradual multi-hop transfer of information, through a weakly connected communication network, between helicopters that are not within the immediate vicinity of each other. The collection of information over a longer time period allows for intermittent information losses between links. Formation control on this level involves low frequency decisions regarding the (re)allocation of helicopters to different parts (either vertices or queues) of a formation.

The interactions between the helicopters are mostly local since the helicopters respond immediately and reactively to data they obtain from others around themselves based on direct communication. This is not equivalent to requiring global information at all times for all decisions. (Re)Allocation based on long term information flows occurs at fixed periods. This information might not be the most current and subjected to time delays. Hence, there is no need for constant global communications between all helicopters. In addition, while information regarding out-of-range helicopters may be available, this is not taken into consideration while making pathing decisions other than for (re)allocation.

### 7.2.3 Elements of the Q-structure

In practical applications, formations usually take the form of geometric shapes, which may be conveniently subdivided into a series of smooth line segments. Here, each of these line segments are referred to as *queues*. The proportion of all the $N_{\text{tot}}$ helicopters in the team to be allocated to each queue for each formation is pre-specified. This increases the flexibility of formations, which scales according to changes in $N_{\text{tot}}$, since proportions are used instead of exact numbers.

**Definition 1 (Formations).** A formation is denoted by $\mathcal{F} = (\mathcal{Q}, \mathcal{G}(N_{\text{tot}}))$, where $\mathcal{Q}$ is the set of all the queues that make up the formation,[3] and $\mathcal{G}(N_{\text{tot}})$ represents the set of formation vertices, $V_i$ $(i = 1, \ldots, N_v)$,[4] around the target.

The positions of the formation vertices may be specified such that they scale proportionally with $N_{\text{tot}}$, and thereby avoid having an arbitrarily densely packed queue. For instance, assuming that $V_2 = (-4, 5)$ in Fig. 7.2a is specified for a team of $N_{\text{tot},0}$ helicopters, $V_2$ may be rewritten as $V_2 = \frac{N_{\text{tot}}}{N_{\text{tot},0}}(-4, 5)$. Note that formation vertices do not uniquely define the appearance of a formation. They merely represent a minimal number of pertinent locations in a formation and pure examination of the formation vertices does not yield complete information about what the formation looks like. As such, two different formations may have the same $\mathcal{G}(N_{\text{tot}})$. This may be seen from the two formations in Fig. 7.2. The actual appearance of the formation is mainly specified by the queues.

**Definition 2 (Queues).** A queue, $\mathcal{Q}_j \in \mathcal{Q}$, is denoted as $\mathcal{Q}_j = (\mathcal{V}_j, \mathcal{S}_j, \mathcal{C}_j, \mathcal{E}_j(N_{\text{tot}}))$. The four elements characterizing a queue are described as follows:

(i) $\mathcal{V}_j \subseteq \mathcal{G}(N_{\text{tot}})$ (Queue Vertices): a list of formation vertices through which $\mathcal{Q}_j$ passes.
(ii) $\mathcal{S}_j$ (Shape): a set of points following an equation in $\mathbb{R}^3$ that describes the spatial appearance of $\mathcal{Q}_j$, and is specified in the coordinate frame of the first formation vertex in the list $\mathcal{V}_j$. In general, this can be the equation of a curve in $\mathbb{R}^3$ that produces a queue like the example shown in Fig. 7.2.
(iii) $\mathcal{C}_j$ (Capacity): a fraction that refers to the proportion of all the helicopters in the formation it can hold, i.e., $\sum_{j=1}^{N_q} \mathcal{C}_j = 1$, where $N_q$ is the total number of queues in the formation.
(iv) $\mathcal{E}_j$ (Encapsulating Region): the set of all the points within a certain distance, $d_{ec}$, of the queue. The region is dependent on the number of helicopters that should reside on the queue, and is hence, related to $N_{\text{tot}}$.[5]

Queues may further be classified into *closed* and *open* queues. This characteristic of queues influences the constraints on the shape of the queues.

---

[3]Note that when the number of helicopters is too small (i.e. $\leq N_v$, the number of formation vertices), the helicopters will all be located at the vertices, and the scheme becomes highly similar to strategies using node-to-helicopter formation structures. However, in such a case, the helicopters are not able to reasonably form up into the desired formation no matter what scheme is used (e.g. two helicopters trying to form a wedge formation).

[4]Each formation vertex is represented by its position relative to the coordinate frame of the target.

[5]In a way, $\mathcal{E}_j$ provides a wrap around each queue, and when the formation reaches its intended form, all helicopters should rightly be within the encapsulating region of their respective queues.

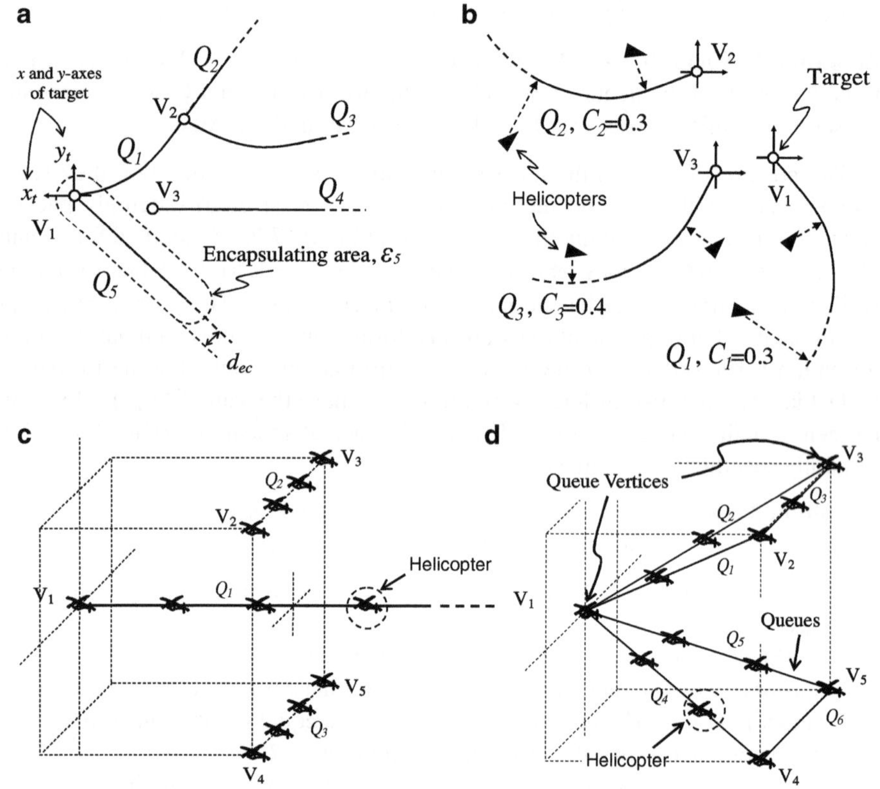


Formation with 5 formation vertices and 3 queues

Formation with 5 formation vertices and 6 queues

**Fig. 7.2** Examples of queues, and formation vertices (*circles*), where $x_t$ and $y_t$ are the axes of the coordinate frame of the target centered at $V_1$. *Open queues* are drawn with *solid (and dashed)* lines, indicating that they extend indefinitely from the vertex. (**a**) Queues and vertices. (**b**) Helicopters (*black triangles*) entering their queues on a 2-D plane. (**c**) and (**d**) Queues and vertices in 3D space, with three and six queues respectively

**Definition 3 (Closed Queues).** Closed queues are those that have two formation vertices in $\mathcal{V}_j$. The curve describing $\mathcal{S}_j$ is constrained to pass through the second vertex in $\mathcal{V}_j$ (e.g., $\mathcal{Q}_1$ in Fig. 7.2a). As the formation reaches steady state, all helicopters residing on a closed queue $\mathcal{Q}_j$ will be on the part of $\mathcal{Q}_j$ that are between the two vertices.

**Definition 4 (Open Queues).** Open queues are able to extend to infinity starting from the formation vertex in $\mathcal{V}_j$, and where $|\mathcal{V}_j| = 1$.

Many commonly used formations, such as the wedge and line formations, involve open queues. Formations such as the diamond and circle, consist of closed queues. Figure 7.2a shows an arbitrary formation that consists of five queues together with the corresponding queue capacities, in the coordinates of the target. The formation vertices (circles, labeled $V_1 - V_3$) are also reflected. In the figure, $\mathcal{Q}_2$ and $\mathcal{Q}_3$ share the same vertex. Queue 1 is a closed queue (with the set of vertices $\mathcal{V}_1 = \{V_1, V_2\}$), starting at $V_1$ and ending at $V_2$, while the rest are classified as open queues. Figure 7.2b shows a formation with three queues, and six helicopters (black triangles) attracted to the nearest point in their respective queues.

### 7.2.4 *Properties of the Q-structure*

The main difference between the proposed approach and other approaches (such as [17, 20]) is the use of the Q-structure. In the following, we derive the graph equivalence of the Q-structure, and use it for comparison with conventional connectivity graph representations based on formation consistency and computational requirements.

#### 7.2.4.1 Graphical Representation of Q-structure

To map the Q-structure into a conventional graph representation, a set of virtual queue vertices, $V_v$ are added to the set $\mathcal{V}_F(N)$, to produce

$$\mathcal{V}_\alpha(N) = \mathcal{V}_F(N) \bigcup V_v \tag{7.5}$$

The virtual Q-vertices are added to impose a limit on the length of queues with only one queue vertex, which would otherwise stretch to infinity. Therefore, each set of queue vertices contains a pair of formation vertices

$$\mathcal{V}_j = \left\{V_i, V_j\right\}, \quad \text{where } i \neq j \text{ and } V_i, V_j \in \mathcal{V}_\alpha \tag{7.6}$$

We then define a function $\Phi_\alpha : \mathcal{F} \to G_N$, such that $\Phi_\alpha(\mathcal{F}_N) = G_F$. Specifically, we have

$$\Phi_\alpha(\mathcal{F}_N) = \left(\mathcal{V}_\alpha(N), \left\{\mathcal{V}_j | j = 1, \ldots, N_q\right\}\right) \tag{7.7}$$

Each queue is represented in $G_F$ by its two queue vertices, which forms an edge of the undirected graph. Formations and their graphical representations are shown in Fig. 7.3. It should be mentioned that these are not connectivity graphs showing sensing/communication links.

#### 7.2.4.2 Consistency in Formation Representation

The Q-structure results in consistent formation representations, independent of the helicopter team size. It does not ascribe specific positions for individual helicopters,

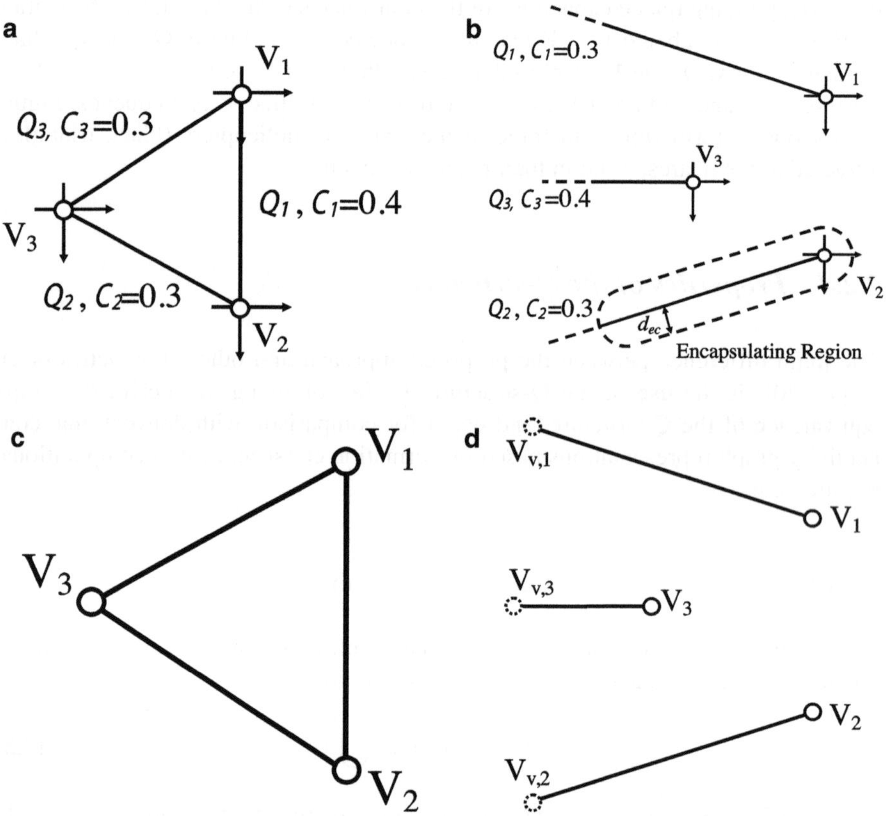


**Fig. 7.3** Graphical representation of Q-structures. *Dotted circles* represent virtual vertices. (**a**) A triangular formation, (**b**) A three column formation, (**c**) Graphical representation of triangular formation, (**d**) Graphical representation of three column formation

and relies on a set of decentralized, self-organizing behaviors to determine the final position of each helicopter. Typical graphical representations of formations rely on exact placement of each helicopter to achieve the final appearance of the formation. Based on graphs defined in [69], the addition or removal of helicopters in formation maintenance schemes in [17, 85] will result in different formations and connectivity graphs. Since the appearance of the formation is the important factor in many applications (such as helicopter convoys or target encirclement), the Q-structure allows formation specification based on appearance, the reverse of what graph-based approaches adopt. The consistent representation dispenses with the additional computation required for the addition/removal of nodes and the calculation of new inter-helicopter relationships. Figure 7.4 shows the two different connectivity graph representations of a triangular formation with three and five helicopters respectively, while the Q-structure for the same formation (Fig. 7.3c) remains unchanged regardless of team size.

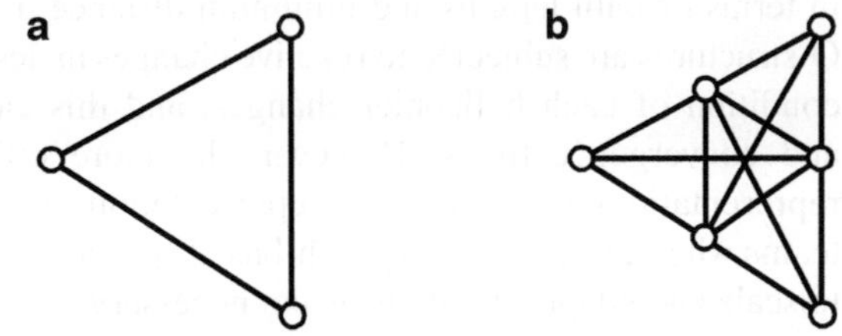


**Fig. 7.4** The triangular formation represented using connectivity graphs. (**a**) Teams with three helicopters, (**b**) Teams with six helicopters

#### 7.2.4.3 Formation Decomposition and Computation

The Q-structure allows a formation to be divided into smaller and simpler formations. Each queue, and its vertices, is a formation, i.e.,

$$\mathcal{F} = \{\mathcal{F}_k \mid k = 1, 2, \ldots, N_q\} \tag{7.8}$$

where $\mathcal{F}_k = (\mathcal{Q}_k, \mathcal{V}_k)$. After the initial allocation phase of helicopters to queues, short term information required by a helicopter can be limited to those within the same queue and others in the immediate vicinity. This reduces the communication load in the system especially for more complex formations.

We measure computational complexity in terms of the frequency at which a helicopter performs a resource expensive computation (e.g., the "comparison" -operation), and thus, how the complexity order scales with $N$. The simplest method for (re)allocating helicopters to either vertices for graph-representations, or to queues, is via greedy allocation.

Assuming that greedy assignment is made based on shortest distance, and that the graph-representation (such as those in [2, 76]) contains the same number of vertices as the number of helicopters, each helicopter compares its distance to $N$ vertices. Therefore, with $N$ helicopters, the computational complexity is $O(N^2)$. For queues, each helicopter makes $2N_q$ comparisons, comparing its distance to the queue and considers also the current capacity of that queue. This results in complexity of $O(NN_q)$ where $N_q \leq N$. Therefore, the Q-structure would potentially result in a lower computational cost, by implicitly decomposing a formation and lumping groups of vertices together.

#### 7.2.4.4 Efficiency and Optimality

As described in the earlier sections, a major difference between the Q-representation and conventional graph-based ones is the flexibility of the individual positioning of each helicopter. This property makes room for easy adaptation and scaling of the formations to changes within a team. However, graph-based representations (intrinsically) produces constant targets, which renders them much more favorable for optimizing the convergence process of helicopters into their desired formations, in terms of path lengths and minimum distance traveled while avoiding obstacles. Q-structures are subjected to reactive changes in desired targets as the neighborhood condition of each helicopter changes, and this can result in longer path lengths and convergence times. However, the more efficient treatment of the teams' representations renders the Q-representations more suitable for large helicopter teams where, in contrast to graph-based approaches, complex computation required to scale and adapt formations is not necessary.

#### 7.2.4.5 Robustness

Related to the issues of scalability and flexibility described above, the Q-structure is robust and more adaptable to team changes compared to existing approaches. Such team changes can encompass helicopter failures which removes subsets of helicopters from the team (resulting in scaling down of formations). The remainder of this chapter examines the adaptation of the system to limitations in communication ranges, under the assumption that helicopters have formed bidirectional links within the ad-hoc network. Due to the reactive nature, each helicopter is highly reliant on neighborhood information when deciding their desired targets. This causes the system to be relatively more sensitive to short term intermittent communication losses compared to graph-based approaches that provides constant targets for each helicopter. What these approaches lack in flexibility, they make up by their constancy and lower sensitivity of the formation framework to environmental changes.

## 7.3 Q-Structure with Perfect Communication

### 7.3.1 *Changing Queues*

Each helicopter changes their queue depending on information gathered via the slow time scale. Let $N_j$ = Nearest Integer($\mathcal{C}_j N_{\text{tot}}$) be the number of helicopters allowed in $\mathcal{Q}_j$, and $\chi_i(t)$ be the queue status[6] of $r_i$ at time $t$. The helicopters are first randomly initialized such that they belong to one of the queues in the current formation. A helicopter $i$ in queue $j$ continually broadcasts its (1) distance ($d_{ij}$) from the first formation vertex in $\mathcal{V}_j$, and (2) queue status, to the other helicopters within the broadcast range. From the data broadcasted by the other helicopters, the following information may be derived by each helicopter:

[6]It may also be interpreted as the information (which may be susceptible to time/ communication lags) regarding the queue status of $r_i$ that another helicopter $r_{i*}$ has at time $t$.

1. The current number of helicopters in $\mathcal{Q}_j$, given by $N_{j,0}$, as well as $N_{\text{tot}}$.
2. The excess length of each queue in the formation, $E_j$. Excess length refers to the number of excess helicopters in the queue, e.g., in a queue with a capacity of 0.5, the excess length $E_j = N_{j,0} - 0.5N_{\text{tot}}$. A negative value of excess length means that the queue is not fully filled up.
3. The last member of each queue, defined by the helicopter in the queue that is the furthest from the corresponding queue vertex.

The queue evaluation process is decentralized and performed individually by each helicopter continuously over time. At every time step $t$, the $r_i$ currently in $\mathcal{Q}_j$ (i.e. $\chi_i(t) = j$) uses the following algorithm, and the most recent information it obtained from the broadcast channel (which may be subjected to network latencies) to arrive at a decision of its queue status for step $t + 1$.

**Algorithm 1** Obtaining the queue status

**for** $k = 1$ to $N_q$ **do**
  Determine:

$$N_{k,0} = \sum_{h=1}^{N_{\text{tot}}} \begin{cases} 1, & \chi_h(t) = k \\ 0, & \text{otherwise} \end{cases} \tag{7.9}$$

  *// Any helicopter that lags too far behind the main team (due perhaps to equipment failure) will eventually move out of the team's broadcast range and be excluded.*
**end for**
**if** $E_j \leq 0$ **then**
  The current queue is either exactly full (and has no extra helicopters), or still has available space for more helicopters. No changes will be made to $r_i$'s queue status.
**else**
  **if** $r_i$ is the last member of its queue **then**
    The queue status will be modified as:

$$\chi_i(t + 1) = \arg\min_{k \in \mathrm{E}^-} (\ell_{k,i,nr}) \tag{7.10}$$

    where $\mathrm{E}^-$ is the set of all the queues with negative excess length, and $\ell_{k,i,nr}$ is the shortest distance between $r_i$ and $\mathcal{Q}_k$. If more than one queue in $\mathrm{E}^-$ are equally near to $r_i$, one will be chosen at random.
  **else**
    $r_i$ will retain its current queue status.
  **end if**
**end if**

Helicopters in queues with (positive) excess length will move towards the nearest queue that has negative $E$. By allowing only the last helicopter in each overpopulated queue to change their queue status, the formation will not experience large reshuffling when many helicopters from an overpopulated queue rush to occupy the extra space in an underpopulated queue. Whenever queue switching occurs, at least one space in all the underpopulated queues will be filled. In situations when an underpopulated queue becomes overpopulated due to a high influx of helicopters from other queues, the algorithm ensures that only the extra helicopters (furthest from the queue vertex) switches queue. The number of helicopters which will potentially change queue will hence be less than that which had entered the queue. As such, the number of helicopters that are eligible for changing their queue will gradually decrease until all the queues reach full capacity.

*Remark 1.* When helicopters join or leave the team, the values of $N_{\text{tot}}$ and $E_j$ will change. The algorithm allows the dynamic redistribution of the helicopters amongst the queues (based on $\mathcal{C}_j$). This scales the formation accordingly.

### 7.3.2 Potential Trench Functions

After a helicopter determines the queue it belongs to, it will be influenced by the artificial potential trench associated with that queue. The artificial potential trench for each queue may be synthesized with respect to the associated formation vertex such that it has the shape of the queue. Helicopters in a potential trench will tend to fall to the bottom of the trench. In other words, these helicopters will be attracted to the line that describes the bottom of the trench, which, in this case is also the equation describing the shape ($\mathcal{S}_j$) of the corresponding queue.

Assuming that $\chi_i = j$, the following analysis is done in the coordinate system of the first formation vertex (at $q_{vj}$ in the world frame) in the list $\mathcal{V}_j$. The x-axis ($x_{(vj)}$) for this coordinate system is taken to be the unit vector of the velocity vector ($\hat{v}_t$), and the z-axis, ($z_{(vj)}$), to be equal to the topside orientation vector $\kappa_t$ of the corresponding formation vertex. This can be seen more clearly from Figs. 7.1 and 7.5b.

In general, let $g_{(vj)}$ define the shape of $\mathcal{Q}_j$, which is continuously differentiable over the range in which the queue exists, and passes through all the formation vertices in the set $\mathcal{V}_\mathcal{Q}$. Furthermore, every point on the curve must be at a different distance from the origin. This ensures that for any point $q_{(vj),i}$ in $\mathbb{R}^3$, there will be a point $q_{(vj),nr}$ on $g_{(vj)}(x, y)$ that is nearest to $q_{(vj),i}$, while maintaining as close a distance from the origin as possible. This is shown in Figs. 7.6 and 7.7. The point $q_{(vj),nr}$ can be obtained from

$$q_{(vj),nr} = \arg \min_{q_{s1} \in Q_n} (\|q_{s1}\|) \tag{7.11}$$

where $Q_n$ is the set of points on the queue that satisfies $\left(\arg \min_{q_s \in \mathcal{S}_j} (\ell_{Q_j}(q_s))\right)$ and

$$\ell_{Q_j}(q_s) = \|q_s - q_{(vj),i}\| \tag{7.12}$$

Note that $g_{(vj)}$ can be any curve that satisfies the conditions listed above and is not restricted to straight lines.

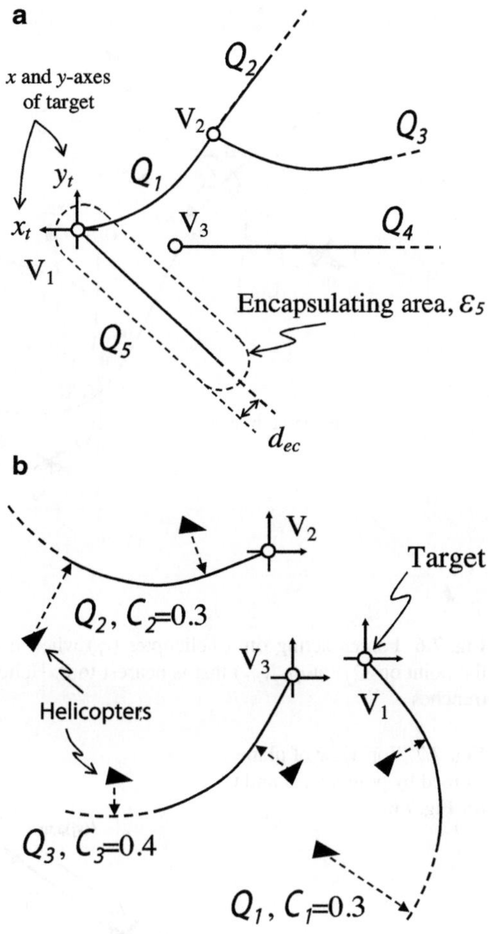


**Fig. 7.5** Examples of queues, and formation vertices (*circles*), where $x_t$ and $y_t$ are the axes of the coordinate frame of the target centered at $V_1$. Open queues are drawn with *solid (and dashed)* lines, indicating that they extend indefinitely from the vertex. (**a**) Queues and vertices. (**b**) Helicopters (*black triangles*) entering their queues on a 2-D plane

Let $U_{\text{cross}}(d)$ be a function that describes the cross section of the potential trench. The potential trench's cross section at any point is taken along the vector $q_{(vj),i,nr} = q_{(vj),nr} - q_{(vj),i}$. The shortest distance between these two points is given by $\ell_{j,i,nr}$. These are shown in Figs. 7.6 and 7.7. Therefore, a helicopter at $q_{(vj),i}$ would be attracted to the nearest point, $q_{(vj),nr}$, on the queue, and the attractive force it experiences may be calculated as

$$\mathbf{F}^{\text{fm}}_{(vj),i} = \left( \nabla_d U_{\text{cross}}(d)|_{\ell_{j,i,nr}} \right) \hat{q}_{(vj),i,nr} \tag{7.13}$$

Note that the force is represented in the coordinate frame of the first vertex in $\mathcal{V}_j$.

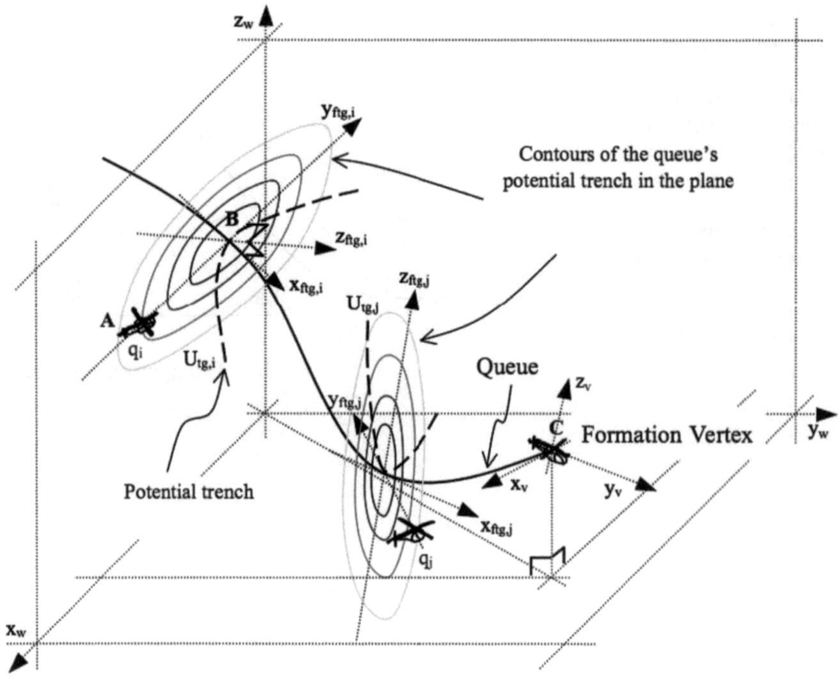


**Fig. 7.6** Forces acting on a helicopter ($r_i$) when it enters a queue. A helicopter is attracted to the point on $\mathcal{Q}_j$ (at $q_{(vj),nr}$) that is nearest to it. Helicopters interacting with queues and potential trenches

**Fig. 7.7** Top view of plane formed by points A, B and C for Fig. 7.6

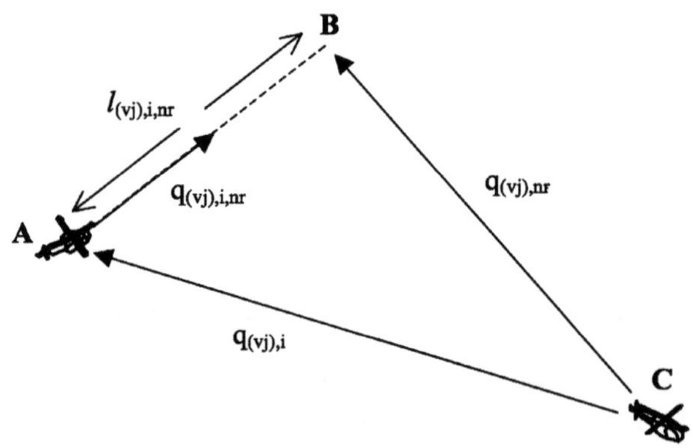


#### 7.3.2.1 Formation Adaptation and Deformation

When an obstacle is detected to be in the direction of $\hat{q}_{(vj),i,nr}$, $r_i$ is attracted to the point that is before the obstacle, but still along the vector $\hat{q}_{(vj),i,nr}$. This can be seen more clearly in Fig. 7.8. In this case, the attractive force is modified to become

$$\mathbf{F}^{\mathrm{fm}}_{(vj),i} = \left( \nabla_d U_{\mathrm{cross}}(d)|_{\ell_{j,i,\mathrm{adp}}} \right) \hat{q}_{(vj),i,nr} \tag{7.14}$$

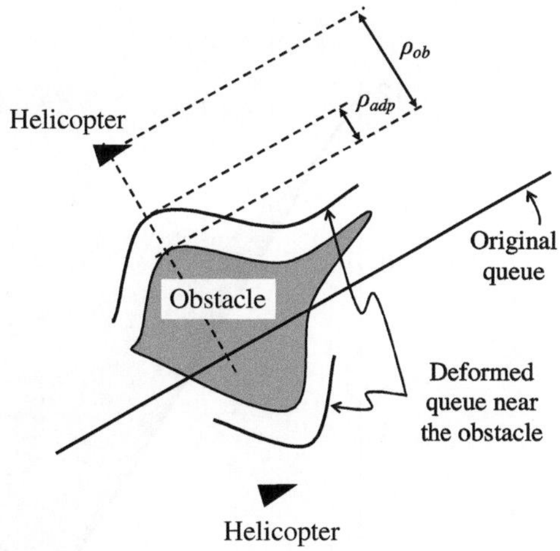


**Fig. 7.8** Instead of the original queue (that passes through the obstacle), the presence of the obstacle causes the helicopters (*triangles*) to be attracted to the deformed queue that hugs the obstacle at a distance of $\rho_{\text{adp}}$

where $\ell_{j,i,\text{adp}}$ is given by

$$\ell_{j,i,\text{adp}} = \begin{cases} \ell_{j,i,nr}, & \text{no obstacles in the direction } \hat{q}_{(vj),i,nr} \\ \rho_{\text{ob}} - \rho_{\text{adp}}, & \text{otherwise} \end{cases} \tag{7.15}$$

and $\rho_{\text{ob}} > 0$ is the distance along $\hat{q}_{(vj),i,nr}$ between $r_i$ and the obstacle, and $\rho_{\text{adp}} > \rho_{\text{sf}}$ is the distance the deformed formation is to be from the obstacle, where $\rho_{\text{sf}}$ is a safety distance between a helicopter and an obstacle.

For the purposes of illustration, consider $\mathcal{Q}_4$ in Fig. 7.5a, with the helicopters moving in $\mathbb{R}^2$. As such, $\kappa_t = [0\ 0\ 1]^{\text{T}}$. The cross section of the queue may be designed to take the form

$$U_{\text{cross}}(d) = a_{\text{fm}} f(d) \tag{7.16}$$

where $f(d) = \sqrt{1+d^2} - 1$, and the user defined parameter $a_{\text{fm}} > 0$ determines the slope of the potential trench. This potential function is similar to that used by Saber and Murray [88]. Forces generated by such potentials have the advantage of being bounded, and will not approach arbitrarily high values when the helicopter is far removed from the zero potential point. Attractive potentials in the rest of chapter will adopt forms similar to $f(d)$. The differential of $f(d)$ with respect to scalar $d$ is given by

$$f'(d) = \frac{\mathrm{d}}{\sqrt{1+d^2}} = \frac{\mathrm{d}}{f(d)+1} \tag{7.17}$$

The entire potential trench function in the 3Dimensional space, for $a = 2$, is shown in Fig. 7.9, with respect to the coordinate space of the vertex $V_3$.

Helicopters belonging to a different queue will be affected by different sets of potential trenches. The repulsive forces (described in Sect. 7.3.3) between

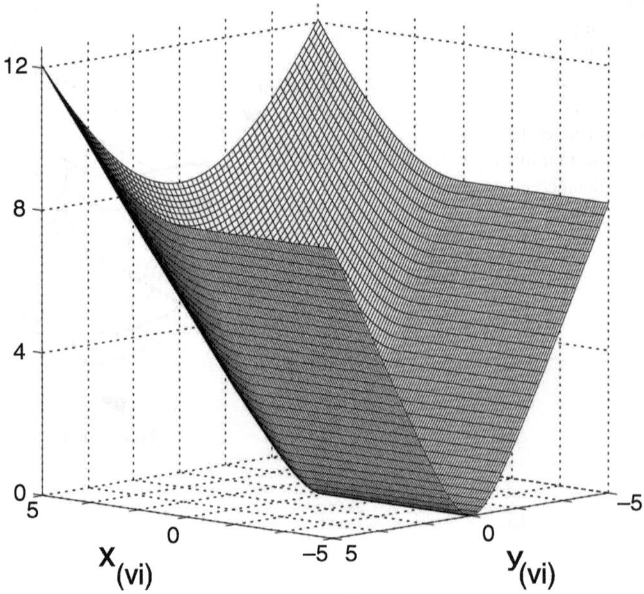


**Fig. 7.9** 3D view of the potential trench function of $Q_4$ in the $(x, y)$-coordinate space of the vertex, $V_3$

helicopters will ensure that the helicopters maintain a desired distance between each other in the potential trench. The force that $r_i$ experiences at $q_{(v3),i}$, in the coordinate system of the queue's vertex, due to the presence of the potential trench is computed as the negative gradient of the potential. The force is calculated using (7.13) as

$$\mathbf{F}^{\mathrm{fm}}_{(v3),i} = a_{\mathrm{fm}} f'(\ell_{4,i,\mathrm{adp}})\hat{q}_{(v3),i,nr} \tag{7.18}$$

with $a_{\mathrm{fm}}$ as defined in (7.16). The forces may then be converted into the world coordinate frame, in which all the forces acting on $r_i$ are calculated, as follows

$$\mathbf{F}^{\mathrm{fm}}_{i} = \mathbf{T}^{(w)}_{(v3)}\mathbf{F}^{\mathrm{fm}}_{(v3),i} \tag{7.19}$$

where

$$\mathbf{T}^{(w)}_{(v3)} = \left(\hat{v}_t \ \kappa_t \otimes \hat{v}_t \ \kappa_t\right) \tag{7.20}$$

## *7.3.3 Helicopter Behaviors*

Besides the formation behavior, helicopters should be equipped with other behaviors, such as target/goal tracking and obstacle avoidance, to navigate effectively. The behavior of $r_i$ is determined by the vector summation of the formation behavior

**Table 7.1** Parameter values for simulations

| Parameters | $a_{\text{fm}}$ | $a_{\text{ob}}$ | $a_{\text{tg}}$ | $a_{\text{ig}}$ | $\rho_0$ | $\rho_{\text{sf}}$ |
|---|---|---|---|---|---|---|
| Value | 10 | 2.5 | 1.5 | 2 | 2.15 m | 0.5 m |

with the target tracking ($\mathbf{F}_i^{\text{tg}}$) and obstacle avoidance ($\mathbf{F}_{i,(j,k)}^{\text{ob}}$) behaviors. This may be written as

$$\mathbf{F}_i^{\text{all}} = \mathbf{F}_i^{\text{fm}} + \sum_{j=1}^{N_\phi}\sum_{k=1}^{N_\theta} \mathbf{F}_{i,(j,k)}^{\text{ob}} + \mathbf{F}_i^{\text{tg}} \tag{7.21}$$

The parameters $a_{\text{fm}}, a_{\text{ob}}, a_{\text{tg}}, a_{\text{ig}} > 0$ are user defined and may be set so as to weight the relative importance of the different behaviors.

### 7.3.4 Simulation Experiments

Simulations are performed using the Player/ Stage platform [36]. Differential drive models of helicopters are used for the simulations. Player and Stage allows the user to set the speed ($v_i$) and turning rate ($\omega_i$) of each differential drive helicopter. Let $\theta_f$ and $\theta_i$ (in degrees) be the angle of $\mathbf{F}_i^{\text{all}}$ and $r_i$ in the world coordinate system respectively. The speed and turning rate is determined with a simple strategy (similar to that used in the experiment section of [28]) as follows

$$v_i = \min\left(K_s F_i^{\text{all}} \cos\left(0.5(\theta_f - \theta_i)\right), v_{\max}\right) \tag{7.22}$$

$$\omega = \omega_{\max}(\theta_f - \theta_i)/180° \tag{7.23}$$

where $K_s$ is a positive constant. For the simulations $K_s = 0.1$, $v_{\max} = 100$ mm/s, and $\omega_{\max} = 30°$/s. Range sensing information is obtained from 32 sonar beams that are equally spaced over 360°. Nine such helicopters are used for the simulations. The parameters used for the simulations are given in Table 7.1.

For the simulations, the four representative formations: (1) Wedge (One Vertex, two open queues), (2) Column (One Vertex, one open queue), (3) Double Column (Two Vertices, two open queues), and (4) Circle (One vertex, two closed queues), are used. In order to determine the closeness of a team of helicopters to a desired formation, we use a distance measure, $\delta$, given by

$$\delta = \frac{1}{N_{\text{tot}}} \sum_{r=1}^{N_{\text{tot}}} (\ell_{\chi_i,i,nr} + \ell_{\chi_i,i,E}) \tag{7.24}$$

where $\ell_{\chi_i,i,E}$ is the distance of the helicopter from the nearest point of the encapsulating area of the queue ($\mathcal{Q}_{\chi_i}$) it belongs to. The distance, $\delta$, may also be viewed as a form of error measure. Note that the distance of helicopters from their queues is, by itself, unable to provide a good measure of how closely the team has formed up into the desired formation. This is because helicopters may have already moved into the potential trench (and $\ell_{\chi_i,i,nr}$ will either be zero or very small) before reaching the vicinity of the formation vertex. As such, the distance from the respective encapsulating area is used together with $\ell_{\chi_i,i,nr}$.

*Remark 2.* Four regular formations are used in the simulations. The concept of the proposed method may also be generalized easily for irregular formations. The difference is only in how the queues (or positions of nodes for NR-approaches) are specified for the two classes of formations. The main advantage of the proposed approach lies in its concise and flexible representation of formations that is independent of team size. This may be more clearly and adequately brought out with commonly used formations (e.g., wedge) without the distractions associated with complicated irregular formations. Consider the case when a team, in a double column formation, increases from 10 to 30 helicopters. The proposed representation will always consist of two vertices and two open queues (regardless of team size), while NR-approaches require 20 nodes to be dynamically added and assigned. This observation is independent of whether the columns are straight or irregular squiggly lines.

#### 7.3.4.1 Convergence to Formations and Scaling

The helicopters are initialized to random positions around a stationary target[7], and the value of $\delta$ against time as the helicopters settle into each of the four formations is shown in Fig. 7.10. It may be observed that for the column and double column formations, the value of $\delta$ decays to almost zero when the formation reaches steady state. On the other hand, for the wedge and circle formations, there is a constant steady state error of approximately 0.15 m. This is due to interference of the obstacle avoidance potential with the potential trench of the formation in the area near the formation vertex, where the helicopters in the two queues are closest to each other. Hence, the helicopters near the front of each queue are pushed a slight distance from the queue due to the repulsion from helicopters at the front of the neighboring queue. Five snapshots of the nine helicopters entering a wedge formation during the simulation are shown in Fig. 7.11. For the circle formation, there are additional, but relatively small, interferences between the potentials at the end of both queues. Due to the presence of uncertainties and imperfection of the position data that each helicopter obtains, as well as the finite reaction times of the helicopters, we observe

[7] We note that the initial positions of the helicopters affect the time of convergence. This is certainly *true for any scheme*, and is also the case for both moving and stationary targets. The main objective of the work presented in this Chapter is not to minimize convergence time, but to investigate how formations may be represented, for greater scalability and flexibility, while achieving convergence in a realistic amount of time.

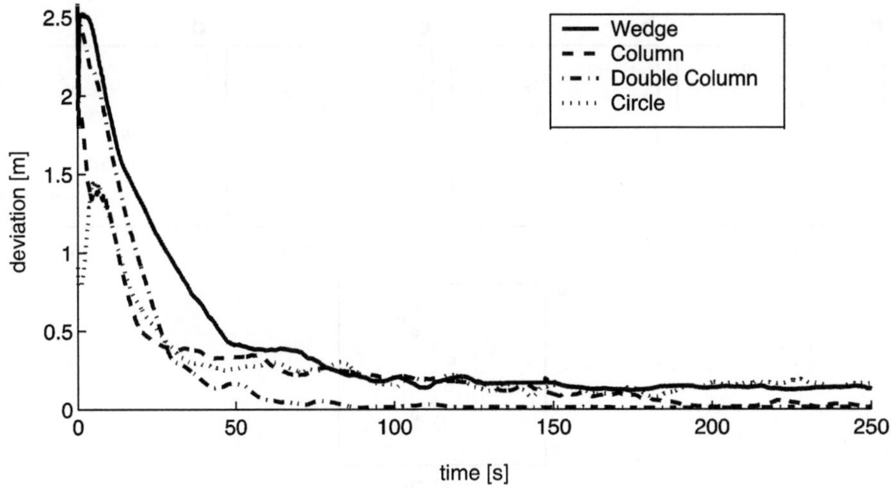


**Fig. 7.10** Convergence of team to desired formation. *Solid*: Wedge, *Dashed*: Column, *Dash-dot*: Double Column, *Dotted*: Circle

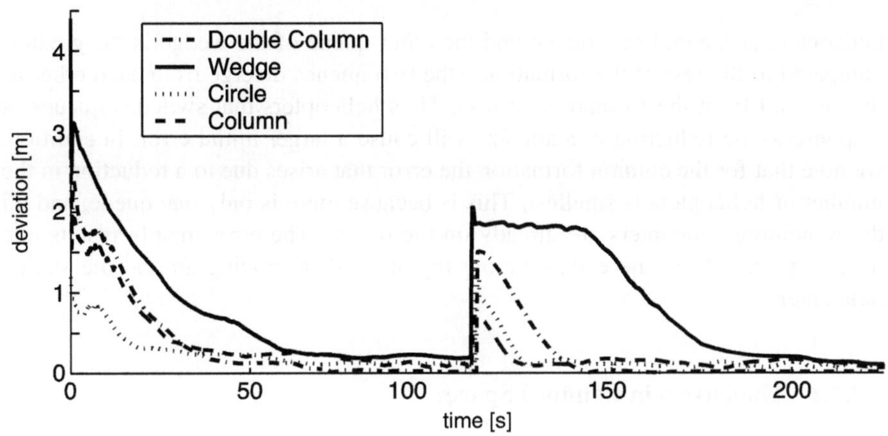


**Fig. 7.11** Snapshots of the team of nine helicopters forming the wedge formation

from the simulations results that the error in the formation never decays to exactly 0. The resulting average error is approximately 0.1–0.15 m. This is relatively small, and does not greatly affect the team's overall formation. The effect of uncertainties is also evident in the experiments carried out in the following sections.

To examine the effect a sudden reduction in $N_{\text{tot}}$, for the simulations, at time $t \simeq 110$ s, we remove half of the helicopters. The simulation results are shown in Fig. 7.12. We observe that for all the four formations, the helicopters are eventually able to scale the formation and form up accordingly. The wedge formation suffers the greatest error. This may be due to the fact that the distance between the

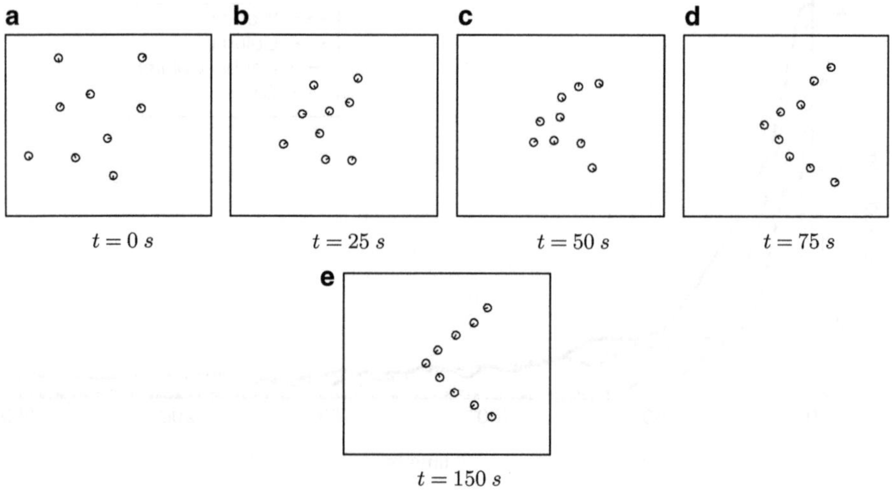


**Fig. 7.12** Scaling of formations. *Solid*: Wedge, *Dashed*: Column, *Dash-dot*: Double Column, *Dotted*: Circle

helicopters at the end of a queue and the other queue of the wedge is the greatest compared to the rest of the formations (the two queues diverge from each other as they extend from the formation vertex). Thus helicopters that switches queues in response to the reduction in team size will cause a larger initial error. In addition, we note that for the column formation the error that arises due to a reduction in the number of helicopters is smallest. This is because there is only one queue, and all the remaining helicopters are already on the queue. The error mainly results due to the distance from the encapsulating region, and in moving around the stalled helicopters.

#### 7.3.4.2 Maneuvers in Confined Spaces

This part of the simulation studies the effect of making turns in confined corridors on the team's formation. The adaptation of the formation to travel into narrow paths is also examined. The team is required to follow a moving target (another helicopter traveling at a speed of 0.6 m/s via a series of waypoints) through a winding corridor shown in Fig. 7.13. The graphs of $\delta$ against time for the column and wedge formations are shown in Fig. 7.14. To observe the effect of having adaptive queues (as described in Sect. 7.3.2), another distance measure

$$\delta_a = \frac{1}{N_{\text{tot}}} \sum_{r=1}^{N_{\text{tot}}} (\ell_{\chi_i,i,\text{adp}} + \ell_{\chi_i,i,E}) \tag{7.25}$$

is also used, and plotted as dashed lines in the graphs of Fig. 7.14. The distance measure in (7.25) is essentially the same as that in (7.24), except that it is dependent

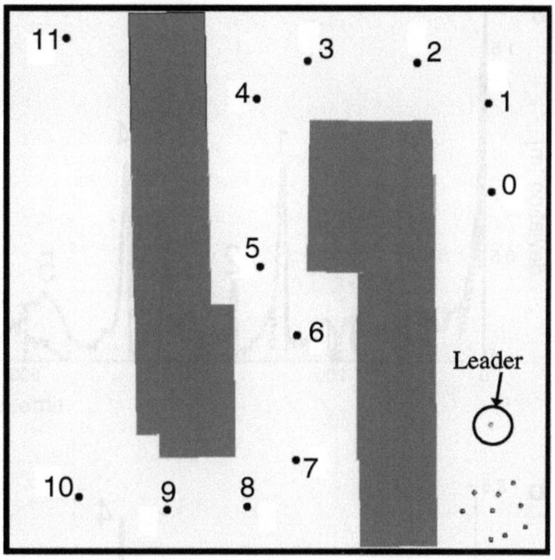


**Fig. 7.13** Snapshot of corridor and waypoints

on the distance of the helicopter from the deformed queue instead of the original queue. As such, it is an indicator of how close the helicopters are to a deformed formation. By comparing the solid and dashed lines, we are able to observe the instances during the team's traversal through the corridor when the formations adapt and deform themselves in response to nearby obstacles. The spikes in the graphs occur when the target (and hence the formation vertex) makes turns around the corners of the corridor. The turns and the corresponding spikes are numbered in Figs. 7.13 and 7.14. As expected, milder turns result in lower spikes in the error graphs. In addition, the wedge formation suffers a greater degree of deformation since it is laterally more spread out, and there are not enough room for the helicopters to spread out in most parts of the narrow corridor. Snapshots of the formations (wedge and column) as the team makes the turn at Point 5 (Fig. 7.13) are shown in Fig. 7.15.

Next, we examine the manner in which a wedge formation may adapt (and deform) itself to suit traversal in two corridors of different widths. For the circle and double column formations, the results are highly similar to that of the wedge. As for the column, travel through a narrow corridor is trivial. Hence, due to space constraints, we shall only present the results for the wedge formation here. The results are shown in Fig. 7.16. For the corridor of width 5 m, the degree of deformation, as can be seen by the difference between the $\delta$ and $\delta_a$ graphs, is smaller and similar to the case above, where the team travels through the twisting corridor. The maximum error is $\delta \simeq 0.8$ m at $t \simeq 360$ s. For the case when the team moves through the 3 m wide corridor, deformation is more severe, with maximum errors $\delta \simeq 1.4$ m and $\delta_a \simeq 1$ m. Due to the lack of space, it was observed that the formation was compressed into a column formation as the team moves through the extremely narrow corridor, although the team is programmed to move in the wedge formation. A snapshot of the deformed wedge is shown in Fig. 7.17.

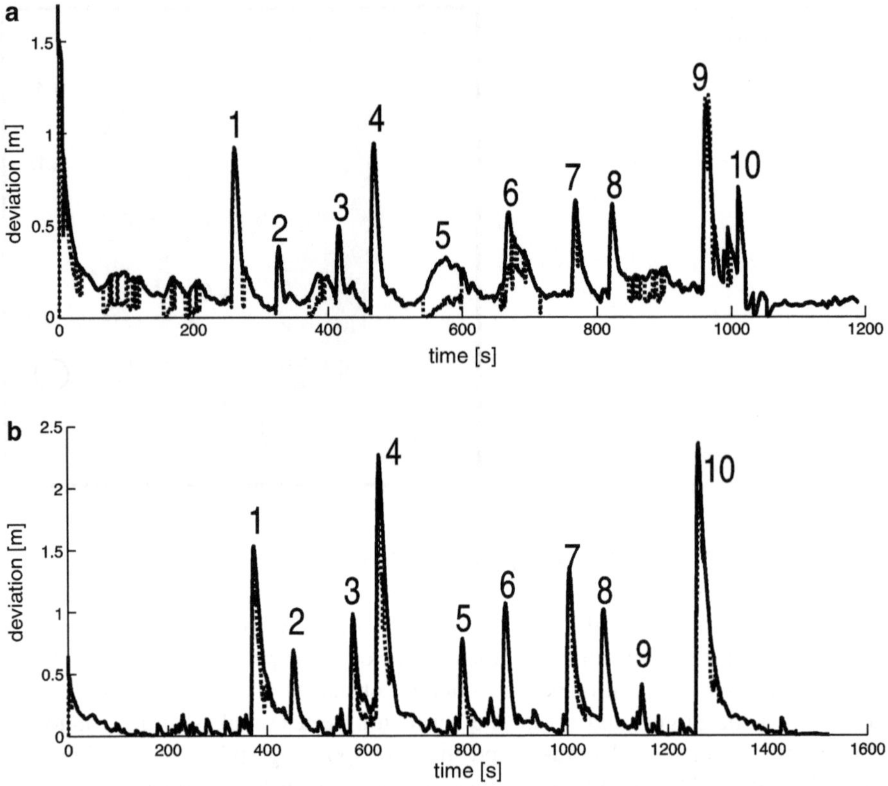


**Fig. 7.14** Team Maneuver through a confined corridor *Solid*: $\delta$, *Dashed*: $\delta_a$. (**a**) Wedge formation movement, (**b**) Column formation movement

#### 7.3.4.3 Reaction of Formations to Obstacles

We examine the effect that the presence of obstacles have on the team formation, as the team follows a virtual target that moves through two different obstacle fields with a speed of 0.6 m from Point A to Point B as shown in Fig. 7.18. The helicopters are positioned at random initial positions near Point A.[8] The obstacles we consider here can mainly be classified into: (1) Type I: Large (more than ten times the radius of each helicopter) and concave and (2) Type II: Small (less than three times of each helicopter's radius) and convex. The effect Type I and Type II obstacles have on the team formation are shown in Fig. 7.19. The wedge formation is used in this part of the simulation studies. For the environment with Type I obstacles in

[8]The helicopters are initialized behind the target so that they require less time to get into formation before encountering any obstacles. This, however, does not detract us from the main objective of this subsection, which is to examine the effects obstacles have on the team formation.

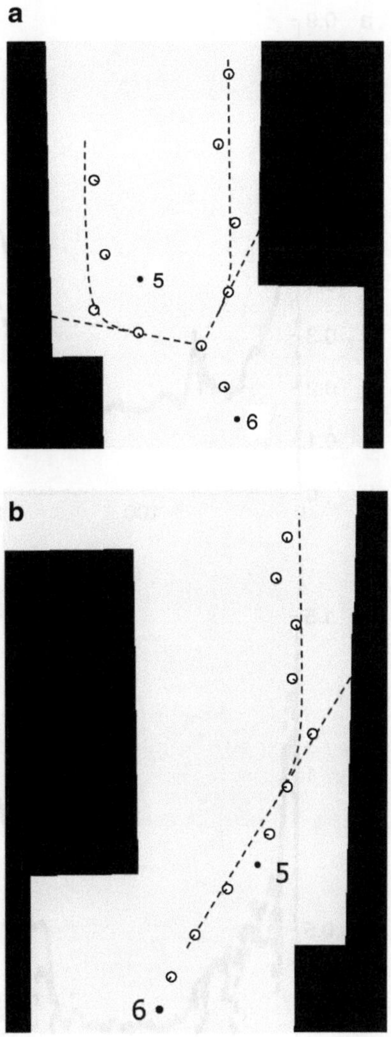


**Fig. 7.15** Formation deformation during a turn. (**a**) Wedge, (**b**) Column

Fig. 7.19a, part of the wedge first encounters obstacle OB1, resulting in the first spike in Fig. 7.19a. The size and shape of the obstacles account for the considerably large error of $\delta \simeq 5.5\,\mathrm{m}$ and amount of time required by the helicopters to maneuver around them. Despite this, the instant goal behavior is able to eventually bring the affected helicopters out of the local minima. Before these helicopters are able to form back into the formation, those at the right of the formation encounter obstacle OB2, causing the second spike in the error function. The helicopters are then able to eventually escape from the local minimums and form back into the wedge formation. In comparison, Type II obstacles produce a smaller effect on the formation, causing only a maximum average error of $\delta \simeq 0.8\,\mathrm{m}$.

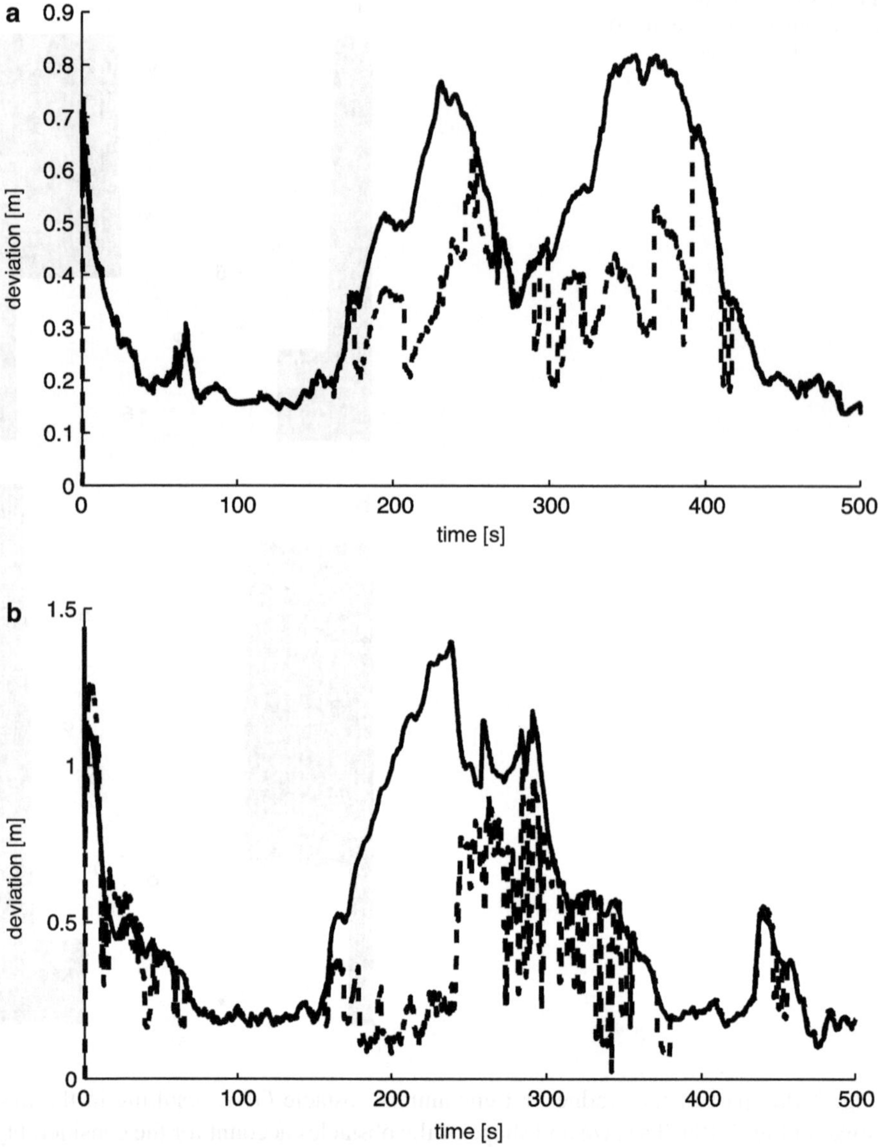


**Fig. 7.16** Team Maneuver through a confined corridor. *Solid*: $\delta$, *Dashed*: $\delta_a$. (**a**) Movement through the 5 m wide corridor, (**b**) Movement through the 3 m wide corridor

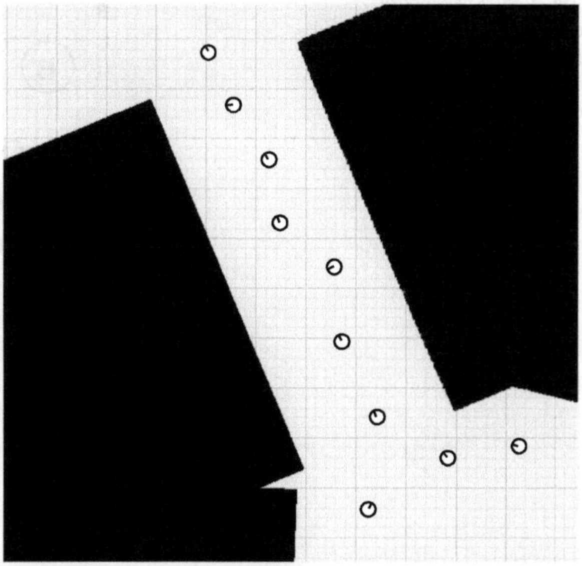

**Fig. 7.17** Snapshot of wedge deformed into a column in a narrow corridor

#### 7.3.4.4 Disruption of Wireless Communications

It is of interest to study the effectiveness of the proposed scheme in the presence of noisy or interrupted communications. Communication in simulated agent societies was studied by MacLennan [65]. Parker [79, 80] investigated how helicopter awareness of the actions of other team members affect the overall performance of the helicopter team. For our simulations, each helicopter loses contact with a random number of team members at random time instants. The double column formation is used since the formation is simple, and can therefore clearly reflect what happens (e.g. queue changes) in the event of communication loss. This formation consists of two open queues, $\mathcal{Q}_1$ and $\mathcal{Q}_2$, with capacities $\mathcal{C}_1 = \mathcal{C}_2 = 0.5$. When a helicopter fails to receive a signal from another helicopter, it assumes that the helicopter is no longer in the team, and performs its calculations for queue status accordingly. Communication links between pairs of helicopters are disrupted with equal probability. We examined the errors ($\delta$) associated with the formations when communications are lost for 5%, 25%, 50%, and 100% of the time, i.e. with probability ($P_{\text{txloss}}$) 0.05, 0.25, 0.5, and 1.0 respectively. When a helicopter experiences breaks in communication links, varying amounts of information ($I_{\text{loss}}$) may be lost. In this section, we study the cases for which the helicopter may lose up to a maximum of (1) all ($I_{\text{loss}} = 1.00$), (2) half ($I_{\text{loss}} = 0.50$), and (3) a quarter $I_{\text{loss}} = 0.25$, of the information in the channel. The plots of the errors for the different frequencies of communication breaks when $I_{\text{loss}} = 0.50$ are shown in Fig. 7.20.

The numerous spikes in the graphs reflect the instants when communication links are lost between at least one pair of helicopters, and the helicopter at the receiving

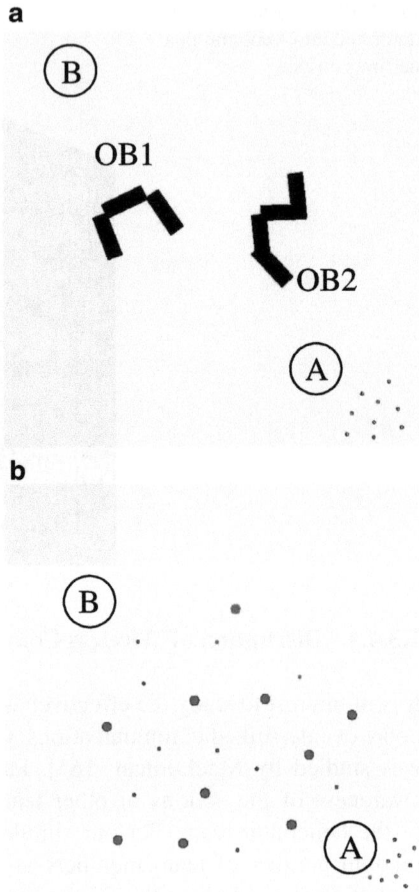


**Fig. 7.18** Snapshot of environment with Types I and II obstacles. (**a**) Type I obstacle field, (**b**) Type II obstacle field

end of the information decides to change queues. This occurs when this helicopter detects that it is the last helicopter in its current queue and that the other queue has negative excess capacity. Therefore, helicopters at the end of both queues have a relatively higher probability of switching queues due to communication breaks. For most cases, most of the team continues in formation, with those at the end of the queues toggling between the queues, and the error $\delta$ is mainly the result of such queue switches. Since all links are not disrupted at the same time, this decision is broadcasted, and will cause the helicopter in the other queue that is furthest from the formation vertex to react by changing queues. For $I_{\text{loss}} = 0.50$ with communication disruptions for 5% of the time, the queue status of the nine helicopters in the team are shown in Fig. 7.21. From Fig. 7.21, we see that only helicopters $r_1$, $r_7$ and $r_8$ change their queue status frequently in response to the breaks in communications. Since the other helicopters are at the front of their queues, they will still detect another helicopter in its queue that is further from

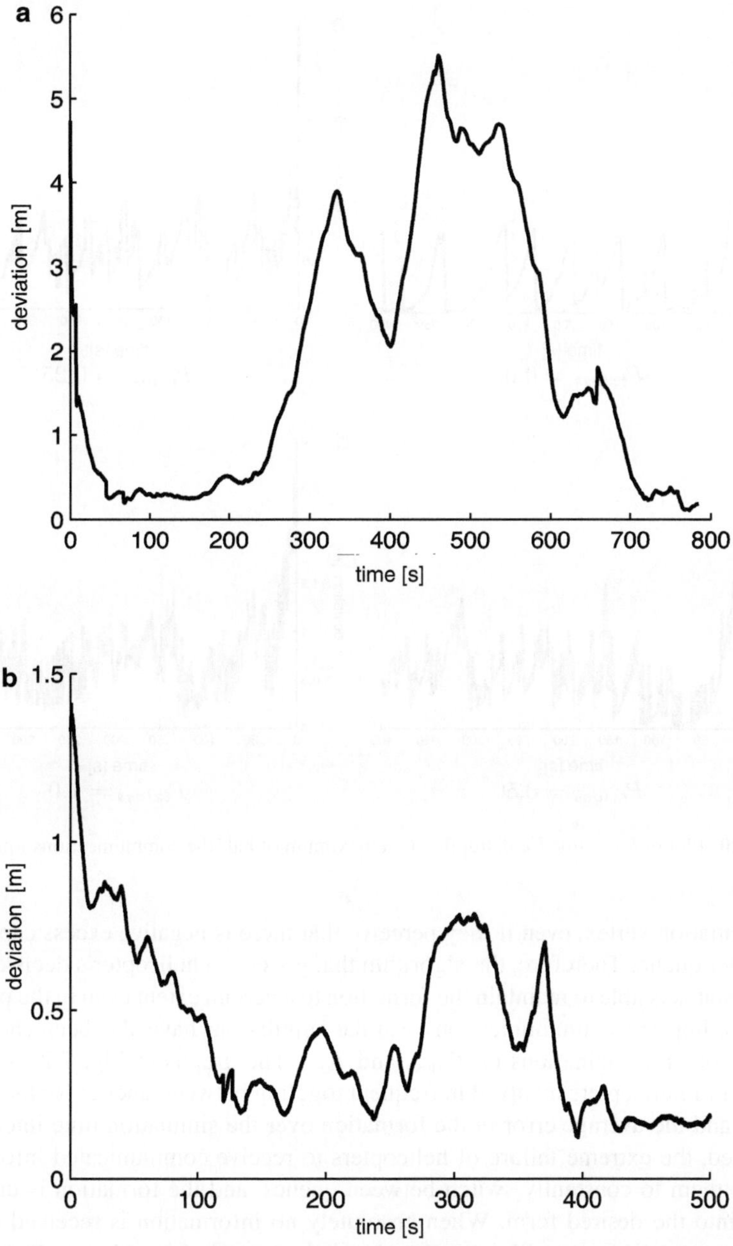


**Fig. 7.19** Plot of $\delta$ vs. time for team traversal through obstacle fields. (**a**) Type I obstacles, (**b**) Type II obstacles

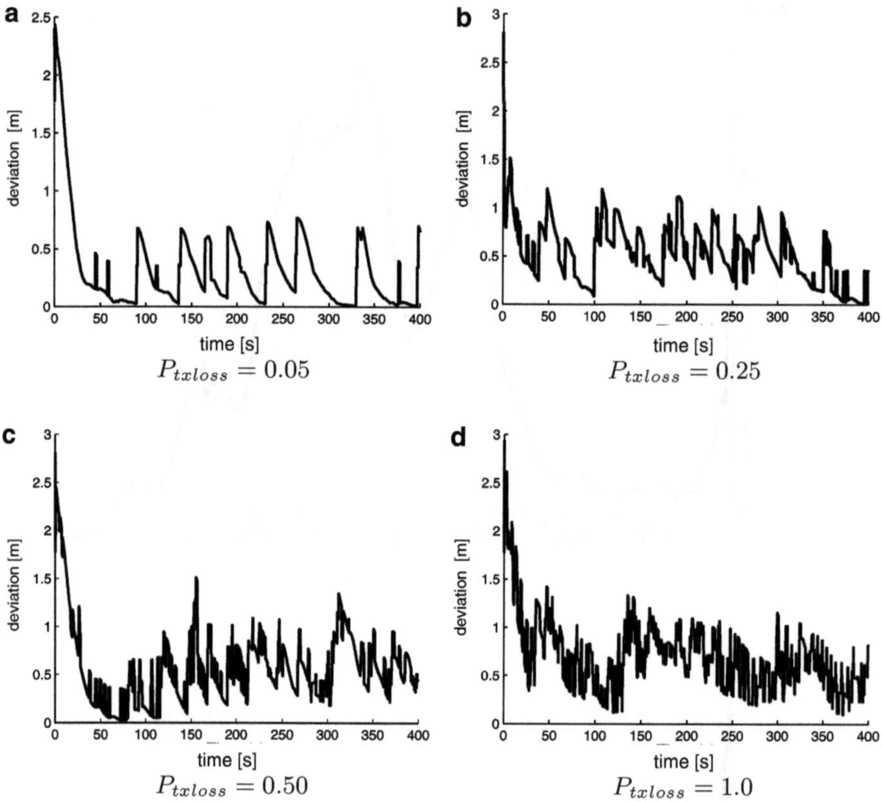


**Fig. 7.20** Plot of $\delta$ vs. time for disruption to a maximum of half the communications links

the formation vertex, even if they perceive that there is negative excess capacity in the other queue. Therefore, the algorithm that governs a helicopter's decision of its queue status is able to maintain the formation to a certain extent even in the presence of some imperfect communications. Similar simulations have also been carried out for the other combinations of $P_{\text{txloss}}$ and $I_{\text{loss}}$. The graphs in Fig. 7.22 show the number of helicopters involved in frequent toggling between queues for the various cases, and the average error in the formation over the simulation time interval. As expected, the extreme failure of helicopters to receive communicated information causes them to constantly switch between queues, and the formation is unable to settle into the desired form. When absolutely no information is received over all times, every helicopter will determine they belong to $\mathcal{Q}_1$ (according to Sect. 7.3.1). As a result, a column formation will form instead of the double column.

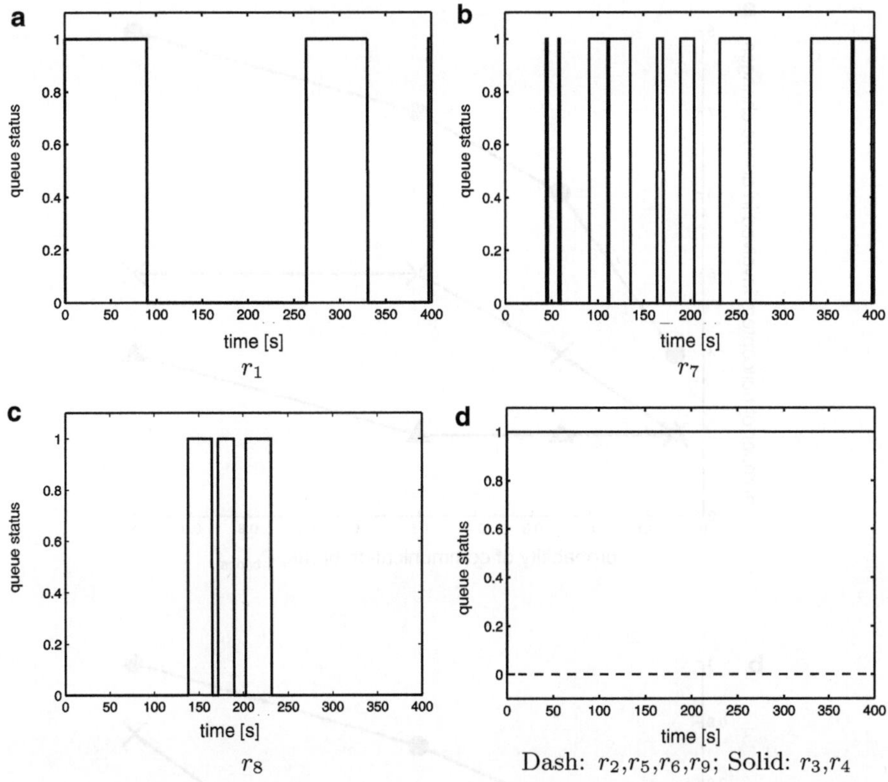


**Fig. 7.21** Queue status of the helicopters for $I_{\text{loss}} = 0.50$, $P_{\text{txloss}} = 0.05$

## 7.4 Q-Structure with Imperfect Communication

As seen from Sect. 7.3.4, over-reliance on inter-robot communication negatively influences the robustness of the team. The use of Q-structures is therefore augmented with an additional target determination mechanism that utilizes only information in the fast time-scale. This allows the system to be robust against limitations in communication ranges.

### *7.4.1 Determination of Target on Queue*

As opposed to formation representations relying on assigning helicopters to specific nodes (targets) within a formation, the helicopters using the Queue-based formation representation do not have fixed targets in their formation. Rather, the helicopters are only constrained by the queue, and can occupy any position in the queue that

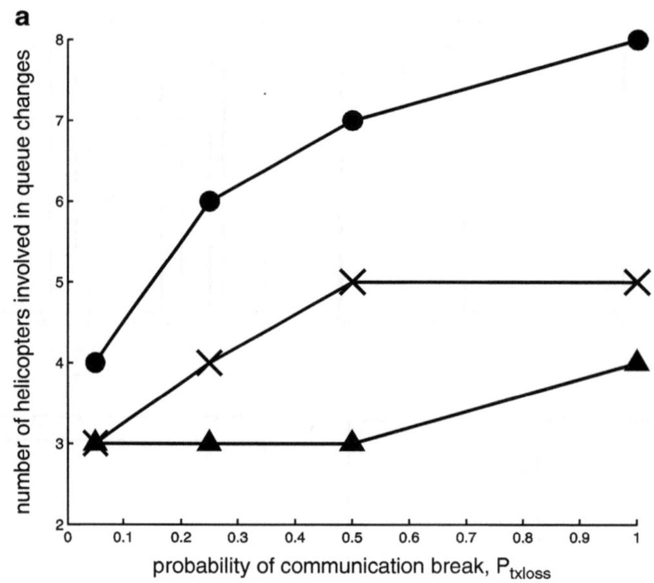


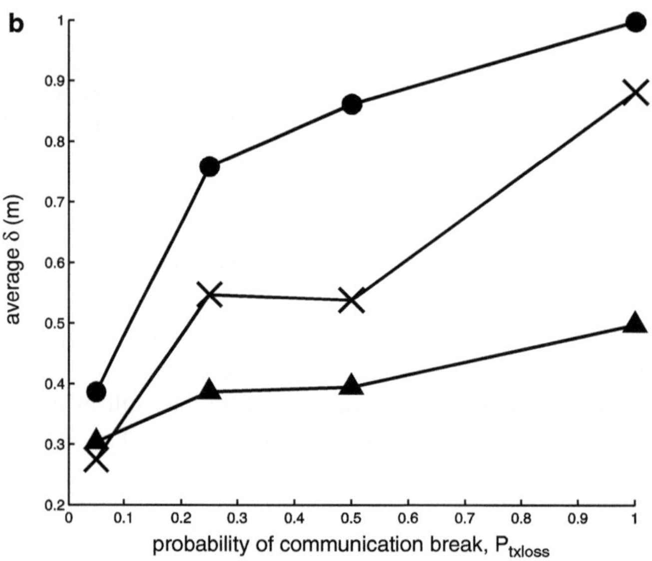


**Fig. 7.22** Effect of different degrees of communication breakdown on the formation. *Circle* (•): $I_{\text{loss}} = 1.00$, *Cross* (×): $I_{\text{loss}} = 0.50$, *Triangle* (▲): $I_{\text{loss}} = 0.25$. (**a**) Number of helicopters involved in queue changes, (**b**) Average error in formation

is the most convenient. This section describes an algorithm that each helicopter $r_i$, associated with a queue $\mathcal{Q}(i)$ uses for target determination.

The algorithm also governs the distance between helicopters within the same queue. Compared to the purely reactive scheme in [29], it improves the scaling of formations through an adaptation of the parameter $d_{ir}$ (acceptable inter-helicopter distance for helicopters on the same queue).

---

**Algorithm 2** Determining target on queue (by helicopter $r_i$)

1: Let $R_{c,i} \in R_N$ be an ordered set of helicopters (according to increasing Euclidean distance from $\mathcal{V}_{\mathcal{Q}(i)}(1)$) within communication range of $r_i$ and belonging to the same queue as $r_i$, i.e., belonging to $\mathcal{Q}(i)$.
2: Suppose $r_i$ is the $n$-th helicopter in the list $R_{c,i}$.
3: **if** n=1 **then**
4: Set $q_{\text{tg},i} = \mathcal{V}_{\mathcal{Q}(i)}(1)$.
5: **else**
6: Let $r_j \in R_{c,i}$ be the $(n-1)$-th helicopter in the list.
7: Set $q_{\text{tg},i} = \arg\min_{q \in Q} \|q - \mathcal{V}_{\mathcal{Q}(i)}(1)\|$ where $Q = \{q \in \mathcal{Q}(i) \mid \|q - q_{\text{tg},j}\| = d_{ir}$ and $\|q - \mathcal{V}_{\mathcal{Q}(i)}(1)\| > \|q_{\text{tg},j} - \mathcal{V}_{\mathcal{Q}(i)}(1)\|\}$.
8: **end if**

---

The algorithm is executed when $R_{c,i}$ changes. It works by considering the helicopters within communication range of $r_i$ which also belong to the same queue as $r_i$. The target of $r_i$ is set to be a point on $\mathcal{Q}(i)$ and at a distance of $d_{ir}$ away from the target of $r_j$. If $r_i$ is the helicopter in $R_{c,i}$ that is closest to the queue vertex $\mathcal{V}_{\mathcal{Q}(i)}(1)$, its target will be set to be the queue vertex.

The target changes in response to the information it has of other helicopters within communication range and which are of the same queue. The common objective ($\mathcal{F}_N$, as mentioned in Sect. II) will result in a weakly connected communication network for each subset of helicopters within the same queue. Although a helicopter may not be in direct communications with some others within the same queue, the decisions of preceding helicopters will be reflected/propagated via the decisions made by others within communication range.

**Lemma 7.4.1.** *Given a set of helicopters and considering only direct communications between a helicopter and those in its neighborhood, Algorithm 2, together with the common objective given in the form of the desired formation $\mathcal{F}_N$, will result in constant targets for each helicopter on each queue.*

*Proof.* Let $r_i$ and $r_j$ be the $n$-th and $(n-1)$-th furthest helicopters in $R_{c,i}$ from the queue vertex $\mathcal{V}_{\mathcal{Q}(i)}(1)$. According to Algorithm 2, if $q_{\text{tg},j}$ is constant, $q_{\text{tg},i}$ will be constant too, and at a distance of $d_{ir}$ along the queue from $q_{\text{tg},j}$.

Consider a queue $\mathcal{Q}_*$ where all helicopters belonging to this queue have converged into a weakly connected net due to the common objective. Let $R_{Q*} = \{r_{q1}, r_{q2}, \ldots, r_{qN_q}\}$ be this set of $N_q$ helicopters, ordered in ascending order according to their distance from the queue vertex $\mathcal{V}_{\mathcal{Q}*}(1)$. For the set $R_{c,q1}$, $r_{q1}$ will be the closest to the vertex, and from Algorithm 2, its target will be constant and locked to $q_{\text{tg},q1} = \mathcal{V}_{Q*}(1)$. From the argument in the preceding paragraph, the target of the second helicopter in $R_{Q*}$, $q_{\text{tg},q2}$ will be constant because $q_{\text{tg},q1}$ is fixed. Therefore, by induction, the target of the $n$-th helicopter will be fixed and constant, once the helicopters have converged into a weakly connected net around their respective queues. □

### 7.4.2 Navigation of Helicopters to Positions in Formations

Direct communication between helicopters in each others' neighborhoods allows target and position information to be transmitted between these helicopters. At any time instant, each helicopter will have their targets determined by their position relative to their related queues as described in Algorithm 2. Upon termination of Algorithm 2, from Lemma 7.4.1, the targets of each helicopter will become constant within finite time, and the control laws presented in this section will first bring each helicopter to converge according to the common objective (queues within formations) and onto their desired targets.

Consider the following potential function:

$$U = U_{\text{tg}} + U_{\text{ob}} \tag{7.26}$$

that consists mainly of two parts:

1. $U_{\text{tg}}$ describes the attractive potentials between the helicopters and their targets, and may be written as:

$$U_{\text{tg}} = \frac{1}{2} \sum_{i=1}^{N} \|q_i - q_{\text{tg},i}\|^2 \tag{7.27}$$

   since it is initially assumed that each helicopter's communications range is large enough to cover the team.
2. $U_{\text{ob}}$ reflects the collision avoidance behavior of the helicopters with their neighbors. It is chosen such that it is equal to infinity in the presence of collisions, and is at its minimum value when the helicopters are at their desired locations. Furthermore, in real life scenarios, the communication range of a helicopter is often limited to a set of helicopters near it. This can be due to power constraints and the presence of obstacles and noise. Let a helicopter, $r_i$, be able to reliably communicate with only $N_i$ helicopters (comprising the set $R_i \in R$). Communication signals that could be received from helicopters outside this range would be heavily attenuated. Each helicopter treats the other helicopters within its communications neighborhood as obstacles and constructs an instantaneous path according to a control law $u_i$. In view of these considerations, we may choose the function to be

$$U_{\text{ob}} = \sum_{i=1}^{N-1} \sum_{j=i+1}^{N} U_{\text{ob},ij} \tag{7.28}$$

where $U_{\text{ob},ij}$ is a function of $U_{ij}$ and $U_{\text{tg},ij}$, which are given by

$$U_{ij} = \frac{1}{2}\|q_i - q_j\|^2 \tag{7.29}$$

$$U_{\text{tg},ij} = \frac{1}{2}\|q_{\text{tg},i} - q_{\text{tg},j}\|^2 \tag{7.30}$$

and $U_{\text{ob},ij}$ is chosen such that it exhibits the following properties:

(a) $U_{\text{ob},ij} = \infty$, if $U_{ij} = 0$
(b) $U_{\text{ob},ij} > 0$, if $U_{ij} \neq 0$
(c) $U'_{\text{ob},ij} = \frac{\partial U_{\text{ob},ij}}{\partial U_{ij}} = 0$, if $U_{ij} = U_{\text{tg},ij}$
(d) $U''_{\text{ob},ij} = \frac{\partial^2 U_{\text{ob},ij}}{\partial U_{ij}^2} \geq 0$, if $U_{ij} = U_{\text{tg},ij}$
(e) $U_{\text{ob},ij} \approx 0$, if $U_{ij} \geq 0.5 d_{ij}^2$

Based on the above properties, $U_{\text{ob},ij}$ is chosen as

$$U_{\text{ob},ij} = f_{ij}\left(\frac{U_{ij}}{U_{\text{tg},ij}^2} + \frac{1}{U_{ij}}\right) \tag{7.31}$$

where

$$f_{ij} = \frac{1}{1 + \exp(a_t (U_{ij} - U_{\text{tg},ij})^3)} \tag{7.32}$$

where $a_t$ is a user-defined constant.

At each time instant, each helicopter moves along the negative gradient of the potential function $U$. In general, the time derivative of the overall potential function $U$ in (7.26) is given by

$$\begin{aligned}
\dot{U} &= \sum_{i=1}^{N} (q_i - q_{\text{tg},i})^{\mathrm{T}} u_i + \sum_{i=1}^{N-1} \sum_{j=i+1}^{N} U'_{\text{ob},ij} (q_i - q_j)^{\mathrm{T}} (u_i - u_j) \\
&= \sum_{i=1}^{N} \left( (q_i - q_{\text{tg},i})^{\mathrm{T}} + \sum_{j \neq i}^{N} U'_{\text{ob},ij} q_{ij}^{\mathrm{T}} \right) u_i \\
&= \sum_{i=1}^{N} \Omega_i^{\mathrm{T}} u_i
\end{aligned} \tag{7.33}$$

where $q_{ij} = q_i - q_j$ and $\Omega_i$ is defined as

$$\Omega_i = (q_i - q_{\text{tg},i}) + \sum_{j \neq i}^{N} U'_{\text{ob},ij} q_{ij} \tag{7.34}$$

This implies that a choice of

$$u_i = -C\Omega_i \tag{7.35}$$

where $C \in \mathbb{R}_{+}^{n_w \times n_w}$ is a symmetric, positive definite matrix, which is chosen as $C = \mathbf{I}_{n_w \times n_w} c$ where $c > 0$, will result in

$$\dot{U} = -\sum_{i=1}^{N} \Omega_i^{\mathrm{T}} C \Omega_i \tag{7.36}$$

and the closed loop dynamics of a single helicopter $r_i$ in the team is then given by

$$\dot{q}_i = -C\Omega_i \tag{7.37}$$

If the helicopters are at different positions (i.e. non-colliding) at an initial time $t_0$, and the target of each helicopter is different as well, these conditions may be written as

$$\|q_i(t_0) - q_j(t_0)\| \geq \epsilon_1 \tag{7.38}$$

where $\epsilon_1$ is a strictly positive constant, and $R$ is the set of helicopters comprising the team. In addition, Algorithm 2 guarantees that if the condition in (7.38) is satisfied, the targets for each cycle do not collide, i.e., $\|q_{\text{tg},i} - q_{\text{tg},j}\| \geq \epsilon_2, \quad \forall i, j \in R$, where $\epsilon_2$ is strictly positive. It is thus desired that, under such conditions, each helicopter will converge toward their targets, and at the same time avoiding collisions, i.e.

$$\begin{aligned} &\lim_{t \to \infty} (q_i(t) - q_{\text{tg},i}) = 0 \\ &\|q_i(t) - q_j(t)\| \geq \epsilon_3, \quad \forall i, j \in R \text{ and } \forall t \geq t_0 \geq 0 \end{aligned} \tag{7.39}$$

where $\epsilon_3$ is a strictly positive number representing the minimum acceptable inter-helicopter distance.

**Theorem 7.4.2.** *Under the conditions stated in (7.38), the common formation objective given by $\mathcal{F}_N$, and Algorithm 2, the control input to each helicopter, given in (7.35), will result in the convergence of each helicopter to their desired targets, such that:*

*(i) The target at $q_{\text{tg}}$ is located at an asymptotically stable equilibrium point of (7.37), and*
*(ii) The critical points of the system other than that at $q_{\text{tg}}$ are unstable equilibrium points.*

*Proof.* The proof is structured into two main parts. It begins with the proof of non-collision between the agents in the team, followed by the examination of the system's behavior around the set of critical points to show that only critical points coinciding with the location of the desired targets are stable. The latter portion of the proof is achieved by splitting the critical points into two non-intersecting sets, the set which, by design, coincides with the set of desired targets, and the set consisting of all other critical points. The behavior of the system around each of these two sets are examined. The first set is shown to be stable equilibrium points, while the second set is shown to be unstable.

To show that there will be no collision between any two agents, (7.36) is integrated on both sides, from $t_0$ to $t$, to obtain

$$U_{\mathrm{tg}}(t) + \sum_{i=1}^{N-1} \sum_{j=i+1}^{N} U_{\mathrm{ob},ij}(t) \leq U_{\mathrm{tg}}(t_0) + \sum_{i=1}^{N-1} \sum_{j=i+1}^{N} U_{\mathrm{ob},ij}(t_0) \tag{7.40}$$

where

$$\begin{aligned} U_{\mathrm{tg}}(t) &= \frac{1}{2} \sum_{i=1}^{N} \|q_i(t) - q_{\mathrm{tg},i}\|^2 \\ U_{\mathrm{ob},ij}(t) &= f_{ij}(t) \left( \frac{U_{ij}(t)}{U_{\mathrm{tg},ij}^2} + \frac{1}{U_{ij}(t)} \right) \end{aligned} \tag{7.41}$$

From the conditions in (7.38), $U_{ij}(t_0)$ and $U_{\mathrm{tg},ij}$ are strictly larger than some positive constants. Furthermore, since $f_{ij}$ is also bounded ($0 < f_{ij} < 1$), the right hand side of (7.40) is bounded by some positive constant (the value of which depends on the initial conditions at $t_0$). Hence, the left hand side is also bounded, which in turn implies that $U_{ij}(t)$ must be strictly larger than some positive constant for all $t \geq t_0 \geq 0$. From (7.41), $\|q_i(t) - q_j(t)\|$ will therefore always be larger than some strictly positive constant, and there will be no collisions. The boundedness of the left hand side of (7.40) also implies that of $\|q_i(t)\|$ for all $t \geq t_0 \geq 0$, and the solutions of the closed loop system in (7.37) exist.

To prove that the system will converge onto the subset of critical points that by design coincides with the set of desired targets, we begin by letting the root sets (critical points) of the system in (7.37) be represented by $q_e$. It consists of points at $q = q_{\mathrm{tg}}$ (due to Property (c) of $U_{\mathrm{ob},ij}$) and $q = q_{\mathrm{c}}$ (representing the remaining critical points), where the overall system for the $N$ helicopters is $\dot{q} = -c\Omega$, with $q = [q_1^{\mathrm{T}}, \ldots, q_N^{\mathrm{T}}]^{\mathrm{T}}$, $\Omega = [\Omega_1^{\mathrm{T}}, \ldots, \Omega_N^{\mathrm{T}}]^{\mathrm{T}}$, $q_{\mathrm{tg}} = [q_{\mathrm{tg},1}^{\mathrm{T}}, \ldots, q_{\mathrm{tg},N}^{\mathrm{T}}]^{\mathrm{T}}$ and $q_{\mathrm{c}} = [q_{\mathrm{c},1}^{\mathrm{T}}, \ldots, q_{\mathrm{c},N}^{\mathrm{T}}]^{\mathrm{T}}$. The equilibrium points are not separated into stable and unstable points at the outset before the following analysis, but rather, the properties of the points which are desired to be stable are examined vs. the rest of the equilibrium points. By construction, $q_{\mathrm{tg}}$ is an equilibrium point, and the main objective is for this to be stable and for the remaining critical points $q_{\mathrm{c}}$, wherever they may be, to be unstable. Furthermore, the targets for each helicopter in the system is determined by Algorithm 2 and are constant for the time period under consideration, and the system, by inspection, is Linear-Time-Invariant (LTI). For the remainder of the proof, the property at $\bar{q}_{\text{tg}}$ is examined first, followed by the properties of $\bar{q}_c$. Linearizing the closed loop system about the equilibrium point $q_e$ gives

$$\dot{q} = -c \left.\frac{\partial \Omega}{\partial q}\right|_{q=q_e} (q - q_e) \tag{7.42}$$

where the general gradient of $\Omega$ with respect to $q$ is

$$\frac{\partial \Omega}{\partial q} = \begin{bmatrix} \frac{\partial \Omega_1}{\partial q_1} & \frac{\partial \Omega_1}{\partial q_2} & \cdots & \cdots & \frac{\partial \Omega_1}{\partial q_N} \\ \vdots & \vdots & \vdots & \vdots & \vdots \\ \frac{\partial \Omega_i}{\partial q_1} & \cdots & \frac{\partial \Omega_i}{\partial q_i} & \cdots & \frac{\partial \Omega_i}{\partial q_N} \\ \vdots & \vdots & \vdots & \vdots & \vdots \\ \frac{\partial \Omega_N}{\partial q_1} & \cdots & \cdots & \cdots & \frac{\partial \Omega_N}{\partial q_N} \end{bmatrix} \tag{7.43}$$

with

$$\frac{\partial \Omega_i}{\partial q_i} = \left(1 + \sum_{j \neq i}^{N} U'_{\text{ob},ij}\right) \mathbf{I} + \sum_{j \neq i}^{N} U''_{\text{ob},ij} q_{ij} q_{ij}^{\text{T}} \tag{7.44}$$

$$\frac{\partial \Omega_i}{\partial q_j} = -U'_{\text{ob},ij} \mathbf{I} - U''_{\text{ob},ij} q_{ij} q_{ij}^{\text{T}} \tag{7.45}$$

At the equilibrium points at $q_e = q_{\text{tg}}$, based on the properties of $U_{\text{ob},ij}$, and letting $q_{\text{tg},ij} = q_{\text{tg},i} - q_{\text{tg},j}$, (7.44) and (7.45) become

$$\frac{\partial \Omega_i}{\partial q_i} = \mathbf{I} + \sum_{j \neq i}^{N} U''_{\text{ob},ij} q_{\text{tg},ij} q_{\text{tg},ij}^{\text{T}} \tag{7.46}$$

$$\frac{\partial \Omega_i}{\partial q_j} = -U''_{\text{ob},ij} q_{\text{tg},ij} q_{\text{tg},ij}^{\text{T}} \tag{7.47}$$

Considering the Lyapunov candidate

$$V_{q_{\text{tg}}} = \frac{1}{2} \|q - q_{\text{tg}}\|^2 \tag{7.48}$$

and using (7.46) and (7.47) we obtain

$$\dot{V}_{q_{tg}} = -c(q - q_{tg})^{T} \left.\frac{\partial \Omega}{\partial q}\right|_{q=q_{tg}} (q - q_{tg}) \tag{7.49}$$

$$= -c \sum_{i=1}^{N} \sum_{j=1}^{N} (q_i - q_{tg,i})^{T} \frac{\partial \Omega_i}{\partial q_i} (q_j - q_{tg,j}) \tag{7.50}$$

$$= -c \sum_{i=1}^{N} \| q_i - q_{tg,i} \|^2 - c \sum_{i=1}^{N} \sum_{j=1, j \neq i}^{N} U''_{ob,ij} (q_{tg,ij}^{T} (q_{ij} - q_{tg,ij}))^2 \tag{7.51}$$

Since $U''_{ob,ij} \geq 0$ at $q = q_{tg}$,

$$\dot{V}_{q_{tg}} \leq -2c V_{q_{tg}} \tag{7.52}$$

indicates the equilibrium points at $q_{tg}$ are asymptotically stable.

To show that the remaining critical points of the system, i.e., $q_c$, are unstable equilibrium points, consider the following.

$$\bar{q}_c^{T} F(\bar{q}_c, \bar{q}_{tg}) = 0 \tag{7.53}$$

$$\Rightarrow \sum_{i=1}^{N-1} \sum_{j=i+1}^{N} \left( q_{c,ij}^{T} (q_{c,ij} - q_{tg,ij}) + N \left. U'_{ob,ij} \right|_{q_{ij}=q_{c,ij}} q_{c,ij}^{T} q_{c,ij} \right) = 0$$

$$\Rightarrow \sum_{i=1}^{N-1} \sum_{j=i+1}^{N} \left( 1 + N \left. U'_{ob,ij} \right|_{q_{ij}=q_{c,ij}} \right) q_{c,ij}^{T} q_{c,ij} = \sum_{i=1}^{N-1} \sum_{j=i+1}^{N} q_{c,ij}^{T} q_{tg,ij} \tag{7.54}$$

where $q_{c,ij} = q_{c,i} - q_{c,j}$, $\Omega_{ij} = \Omega_i - \Omega_j$ and

$$\bar{q} = [q_{12}^{T}, q_{13}^{T}, \ldots, q_{ij}^{T}, \ldots, q_{N-1N}^{T}]^{T} \tag{7.55}$$

$$\bar{q}_{tg} = [q_{tg,12}^{T}, q_{tg,13}^{T}, \ldots, q_{tg,ij}^{T}, \ldots, q_{tg,N-1N}^{T}]^{T} \tag{7.56}$$

$$\bar{q}_c = [q_{c,12}^{T}, q_{c,13}^{T}, \ldots, q_{c,ij}^{T}, \ldots, q_{c,(N-1)(N)}^{T}]^{T} \tag{7.57}$$

$$F(\bar{q}, \bar{q}_{tg}) = [\Omega_{12}^{T}, \Omega_{13}^{T}, \ldots, \Omega_{ij}^{T}, \ldots, \Omega_{N-1N}^{T}]^{T} \tag{7.58}$$

Consider the term $q_{c,ij}^{T} q_{tg,ij}$ and the helicopters $i$ and $j$. The helicopter $j$ can be seen as an obstacle situated at $q_{ij} = 0$. Similarly, helicopter $i$ is an obstacle with respect to $j$ at $q_{ji} = 0$. At $q_{ij} = q_{c,ij}$, both helicopters are at their critical points. For this to hold, both critical points must lie along a straight line along the vector $q_{tg,ij}$ and between $q_{tg,i}$ and $q_{tg,j}$. That is, the point $q_{ij} = 0$ must lie between the points $q_{ij} = q_{tg,ij}$ and $q_{ij} = q_{c,ij}$, and such that these three points are colinear. Thus, the term $\sum_{i=1}^{N-1} \sum_{j=i+1}^{N} q_{c,ij}^{T} q_{tg,ij}$ is strictly negative and there exists at least one pair $(i, j)$ denoted by $(i^*, j^*) \in R^*$ such that

$$1 + N\, U'_{\text{ob},i^* j^*}\Big|_{q_{i^* j^*} = q_{\text{c},i^* j^*}} \leq -b \tag{7.59}$$

where $b$ is a strictly positive constant. For the system under consideration, the inter-helicopter repulsive forces are dependent on the relative distances between individual helicopters. For a helicopter $i$, the other helicopters can be treated as obstacles, and the equilibrium points are a direct result of the relative positions. Therefore, instead of considering the function $V_c = \|q - q_\text{c}\|^2$, the behavior of the equilibrium points in the system are examined first by considering the Lyapunov function based on the relative distances (i.e., $\bar{q}$ and $\bar{q}_c$), and the result is then linked to stability of the points $q_\text{c}$ in the last part of the proof. Consider the Lyapunov function candidate

$$V_{\bar{q}_c} = \|\bar{q} - \bar{q}_c\|^2 \tag{7.60}$$

whose derivative along the solution of (7.60) gives

$$\begin{aligned}
\dot{V}_{\bar{q}_c} &= -2c \sum_{i=1}^{N-1} \sum_{j=i+1}^{N} (q_{ij} - q_{\text{c},ij})^\text{T} \Bigg( \mathbf{I}_{n_w \times n_w} + N \mathbf{I}_{n_w \times n_w}\, U'_{\text{ob},ij}\Big|_{q_{ij} = q_{\text{c},ij}} \\
&\quad + N\, U''_{\text{ob},ij}\Big|_{q_{ij} = q_{\text{c},ij}} q_{\text{c},ij} q_{\text{c},ij}^\text{T} \Bigg) (q_{ij} - q_{\text{c},ij}) \\
&\geq 2cb(q_{i^* j^*} - q_{\text{c},i^* j^*})^\text{T}(q_{i^* j^*} - q_{\text{c},i^* j^*}) - 2c \sum_{i=1, i \neq i^*}^{N-1} \; \sum_{j=i+1, j \neq j^*}^{N} (q_{ij} - q_{\text{c},ij})^\text{T} \\
&\quad \left( \mathbf{I}_{n_w \times n_w} + N \mathbf{I}_{n_w \times n_w}\, U'_{\text{ob},ij}\Big|_{q_{ij} = q_{\text{c},ij}} \right) (q_{ij} - q_{\text{c},ij}) \\
&\quad - 2c \sum_{i=1}^{N-1} \sum_{j=i+1}^{N} (q_{ij} - q_{\text{c},ij})^\text{T} \left( N\, U''_{\text{ob},ij}\Big|_{q_{ij} = q_{\text{c},ij}} q_{\text{c},ij} q_{\text{c},ij}^\text{T} \right) (q_{ij} - q_{\text{c},ij})
\end{aligned} \tag{7.61}$$

Consider a subspace such that $q_{ij} = q_{\text{c},ij}\ \forall (i, j) \in \{1, \ldots, N\}, (i, j) \neq (i^*, j^*)$ and $(q_{ij} - q_{\text{c},ij})^\text{T} q_{\text{c},ij} q_{\text{c},ij}^\text{T} (q_{ij} - q_{\text{c},ij}) = 0,\ \forall (i, j) \in \{1, \ldots, N\}$. In this subspace, the following holds

$$V_{\bar{q}_c} = \sum_{(i,j) \in R^*} \|q_{ij} - q_{\text{c},ij}\|^2 \tag{7.62}$$

$$\dot{V}_{\bar{q}_c} \geq 2bc V_{\bar{q}_c} \tag{7.63}$$

which implies that

$$\sum_{(i,j)\in R^*} \|q_{ij}(t) - q_{c,ij}\| \geq \sum_{(i,j)\in R^*} \|q_{ij}(t_0) - q_{c,ij}\| e^{bc(t-t_0)} \tag{7.64}$$

where $t \geq t_0 \geq 0$. Assume that $q_c$ is a stable equilibrium with $\lim_{t\to\infty} \|q_i(t) - q_{c,i}\| = a_i$ where $a_i$ is a positive constant. This further implies that $\lim_{t\to\infty} \sum_{(i,j)\in R^*} \|q_{ij}(t) - q_{c,ij}\| = a_*$, $\forall(i,i) \in R^*$ and $a_*$ being a positive constant, which contradicts the result obtained in (7.64), and $q_c$ is an unstable equilibrium point of the closed loop system. □

*Remark 3.* In the above proof of $q_c$ being unstable equilibrium points, an exception occurs when all the helicopters start at positions that coincides exactly with their critical points at $q_c$ (i.e., $\sum_{(i,j)\in R^*} \|q_{ij}(t_0) - q_{c,ij}\| = 0$), in this case $q_c$ will be marginally stable (similar to a linear system where the real part of one or more eigenvalues equals zero). However, for practical systems that are considered here, noise and other disturbances will cause $\sum_{(i,j)\in R^*} \|q_{ij}(t_0^*) - q_{c,ij}\| \neq 0$ for some finite $t_0^* > t_0$. Therefore instability of $q_c$ can be analyzed in the same way as above with $t_0$ replaced by $t_0^*$.

For a practical implementation, a helicopter $r_i$ may only be able to compute an approximate value of $\Omega_i$ in (7.34) since it may not receive any information at all from helicopters outside the communications radius $d_i$. The approximation of $\Omega$ is given by

$$\hat{\Omega}_i = (q_i - q_{tg,i}) + \sum_{j\neq i, j\in R_i} U'_{ob,ij} q_{ij} \tag{7.65}$$

where $R_i$ is the set of helicopters within the $d_i$-neighborhood of $r_i$, and the control law becomes

$$\dot{\hat{u}} = -C\hat{\Omega} \tag{7.66}$$

The approximation error for each helicopter may thus be written as

$$\begin{aligned} e_\Omega &= \Omega - \hat{\Omega} \\ &= \sum_{j\neq i, j\in R_{ni}} U'_{ob,ij} q_{ij} \end{aligned} \tag{7.67}$$

where $R_{ni} = R \setminus R_i$ is the set of helicopters that $r_i$ cannot communicate with, and "\" denotes the set subtraction operation. From property (e) of $U_{ob,ij}$, we know that for $j \in R_{ni}$, $U_{ob,ij}$, $U_{ob,ij} \approx 0$. In addition, assuming that $\|q_{ij}(0)\|$ is bounded, since the helicopters converge to their targets on the queues and $\|q_{tg,ij}\|$ is also bounded, the value of $e_\Omega$ is bounded by some small positive real value, and the error that arises due to incomplete information from helicopters out of communication range can be kept small through the use of $f_{ij}$ to weight the importance of repulsive forces between helicopters. Therefore, for the control law described in (7.34) and (7.35) for an helicopter $i$, the use of $f_{ij}$ heavily attenuates the contribution of any helicopter $j$ that is out of range to approximate the control when global communications is present, to facilitate practical implementation with limited communication ranges and scaling.

### 7.4.3 *Simulation Studies*

As in the previous section, simulations were conducted using Player and Stage [103]. The simulated system consists of five circular, omni directional helicopters, of diameter 0.3 m. Each of these helicopters acts based on commands to their speed, which in this case is determined by the control input $u_i$ in (7.66) with the estimated $\hat{\Omega}_i$ in (7.65). The parameters $a_t$, $d_{ir}$ and $C$ are chosen to be 10, 2 and the identity matrix respectively. It is assumed that each helicopter is able to localize itself in the global frame. Furthermore, each helicopter is equipped with a laser scanner (180°) and 16 sonar range sensors arranged in a ring around the circular helicopters for obstacle avoidance. The sensor noise introduced into the range sensing has a normal distribution of 0.2 variance. The communication range of the helicopters is set to 3 m, and there will be connectivity between each helicopter on each queue as they converge into position as long as $d_{ir}$ is set to be less than the communication range.

#### 7.4.3.1 Formation Convergence and Scaling

The first part of the simulations consists of examining the convergence of the helicopters to a given wedge formation, and how it scales when two helicopters are removed (deactivated) at $t = 10$ s. In the final formation, helicopters are to be a minimum of 2 m from others. The helicopters are initialized at random (non-colliding) positions in a 20 m×20 m square around the point (10 m,10 m) in the workspace. Figure 7.23 shows how the distance of the helicopters from their targets vary over time. The targets evolve according to Algorithm 2.

From the graphs, we can see that the helicopters are able to converge to the formation in a relatively short time of 6–8 s, and in approximately 3 s after scaling. Figure 7.24a shows the minimum center-to-center distance that exists between any two helicopters in the team at each time. It can be seen that the minimum distance between any two helicopters is always greater than 0.5 m at all times, and hence, no collisions occur. Figure 7.24b shows the control signals applied by each helicopter over time. It should be noted that Fig. 7.23 shows the distance of each helicopter from their target at each time instant. The spikes in the graphs are the result of changes in the targets for each helicopter (according to Algorithm 2) as they interact with others within communication range. It can also be noted that these spikes, however, cease to appear when the helicopters get within communication range of each other and their targets reach a constant state. This is further evidenced by the absence of spikes when scaling occurs at $t = 10$ s, and the helicopters converge to their new targets. This observation applies also to the subsequent subsections.

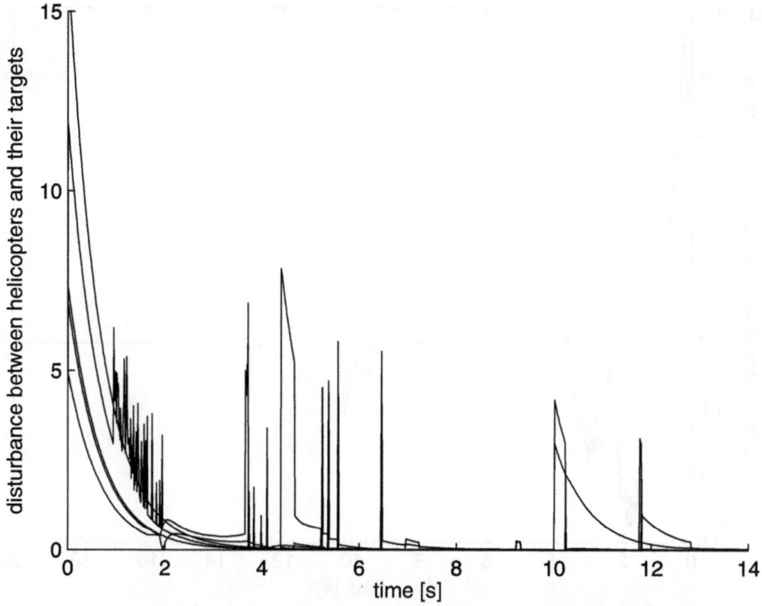


**Fig. 7.23** Helicopter convergence to formation with helicopter deactivation/removal at $t = 10$s

For comparison, the convergence of helicopters using the purely reactive technique presented in [29], with limited communication ranges between helicopters and for a wedge formation, is shown in Fig. 7.25. It can be seen that convergence is adversely affected by limited communication ranges, since the helicopters frequently reallocate themselves to different queues depending on the helicopters within their own neighborhoods, which causes constant shuttling between queues. This is an effect that has been removed by the current proposed scheme.

#### 7.4.3.2 Moving Formations

In order to verify the effectiveness of the proposed method in enforcing formation maintenance relative to a moving target, simulations are run on the same team of helicopters with a separate sixth helicopter acting as a moving target. The formation vertex of the wedge formation is set to be at a distance of 2 m along the negative $x$-axis of the target helicopter as shown in Fig. 7.26a. The moving target is programmed to start moving at time $t \approx 3$ from its initial point at (11.4 m, 11.4 m) (such that the initial formation vertex is approximately at (10 m, 10 m)) at a constant velocity of $[0.2\ 0.2]^{\mathrm{T}}$ m/s. The helicopters' task is to form a straight line formation (2 m apart from each other) behind the target when it is not moving and to follow it in a wedge formation when it starts to move. The convergence of the helicopters to the formation is shown in Fig. 7.26b. It can be observed that when the target begins

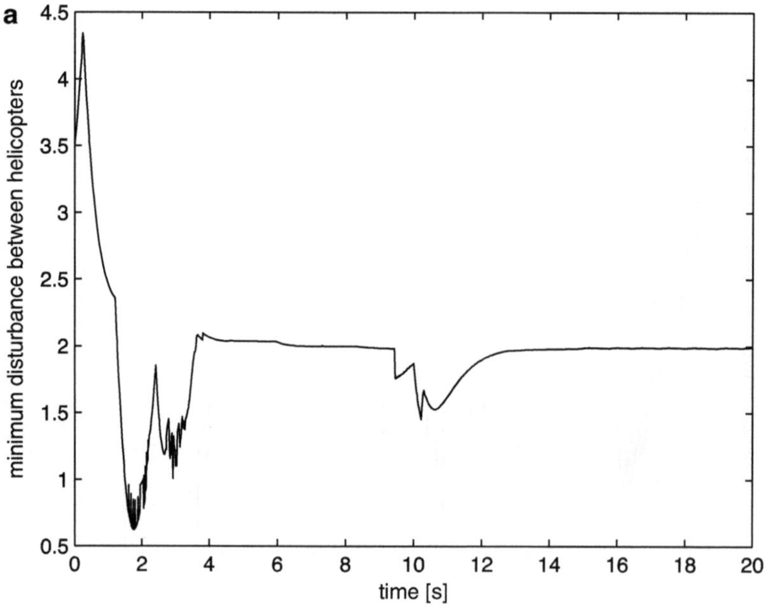


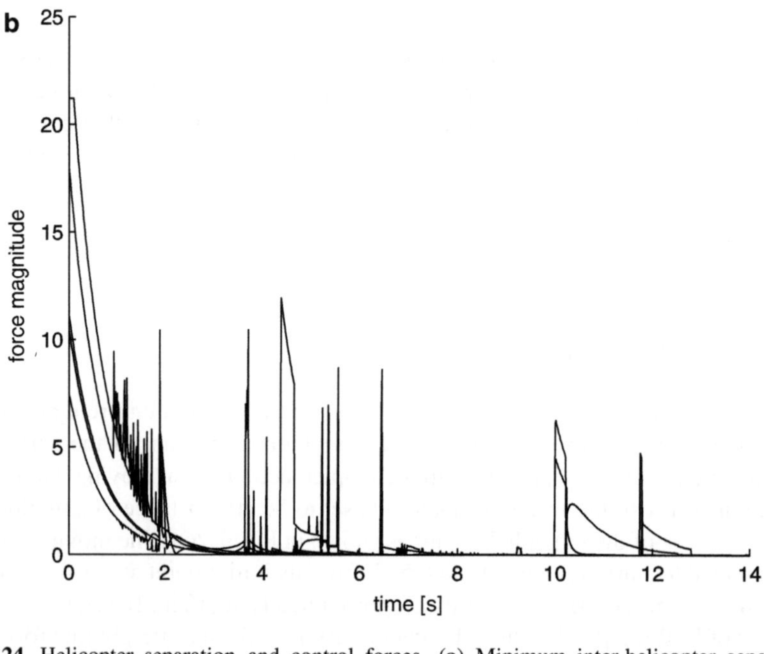


**Fig. 7.24** Helicopter separation and control forces. (**a**) Minimum inter-helicopter separation, (**b**) Forces due to potential field

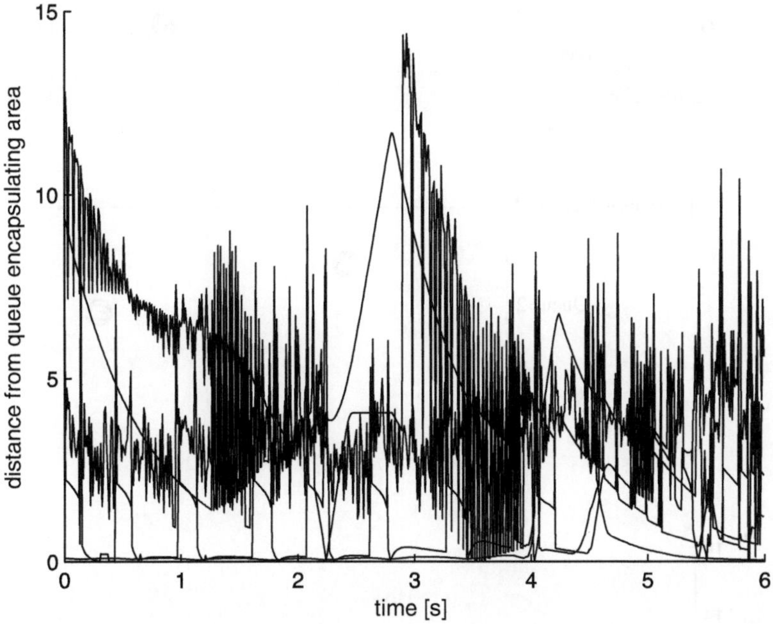


**Fig. 7.25** Distance of helicopters from related queue's encapsulating area

to move, the line formation that the helicopters are originally in is disrupted as they attempt to form a wedge. Convergence into the wedge formation is subsequently observed after approximately 4 s. The minimal inter-helicopter distance and the control signals are shown in Fig. 7.27.

#### 7.4.3.3 Changing Formations

To further investigate the proposed method when the formation changes, we conduct similar experiments for the case when the formation changes at predefined times from a wedge to a column (perpendicular to the orientation of the target), and finally to a line (parallel to the target's orientation). The results are shown in Figs. 7.27 and 7.28. We can observe spikes in the graphs at the times when formation changes are initiated, occurring due to the abrupt change in targets. Furthermore, comparing the second the third clusters of spikes, it can be seen that, as expected, the transition from a column to a line is more disruptive compared to the transition from a wedge to a column, due to the further distances to the new targets. On the whole, the team requires an average of 4–6 s to transition between formations and settle stably into the new formation (Fig. 7.29).

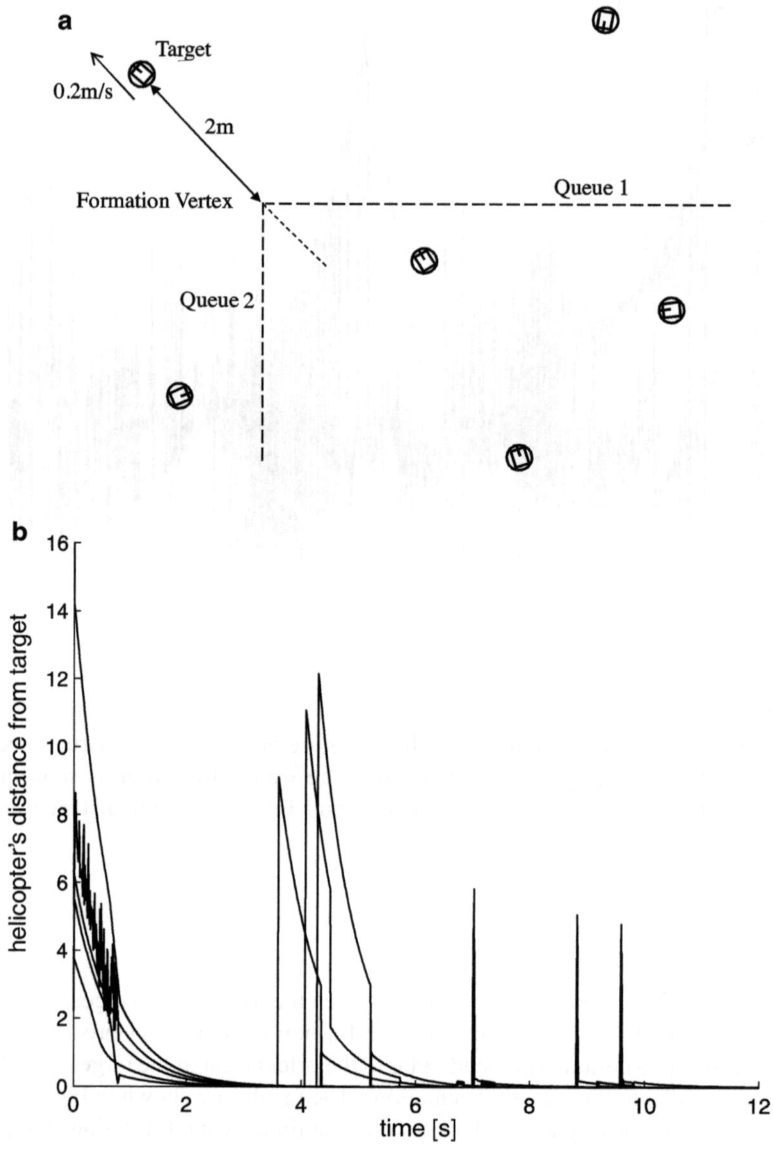


**Fig. 7.26** Formation convergence with a moving target. (**a**) Moving target, the queues and virtual formation vertex, (**b**) Helicopter convergence to formation

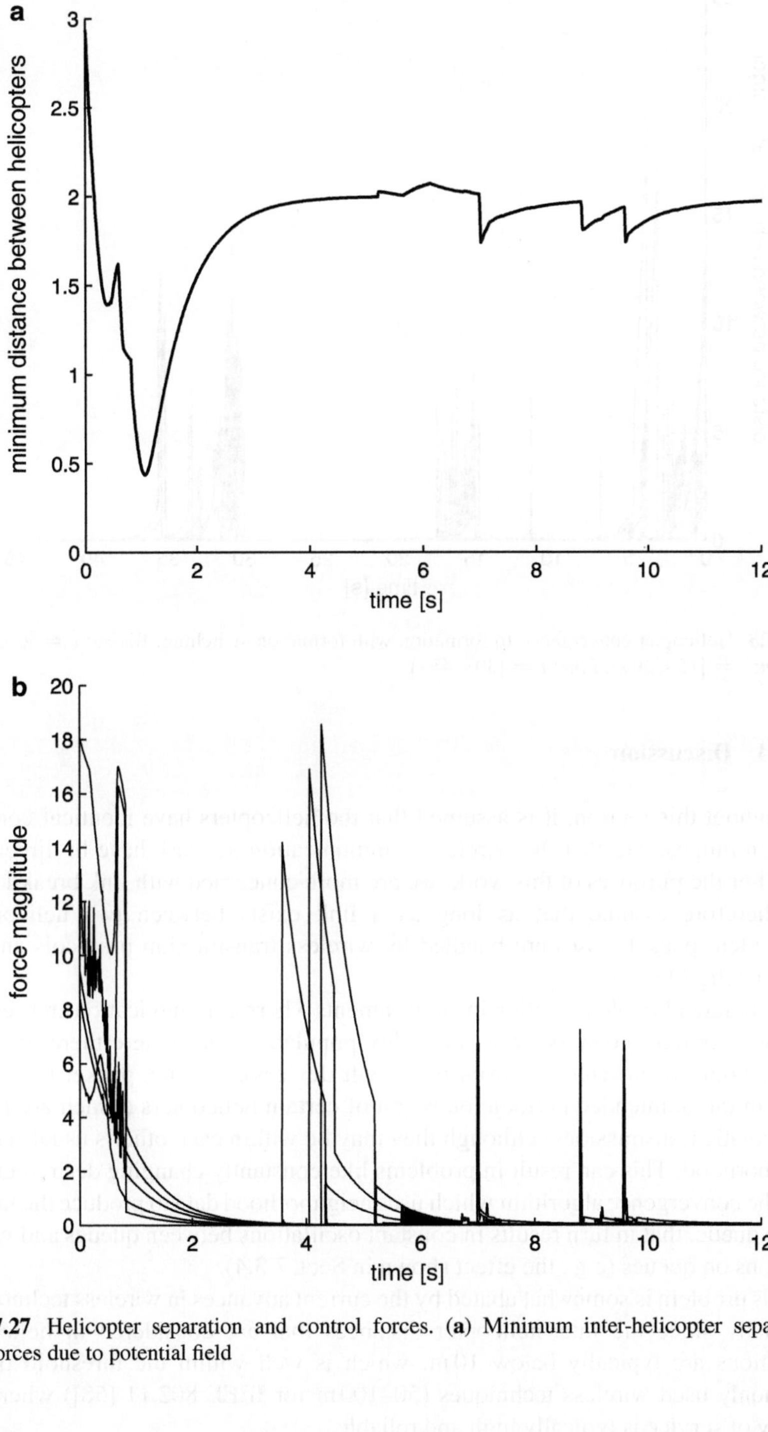


**Fig. 7.27** Helicopter separation and control forces. (**a**) Minimum inter-helicopter separation, (**b**) Forces due to potential field

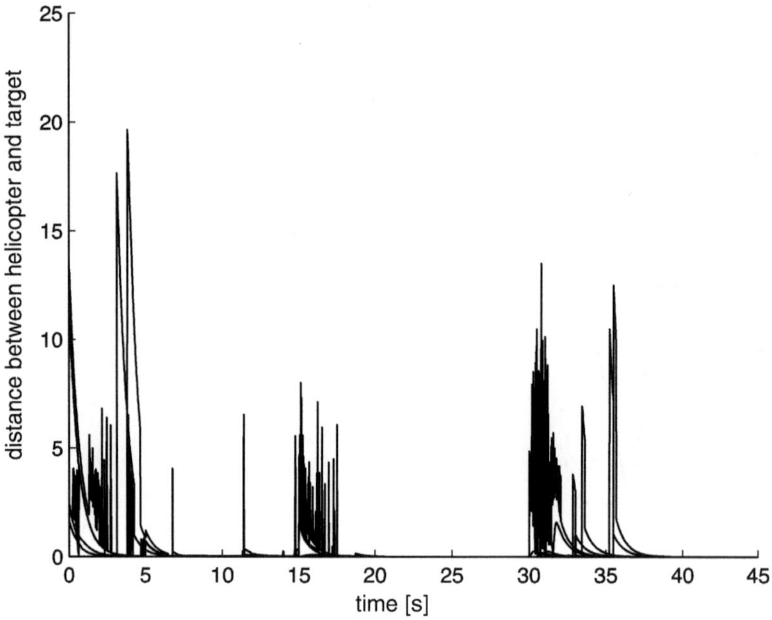


**Fig. 7.28** Helicopter convergence to formation with formation switching. *Wedge*: $t = [0\,\text{s}, 15\,\text{s})$, *Column*: $t = [15\,\text{s}, 30\,\text{s})$, *Line*: $t = [30\,\text{s}, 45\,\text{s})$

### 7.4.3.4 Discussion

Throughout this section, it is assumed that the helicopters have identical communication ranges, and that the wireless communication network have bi-directional links. For the purposes of this work, we are more concerned with link breakdowns, and therefore assume that as long as a link exists between two helicopters, intermittent packet losses are handled by wireless transmission protocols and are hence negligible.

For practical implementation in environments where communication and sensing can be extremely noisy (such as in highly populated areas where there can be a large amount of interference from other wifi devices), extreme packet losses can result in the unintended periodic omission of certain helicopters (which are facing problematic transmissions) although they may be within each other's usual sensing neighborhood. This can result in problems like constantly changing desired targets with the convergence algorithm which uses neighborhood data to produce the targets on the queue, that in turn results in constant oscillations between queues and within positions on queues (e.g., the effect shown in Sect. 7.3.4).

This problem is somewhat abated by the current advances in wireless technology, especially since the inter-helicopter distances that are considered in helicopter formations are typically below 10 m, which is well within the threshold of the commonly used wireless techniques (50–100 m for IEEE 802.11 [58]) where the quality of service is typically high and reliable.

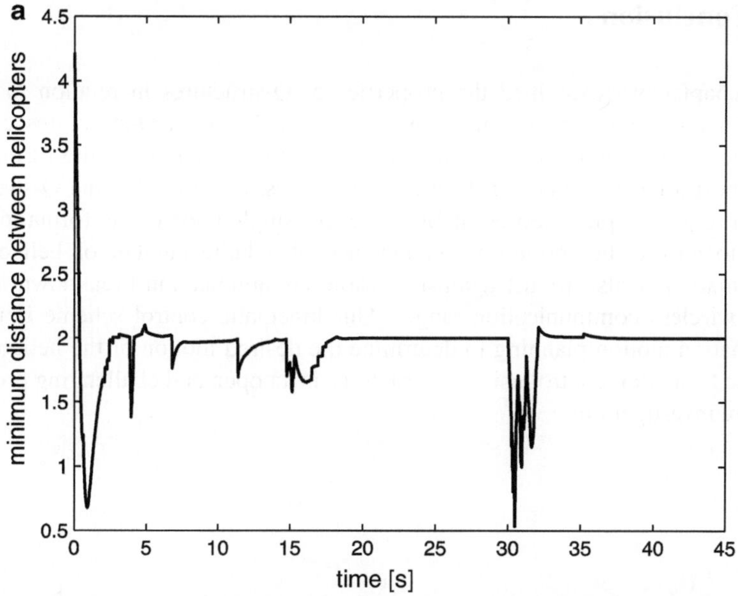


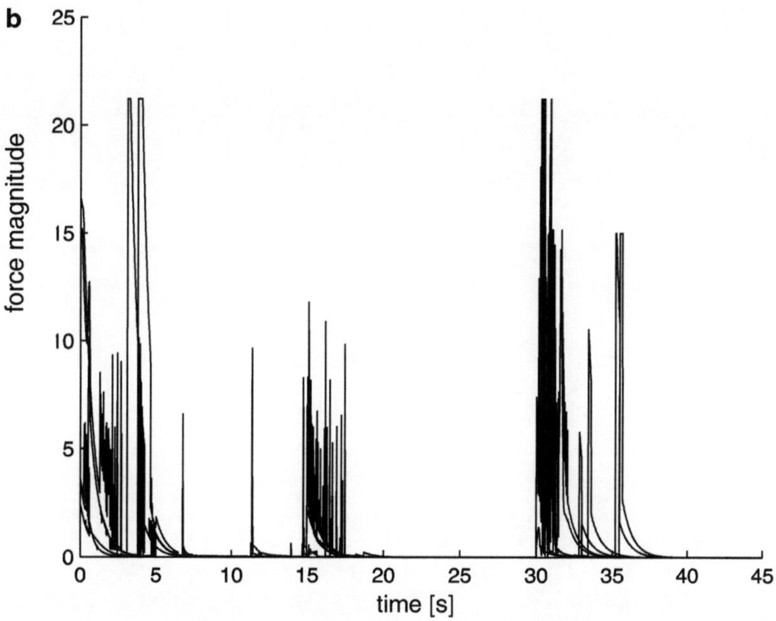


**Fig. 7.29** Helicopter separation and control forces. (**a**) Minimum inter-helicopter separation, (**b**) Forces due to potential field

## 7.5 Conclusion

In this chapter, we examined the properties of Q-structures in relation to other formation representation schemes, and looked at the ways Q-structures can be used with artificial potential trenches to improve the scalability of the formations and support a large number of different formations. In particular, the Q-structure does not require explicit representation of every single node of the formation and is able to ensure the formation maintenance of a large number of helicopters. The formation is also robust against possible communication breakdown and/or limited wireless communication ranges. Our kinematic control scheme is useful for formation motion planning to determine the desired motion of the helicopters. Dynamic formation control using Q-structures is an open and challenging problem for future investigations.

# Chapter 8
# Dynamic Altitude Synchronization Using Graph Theory

## 8.1 Introduction

The Q-structure proposed in Chap. 7 provides a promising avenue for formation control of helicopters under a flexible and scalable framework. We presented a kinematic control scheme that does not consider the dynamics of the helicopters, and is useful for formation motion planning to determine the desired motion of the helicopters. Dynamic formation control using Q-structures in which the helicopter dynamics are directly taken into account in the formation control design can yield better flight performance, but it is an open and challenging problem. In this chapter, we take a different approach to solving the dynamic formation control problem, by combining graph theory with adaptive neural networks.

We focus on the synchronized tracking problem of helicopters in vertical flight, in which multiple helicopters track the same desired trajectory while the desired trajectory is not accessible to all the helicopters in the team. The vertical fight mode starts when the helicopter is at rest on the ground IGE (in ground effect). Then take-off is started and the helicopter climbs. Vertical descent precedes landing. Since the coupling between longitudinal and lateral-directional equations in this flight regime is weak, it can be presented by single-input–single-output (SISO) models with zero-dynamics to yield useful results [83, 102]. In the formation group, the desired trajectory is not available to all the helicopters in the team, synchronized tracking control is designed for each helicopter by using the information exchange with its neighbors. The main contributions of the work are as follows:

1. The extended formation graph Laplacian, which contains a spanning tree which the root helicopter can access for the desired trajectory, is proved to be positive definite.
2. The neural approximation based control is designed for the purpose of synchronized tracking of each helicopter by using the weighted average of its neighbors' states. All signals are proved to be bounded and the tracking errors of all helicopters will converge to a neighborhood of the origin.


DOI 10.1007/978-1-4614-1563-3_8, 

3. A high gain observer is employed for each helicopter to estimate the unaccessible derivation of the states of both itself and its neighbors. It is shown that in this case, the boundness of all the closed-loop signals are guaranteed.

## 8.2 Problem Formulation

### 8.2.1 Helicopter Dynamics

Consider the class of SISO helicopter systems described by

$$\begin{aligned}
\dot{x}_j &= x_{j+1}, \quad j = 1, \ldots, \rho - 1 \\
\dot{x}_\rho &= f(\eta, x) + g(x, \eta)(u + d) \\
\dot{\eta} &= q(x, \eta) \\
y &= x_1
\end{aligned} \tag{8.1}$$

where $x = [x_1, \ldots, x_\rho]^{\mathrm{T}} \in \mathbb{R}^\rho$ and $\eta \in \mathbb{R}^{n-\rho}$ are the states of the system, $u,\ y \in \mathbb{R}$ the input and output, respectively, $f : \mathbb{R}^n \to \mathbb{R}$ an unknown smooth function, and $q : \mathbb{R}^n \to \mathbb{R}$ is a partially unknown vector field satisfying certain properties, which will be described shortly, $g : \mathbb{R}^n \to \mathbb{R}$ is an unknown function with certain properties, and $d$ is the external disturbance in the input channel.

**Assumption 8.1.** The zero dynamics of system (8.1), given by $\dot{\eta} = q(x, \eta)$, are exponentially stable. In addition, $q(\xi, \eta)$ is Lipschitz in $x$, i.e., there exists positive constants $a_q$ and $a_x$ such that

$$\|q(x, \eta) - q(0, \eta)\| \le a_x \|x\| + a_q \quad \forall (x, \eta) \in \mathbb{R}^n \tag{8.2}$$

Under the assumption that the zero dynamics are stable, by the converse Lyapunov theorem, there exists a Lyapunov function $V_0(\eta)$ which satisfies the following Lyapunov inequalities for $(x, \eta) \in \mathbb{R}^n$:

$$\gamma_1 \|\eta\|^2 \le V_0(\eta) \le \gamma_2 \|\eta\| \tag{8.3}$$

$$\frac{\partial V_0}{\partial \eta} q(0, \eta) \le -\lambda_a \|\eta\|^2 \tag{8.4}$$

$$\left\| \frac{\partial V_0}{\partial \eta} \right\| \le \lambda_b \|\eta\| \tag{8.5}$$

where $\gamma_1$ , $\gamma_2$, $\lambda_a$, and $\lambda_b$ are positive constants.

**Assumption 8.2.** The external disturbance $d$ is an uncertain bounded function $d \in L_\infty$. That is, there exists unknown positive constants $\varrho$ such that $|d(t)| \leq \varrho < \infty$ where $\varrho$ can be arbitrarily large.

**Assumption 8.3.** There exist smooth functions $\bar{g}(x,\eta)$ and a positive constant $\underline{g} > 0$, such that $\bar{g}(x,\eta) \geq g(x,\eta,u) > \underline{g} > 0$, $\forall (x,\eta) \in \bar{U}$. Without loss of generality, it is further assumed that the sign of $g(x,\eta,u)$ is positive $\forall (x,\eta) \in \bar{U}$.

**Assumption 8.4.** There exists a positive function $g_0(x,\eta)$ satisfying $|\dot{g}(x,\eta)/2g(x,\eta)| \leq g_0(x,\eta)$, $\forall (x,\eta) \in \bar{U}$.

*Remark 8.1.* The SISO representation considered in this chapter is valid for simple operations involving the regulation or tracking of a single degree of freedom, such as altitude tracking and pitch regulation, among others. The general nonlinear SISO helicopter model can be described in [48]

$$\begin{aligned} \dot{x} &= f(x,u) \\ y &= h(x) \end{aligned} \tag{8.6}$$

with some assumptions such as it can be input–output linearizable with strong relative degree $\rho < n$, which can be described as (8.1). In addition, we will show that the helicopter given in Sect. 4.5.2, which will be used in the subsequent simulation section, can be changed to (8.1) and satisfies the above assumptions.

### 8.2.2 Formation Control of Helicopters

We associate the helicopters with nodes in a graph and information exchange with the graph edges. The Following definitions are useful for describing the formation.

**Definition 8.2.** [24] A directed graph $\mathcal{G}'$ consists of a non-empty finite set $\mathcal{V}'$ of elements called nodes and a finite set $\mathcal{E}' \subset \mathcal{V}'^2$ of ordered pairs of nodes called arcs, where $e = (v_i, v_j) \in \mathcal{E}'$ and $v_i, v_j \in \mathcal{V}'$. The neighbors set of vertical $v_i$ is defined as $\mathcal{N}_i' = \{v_j \in \mathcal{V}' | \, (v_j, v_i) \in \mathcal{E}'\}$.

For the multiple agents tracking problem, we introduce a virtual agent $v_0$, whose motion follows the desired trajectory restrictively. And we define a non-empty set $\mathcal{V}_0 \subset \mathcal{V}'$, in which the elements can access the desired trajectory, i.e., $v_0 \in \mathcal{N}_j$, iff $v_j \in \mathcal{V}_0$. Then the extended formation graph can be described as $\mathcal{G} = \{\mathcal{V}, \mathcal{E}\}$, where $\mathcal{V} = \mathcal{V}' \cup \{v_0\}$, and $\mathcal{E} = \mathcal{E}' \cup \{(v_0, v_j) | v_0 \in \mathcal{N}_j\}$. For all agents $v_j \in \mathcal{V}_0$, $\mathcal{N}_j = \mathcal{N}_j' \cup \{v_0\}$.

**Definition 8.3.** The weighted adjacency matrix of the extended formation graph $\mathcal{G}$, denoted as $A^*(\mathcal{G})$, is a square matrix of size $|\mathcal{V}|$, with its elements $A_{ij}^* > 0$ if $(v_j, v_i) \in \mathcal{G}$, and is zero otherwise. Define a diagonal matrix $\Delta(\mathcal{G})$ with its elements $\Delta_{jj} = \sum_k A_{jk}^*$, and the normalized Laplacian of the graph is defined as $L = I - A$ with $A = \Delta^{-1} A^*$.

**Definition 8.4.** A spanning tree of a directed graph $\mathcal{G}'$ is a directed tree formed by graph edges that connect all the nodes of the graph. We say that a graph has (or contains) a spanning tree if a subset of the edges forms a spanning tree.

**Definition 8.5.** A substochastic matrix is a square matrix with nonnegative entries such that every row adds up to at most 1.

**Definition 8.6.** A directed graph is called weakly connected if there exits a node which is globally reachable.

**Definition 8.7.** [92] If matrix $L = (\ell_{ij}) \in \mathbb{R}^{(n+1)\times(n+1)}$ satisfies the following three conditions:

1. $|\ell_{ii}| \geq \sum_{j\neq i} |\ell_{i,j}|, \quad (i = 0, 1, \ldots, n)$;
2. $J = \{k \in N| \, |\ell_{kk}| > \sum_{j=1, j\neq k}^{n} |\ell_{kj}|\} \neq \varnothing$, where $N = \{0, 1, \ldots, n\}$, it also means that there at least exists an $i$ that satisfies $|\ell_{ii}| > \sum_{j\neq i} |\ell_{ij}|$; and
3. For each $i \notin J$ there exists a sequence of nonzero elements of $L$ with the form $\ell_{ii_1}, \ell_{i_1 i_2}, \ldots, \ell_{i_s k}$ with $k \in J$.

Then we say $L$ is a diagonally dominant matrix with nonzero elements chain.

*Property 8.8.* [92] For a diagonally dominant matrix with nonzero elements chain $L = (\ell_{ij})$, we have the following properties:

1. $L$ is a nonsingular matrix;
2. If $B = I - D^{-1}L$, where $D = \text{diag}\{\ell_{11}, \ldots, \ell_{nn}\}$, $\ell_{ii} \neq 0$, then $\rho(B) < 1$, where $\rho(B)$ is the spectrum of $B$; and
3. If $L$ is real and $\ell_{ij} \leq 0$, $\ell_{ii} > 0$, then $L$ is an $M$-matrix.

**Theorem 8.9.** *Consider the multiple agent synchronized tracking problem, if the formation graph $\mathcal{G}'$ contains a spanning tree with its root $v_j \in \mathcal{V}_0$. Then the normalized adjacent matrix $A$ of the extended formation graph $\mathcal{G}$ is sub-stochastic, and $L = I - A$ is positive definite, which inverse is given by $L^{-1} = \sum_{l=0}^{\infty} A^l$.*

*Proof.* By introducing the virtual agent $v_0$, we know that $\mathcal{N}_0 = \varnothing$ in the extended formation graph $\mathcal{G}$, as it does not accept any other agents' information and follows the desired trajectory strictly, and we also know that all the elements of the first row of $A$ are zero. Since $\mathcal{G}'$ has a spanning tree and $v_j \in \{\mathcal{V}_0\}$ is the root, this means that each agent has at least one neighbor, therefore the sum of any other row of $A$ equals to 1. According to Definition 8.5, we know that $A$ is a sub-stochastic matrix.

It is clear that all the diagonal elements of $L$ are 1, and all the row sums of $A$ are 1 except the first row, this means $L$ is a diagonal dominant matrix with the $J = \{0\}$ (Since the virtual agent is added, we start the row number from 0 corresponding to the label of agents). Let us revisit that $\mathcal{G}'$ has a spanning tree with $v_j \in \mathcal{V}_0$ as the root, this also means that there is a path from $v_0$ to any agent $v_i \in \mathcal{V}$; therefore, in the matrix $L$, for every element $i \neq 0$, there exists a sequence of nonzero elements form $\ell_{ii_1}, \ell_{i_1 i_2}, \ldots, \ell_{i_s 0}$. Then $L$ satisfies all the conditions of Definition 8.7. Since $L$ is real and $\ell_{ij} < 0, i \neq j$, $\ell_{ii} = 1$, According to Property 8.8, $L$ is a nonsingular $M$-matrix [92]. By using Gerschgorin disc theory, we also know that all

**Fig. 8.1** Eigenvalue distribution of $L = (I - A)$

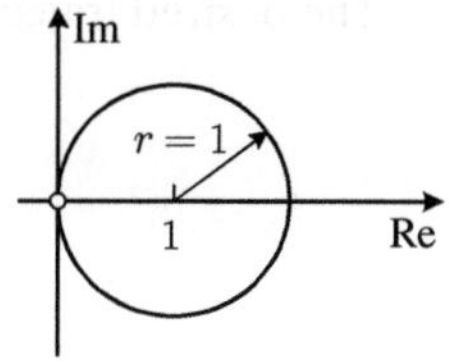


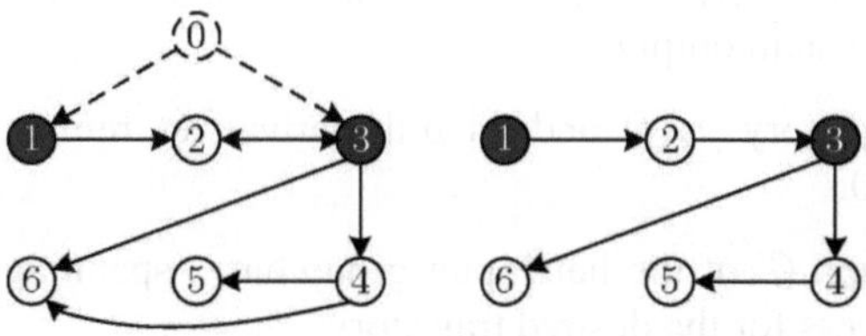


$$L = \left[\begin{array}{ccccccc|c} 1 & 0 & 0 & 0 & 0 & 0 & 0 & 0 \\ -1 & 1 & 0 & 0 & 0 & 0 & 0 & 1 \\ 0 & -2/3 & 1 & -1/3 & 0 & 0 & 0 & 2 \\ -3/4 & 0 & -1/4 & 1 & 0 & 0 & 0 & 3 \\ 0 & 0 & 0 & -1 & 1 & 0 & 0 & 4 \\ 0 & 0 & 0 & 0 & -1 & 1 & 0 & 5 \\ 0 & 0 & 0 & -1/2 & -1/2 & 0 & 1 & 6 \end{array}\right]$$

Matrix $L$

**Fig. 8.2** Sample graph and its Laplacian

the eigenvalues of $L$ lie in the right part of the complex plane as shown in Fig. 8.1; therefore, we can conclude that $L$ is positive definite. Furthermore, it follows from $\rho(A) < 1$ that $\lim_{l\to\infty} A^l = 0$. Then,

$$(I - A)(I + A + A^2 + \cdots) = (I + A + A^2 + \cdots) - (A + A^2 + A^3 + \cdots) = I \quad (8.7)$$

We obtain $L^{-1} = \sum_{l=0}^{\infty} A^l$. This completes the proof. □

*Example 8.10.* To demonstrate Theorem 8.9 clearly, take the sample graph shown in Fig. 8.2 for example. Both $v_1$ and $v_3$ can access the desired trajectory and $\mathcal{G}'$ contains a spanning tree with 1 as its root. Take the node 5 for example, we can find that in the Laplacian matrix $L$, there exists a sequence $\ell_{54}, \ell_{43}, \ell_{32}, \ell_{21}, \ell_{10} \neq 0$.

In this chapter, we studied the synchronized tracking problem of multiple unmanned helicopters as follows:

Considering a group of helicopters, the desired trajectory of the team $y_d(t)$ and its derivations up to $\rho$-th order are bounded, and are only available to the helicopters $v_j \in \mathcal{V}_0$. For each helicopter, design a control, (1) using its own full states and its neighbors' full states and (2) using its outputs and its neighbors' outputs, such that

$$\lim_{t\to\infty} |y_i(t) - y_d(t)| = \bar{\varepsilon}, \quad i = 1, \ldots, N \quad (8.8)$$

where $\bar{\varepsilon}$ is a small positive constant.

The desired trajectory $y_d(t)$ is generated by the following reference model:

$$\begin{aligned}\dot{x}_{dj} &= \dot{x}_{di+1}, \quad i = 1, \ldots, \rho - 1 \\ \dot{x}_{d\rho} &= f_d(x_d, t) \\ y_d &= x_{d1}\end{aligned} \tag{8.9}$$

where $\rho \geq 2$ is a constant index, $x_d = [x_{d1}, \ldots, x_{d\rho}]^{\mathrm{T}} \in \mathbb{R}^\rho$ are the states of the reference system, and $y_d \in \mathbb{R}$ is the system output.

**Assumption 8.5.** The reference trajectory $y_d(t)$ and its $\rho$-th derivatives remain bounded, i.e., $x_d \in \Omega_d \subset \mathbb{R}^\rho$, $\forall t \geq 0$.

**Assumption 8.6.** The formation graph $\mathcal{G}'$ of the helicopter group has a spanning tree which the root helicopter can access for the desired trajectory.

The following lemma is useful for analysis of the internal dynamics of the helicopter.

**Lemma 8.11.** *[35] Denote positive constants* $a_1 = (\lambda_b a_x)/\lambda_a$ *and* $a_2 = (\lambda_b a_q)/\lambda_a$. *If Assumptions 8.1 and 8.5 satisfied, there exists a positive constant* $T_0$ *such that the trajectories* $\eta(t)$ *of the internal dynamics satisfy*

$$\|\eta\| \leq a_1 \|x(t)\| + a_2 \tag{8.10}$$

## 8.3 Control with Full Information

In this section, we design the tracking control for each helicopter using the full information of itself and its neighbors. The adaptive NN control scheme is constructed for the synchronized tracking control. Since not all the helicopters can access the information of the desired trajectory, the tracking control is designed based on the relative states with its neighbors. Define the following error variables for the helicopters:

$$z_{i,1} = y_{i,1} - y_{ir}, \quad z_{i,2} := \dot{z}_{i,1} = x_{i,2} - \dot{y}_{ir}, \ldots, \quad z_{i,\rho} := z_{i,1}^{(\rho)} = x_{i,\rho} - y_{ir}^{(\rho)} \tag{8.11}$$

with

$$y_{ir}(t) = \sum_{j \in \mathcal{N}_i} a_{ij} y_j(t), \qquad y_{ir}^{(k)}(t) = \sum_{j \in \mathcal{N}_i} a_{ij} y_j^{(k)}(t), \quad k = 1, \ldots, \rho - 1 \tag{8.12}$$

where $a_{ij}$ is the element of the normalized adjacent matrix $A$ of the extended formation graph $\mathcal{G}$.

*Remark 8.12.* In (8.12), we defined that the reference state of each helicopter is the weighted average of its neighbors' states. If the helicopter $v_i$ can access the desired trajectory, the virtual agent $v_0$ is viewed as one of its neighbors, and $1 \geq a_{i0} > 0$. While considering the pure tracking, we may choose $a_{i0} = 1$ for better tracking performance. In the synchronized tracking problem, if $v_i$ has other neighbors $v_j \in \mathcal{N}_i$, we prefer to choose $a_{i0} < 1$ for better synchronization with its neighbors.

For each helicopter, we define vectors $\bar{z}_i$, and $\mathcal{Z}_i$ as

$$\bar{z}_i = [z_{i,1}, \ldots, z_{i,\rho}]^{\mathrm{T}} \in \mathbb{R}^{\rho}$$
$$\mathcal{Z}_i = [z_{i,1}, \ldots, z_{i,\rho-1}]^{\mathrm{T}} \in \mathbb{R}^{\rho-1}$$

and the filtered tracking error as

$$s_i = [\Lambda^{\mathrm{T}} \;\; 1]\bar{z}_i \tag{8.13}$$

where $\Lambda = [\lambda_1, \lambda_2, \ldots, \lambda_{\rho-1}]^{\mathrm{T}}$ is an appropriately chosen coefficient vector so that $z_{i,\rho} \to 0$ as $s_i \to 0$, i.e., $p^{\rho-1} + \lambda_{\rho-1}p^{\rho-2} + \cdots + \lambda_1$ is Hurwitz. Then we have

$$\dot{\mathcal{Z}}_i = A_p \mathcal{Z}_i + b s_i \tag{8.14}$$

where $A_p = \begin{bmatrix} 0 & 1 & \cdots & 0 \\ \vdots & \vdots & \ddots & \vdots \\ 0 & 0 & \ldots & 1 \\ -\lambda_1 & -\lambda_2 & \cdots & -\lambda_{\rho-1} \end{bmatrix}$, and $b = [\underbrace{0, \ldots, 0}_{\rho-2}, 1]^{\mathrm{T}}$.

The dynamics of $s_i$ are written as

$$\dot{s}_i = f_i(x_i, \eta_i) + g_i(u_i + d_i) + [0 \;\; \Lambda^{\mathrm{T}}]\bar{z}_i - y_{ir}^{(\rho)} \tag{8.15}$$

Consider the Lyapunov function candidate

$$V_i = \frac{1}{2g_i}s_i^2 + \frac{1}{2\gamma_2}\tilde{\theta}_i^{\mathrm{T}}\tilde{\theta}_i + \frac{1}{2\gamma_1}\tilde{\varphi}_i^2 \tag{8.16}$$

where $\tilde{\theta}_i = \hat{\theta}_i - \theta_i^*$, and $\tilde{\varphi}_i = \hat{\varphi} - \varphi_i^*$ are the estimated errors of parameters and the error bounded, respectively.

Then,

$$\begin{aligned}
\dot{V}_i &= -\frac{\dot{g}_i}{2g_i^2}s_i^2 + \frac{1}{g_i}s_i\dot{s}_i + \frac{1}{\gamma_2}\tilde{\theta}_i\dot{\tilde{\theta}}_i + \frac{1}{\gamma_1}\tilde{\varphi}_i\dot{\tilde{\varphi}}_i \\
&= -\left(g_0 + \frac{\dot{g}_i}{2g_i^2}\right)s_i^2 + s_i(u_i + d_i) + s_i\frac{f_i(x_i,\eta_i) + [0\ \ \Lambda^{\mathrm{T}}]\bar{z}_i - y_{ir}^{(\rho)} + g_i g_0 s_i}{g_i} \\
&\quad + \frac{1}{\gamma_2}\tilde{\theta}_i\dot{\tilde{\theta}}_i + \frac{1}{\gamma_1}\tilde{\varphi}_i\dot{\tilde{\varphi}}_i
\end{aligned} \tag{8.17}$$

We use the parameter linearized NN to approximate the unknown nonlinear function $\bar{f}_i(x_i,\eta_i,\bar{z}_i,y_{ir}^{(\rho)}) = \frac{f_i(x_i,\eta_i)+[0\ \ \Lambda^{\mathrm{T}}]\bar{z}_i - y_{ir}^{(\rho)} + g_i g_0 s_i}{g_i}$, which can be described as

$$\bar{f}_i(Z_i) = \theta_i^{*\mathrm{T}}\varphi_i(Z_i) + \bar{\varepsilon}_i \tag{8.18}$$

where $Z_i = [x_i,\eta_i,\bar{z}_i,y_{ir}^{(\rho)}]^{\mathrm{T}}$.

*Remark 8.13.* The NN is constructed to approximate $\bar{f}_i(x_i,\eta_i,\bar{z}_i,y_{ir}^{(\rho)}) = \frac{f_i(x_i,\eta_i)+[0\ \ \Lambda^{\mathrm{T}}]\bar{z}_i - y_{ir}^{(\rho)} + g_i g_0 s_i}{g_i}$ on a whole, which avoids the possible singularity of the direct approximation of $g_i$.

Select the following control $u_i$ for each helicopter

$$u_i = -\hat{\theta}_i^{\mathrm{T}}\psi_i - k_i s_i - \frac{1}{2}\hat{\varphi}_i s_i, \quad i = 1,\ldots,N \tag{8.19}$$

where $\hat{\varphi}_i$ and $\hat{\theta}_i$ denote the estimate of $\varphi_i^* = (\varrho_i + \bar{\varepsilon}_i)^2$ and $\theta_i^*$, respectively.

The update law of parameters are designed as

$$\begin{aligned}
\dot{\hat{\varphi}}_i &= -\gamma_1\left[-\frac{1}{2}(1-\varpi_\varphi)s_i^2 + \sigma_1\hat{\varphi}_i\right] \\
\dot{\hat{\theta}}_i &= -\gamma_2(-\psi_i s_i + \sigma_2\hat{\theta}_i)
\end{aligned} \tag{8.20}$$

By using the Using Young's inequality, we have

$$\begin{aligned}
-\sigma_2\tilde{\theta}_i^{\mathrm{T}}\hat{\theta}_i &\le -\frac{\sigma_2}{2}\|\tilde{\theta}_i\|^2 + \frac{\sigma_2}{2}\|\theta_i^*\|^2 \\
-\sigma_1\tilde{\varphi}_i\hat{\varphi}_i &\le -\frac{\sigma_1}{2}\tilde{\varphi}_i^2 + \frac{\sigma_1}{2}\varphi_i^{*2} \\
(\varrho_i + \bar{\varepsilon}_i)s_i &\le \frac{1}{2} + \frac{1}{2}s_i^2\varphi_i^*
\end{aligned}$$

Considering (8.19) and (8.20), the time derivation of $V_i$ in the closed-loop trajectory can be written as

$$\begin{aligned}
\dot{V}_i &= -\left(g_0 + \frac{\dot{g}_i}{2g_i^2}\right)s_i^2 - k_i s_i^2 + s_i\left(\bar{\varepsilon}_i + d_i - \frac{1}{2}\hat{\varphi}_i\right) - \sigma_2\tilde{\theta}_i^{\mathrm{T}}\hat{\theta}_i + \frac{1}{2}(1-\varpi_{\varphi})s_i\tilde{\varphi}_i s_i \\
&\quad -\sigma_1\tilde{\varphi}_i\hat{\varphi}_i \\
&\le -k_i s_i^2 + \frac{1}{2} + \frac{1}{2}s_i^2\varphi_i^* - \frac{1}{2}s_i^2\hat{\varphi}_i - \frac{\sigma_2}{2}\|\tilde{\theta}_i\|^2 + \frac{\sigma_2}{2}\|\theta_i^*\|^2 + \frac{1}{2}\tilde{\varphi}_i s_i^2 - \frac{\sigma_1}{2}\tilde{\varphi}_i^2 \\
&\quad +\frac{\sigma_1}{2}\varphi_i^{*2} \\
&= -k_i s_i^2 - \frac{\sigma_1}{2}\tilde{\varphi}_i^2 - \frac{\sigma_2}{2}\|\tilde{\theta}_i\|^2 + \frac{\sigma_2}{2}\|\theta_i^*\|^2 + \frac{\sigma_1}{2}\varphi_i^{*2} + \frac{1}{2}
\end{aligned} \tag{8.21}$$

Then,

$$\dot{V}_i \le -c_{1i}V_i + c_{2i} \tag{8.22}$$

$$c_{1i} = \min\{k_i, \gamma_2\sigma_2, \gamma_1\sigma_1\} \tag{8.23}$$

$$c_{2i} = \frac{\sigma_2}{2}\|\theta_i^*\|^2 + \frac{\sigma_1}{2}\|\varphi_i^*\|^2 + \frac{1}{2} \tag{8.24}$$

Now define

$$\Omega_{si} = \left\{ s_i \,\middle|\, |s_i| \le \sqrt{\frac{2c_{2i}}{c_{1i}}} \right\} \tag{8.25}$$

$$\Omega_{\theta_i} = \left\{ (\tilde{\theta}_i, \tilde{\varphi}_i) \,\middle|\, \|\tilde{\theta}_i\| \le \sqrt{\frac{2c_{2i}}{\sigma_2}}, |\tilde{\varphi}_i| \le \sqrt{\frac{2c_{2i}}{\sigma_1}} \right\} \tag{8.26}$$

$$\Omega_{ei} = \left\{ (s_i, \tilde{\theta}_i, \tilde{\varphi}_i) \,\middle|\, k_i s_i^2 + \frac{\sigma_2}{2}\tilde{\theta}_i^{\mathrm{T}}\tilde{\theta}_i + \frac{\sigma_1}{2}\tilde{\varphi}_i^2 \le c_{2i} \right\} \tag{8.27}$$

Since $c_{1i}$, $\sigma_1$, $\sigma_2$, and $k_i$ are positive constants, we know that $\Omega_{si}$, $\Omega_{\theta_i}$ and $\Omega_{ei}$ are compact sets. Equation (8.22) shows that $\dot{V}_i \le 0$ once the errors are outside the compact set $\Omega_{ei}$. According to the standard Lyapunov theorem, we conclude that $s_i$, $\tilde{\theta}_i$, and $\tilde{\varphi}_i$ are bounded. From (8.22) and (8.25), it can be seen that $V_i$ is strictly negative as long as $s_i$ is outside the compact set $\Omega_{si}$. Therefore, there exists a constant $T_1$ such that for $t > T_1$, the filtered tracking error $s_i$ converges to $\Omega_{si}$, that is to say, $s_i \le \beta_{si}(k_i, \gamma_1, \gamma_2, \sigma_1, \sigma_2, \theta_i^*, \varphi_i^*, \varepsilon_i^*) = \sqrt{2c_{2i}/c_{1i}}$.

Now we will show that all the helicopters will track the desired trajectory although only some of them can access the desired trajectory. Define the error between $i$-th helicopter and the desired trajectory as $\tilde{y}_i(t) = y_i(t) - y_d(t) = y_i(t) - y_0(t)$, and the auxiliary states of each helicopter $\xi_i(t) = [\Lambda^{\mathrm{T}} \;\; 1]Y_i$ with

$Y_i = [y_i, y_i^{(1)}, \ldots, y_i^{(\rho-1)}]^{\mathrm{T}}$. The filtered error is denoted as $\tilde{\xi}_i(t) = \xi_i(t) - \xi_d(t) = \xi_i(t) - \xi_0(t)$.

Using the fact that $s_i(t) = \xi_i(t) - \sum_{j\in\mathcal{N}_i} a_{ij}\xi_j(t)$, we have

$$\begin{aligned}\tilde{\xi}_i &= \xi_i - \xi_0 \\ &= \sum_{j\in\mathcal{N}_i} a_{ij}\xi_j + s_i - \xi_0, \quad i = 1, \ldots, N\end{aligned}$$

and in the vector form

$$\tilde{\xi} = A\xi + s - \xi_0\mathbf{1} \tag{8.28}$$

where $\mathbf{1} = [1, \ldots, 1]^{\mathrm{T}}$, $s = [s_0, s_1, \ldots, s_N]^{\mathrm{T}}$, and $A$ is the normalized adjacency matrix of the extended formation graph. Note that the elements in the first row of $A$ are all equal to 0, and the other row summations of the matrix $A$ are 1, and we have $[0, 1, \ldots, 1]^{\mathrm{T}} = A[0, 1, \ldots, 1]^{\mathrm{T}}$. Then,

$$\begin{aligned}\tilde{\xi} &= A(\tilde{\xi} + \xi_0\mathbf{1}) + s + [1, 0 \ldots, 0]^{\mathrm{T}}\xi_0 - \xi_0\mathbf{1} \\ &= A\tilde{\xi} + [0, 1, \ldots, 1]^{\mathrm{T}}\xi_0 + s + [1, 0 \ldots, 0]^{\mathrm{T}}\xi_0 - \xi_0\mathbf{1} \\ &= A\tilde{\xi} + s\end{aligned} \tag{8.29}$$

Under the Assumption 8.6, we know that $L = (I - A)$ is an invertible matrix, and we have

$$\tilde{\xi} = L^{-1}s \tag{8.30}$$

Define vectors

$$\begin{aligned}\mathcal{Y} &= [Y_0^{\mathrm{T}}, Y_1^{\mathrm{T}}, \ldots, Y_N^{\mathrm{T}}]^{\mathrm{T}} \\ \tilde{\mathcal{Y}} &= [\tilde{Y}_0^{\mathrm{T}}, \tilde{Y}_1^{\mathrm{T}}, \ldots, \tilde{Y}_N^{\mathrm{T}}]^{\mathrm{T}} \\ X &= [X_0^{\mathrm{T}}, X_1^{\mathrm{T}}, \ldots, X_{\rho-1}^{\mathrm{T}}]^{\mathrm{T}} \\ \tilde{X} &= [\tilde{X}_0, \tilde{X}_1, \ldots, \tilde{X}_{\rho-1}]^{\mathrm{T}}\end{aligned}$$

where $X_j = [X_{0,j}, X_{1,j}, \ldots, X_{N,j}]^{\mathrm{T}}$, $\tilde{X}_j = X_j - X_{jd} = X_j - y_0^{(j)}\mathbf{1}$, $\tilde{Y}_i = Y_i - Y_d = Y_i - Y_0$. Then we have

$$\dot{\tilde{\mathcal{Y}}} = \bar{A}_p\tilde{\mathcal{Y}} + \bar{b}\tilde{\xi} \tag{8.31}$$

where $\bar{A}_p = I_{N+1} \otimes A_p$ and $\bar{b} = I_{N+1} \otimes b$.

Considering (8.29), the error dynamics can be written as

$$\dot{\tilde{\mathcal{Y}}} = \bar{A}_p\tilde{\mathcal{Y}} + \bar{b}\tilde{\xi} = \bar{A}_p\tilde{\mathcal{Y}} + \bar{b}L^{-1}s \tag{8.32}$$

**Lemma 8.14.** *Define* $s_{i,\max} = \sup_{0\le\tau\le t}|s_i(t)|$, $\beta_{s_i} = \sup_{t>T_1|s_i(t)|}$, *and* $s_{\max,i}(t) = \max_i \sup_{0\le\tau\le t}|s_i(t)|$, *then the following equations hold:*

$$\|\tilde{\mathcal{Y}}(t)\| \le k_0 \mathrm{e}^{-\lambda_0 t}\|\tilde{\mathcal{Y}}(0)\| + \frac{k_0}{\lambda_0}\left[N\lambda_{\max}(L^{-1}) + N - 1\right] s_{\max,i}(t)$$

$$\|\tilde{\mathcal{Y}}(t)\| \le k_0 \mathrm{e}^{-\lambda_0 t}\left(\|\tilde{\mathcal{Y}}(0)\| + \frac{\mathrm{e}^{\lambda_0 T_1}}{\lambda_0}\beta_s(T_1)\right) + \frac{k_0}{\lambda_0}\beta_{sT}$$

*where* $\beta_s(t) = N\lambda_{\max}(L^{-1})s_{\max,i}(t)$ *and* $\beta_{sT} = N\lambda_{\max}(L^{-1})\sup_{T_1\le t} s_{\max,i}(t)$ *with constants* $\lambda_0 > 0$ *and* $k_0 > 0$.

*Proof.* From (8.32) and the fact that $A_p$ is Hurwitz, we have

$$\tilde{\mathcal{Y}}(t) = \tilde{\mathcal{Y}}(0)\mathrm{e}^{\bar{A}_p t} + \int_0^t \mathrm{e}^{\bar{A}_p(t-\tau)}\bar{b}L^{-1}s\,\mathrm{d}\tau$$

$$\|\mathrm{e}^{\bar{A}_p t}\| \le k_0\mathrm{e}^{-\lambda_0 t}$$

Then,

$$\begin{aligned}
\|\tilde{\mathcal{Y}}(t)\| &\le k_0\mathrm{e}^{-\lambda_0 t}\|\tilde{\mathcal{Y}}(0)\| + \int_0^t \mathrm{e}^{-\lambda_0(t-\tau)}\left\|\bar{b}L^{-1}s\right\|\mathrm{d}\tau \\
&\le k_0\mathrm{e}^{-\lambda_0 t}\|\tilde{\mathcal{Y}}(0)\| + k_0\mathrm{e}^{-\lambda_0 t}\left[N\lambda_{\max}(L^{-1})s_{\max,i}(t)\right]\int_0^t \mathrm{e}^{\lambda_0\tau}\,\mathrm{d}\tau \\
&\le k_0\mathrm{e}^{-\lambda_0 t}\|\tilde{\mathcal{Y}}(0)\| + k_0\mathrm{e}^{-\lambda_0 t}\left[N\lambda_{\max}(L^{-1})\right]s_{\max,i}(t)\frac{\mathrm{e}^{\lambda_0 t}-1}{\lambda_0} \\
&\le k_0\mathrm{e}^{-\lambda_0 t}\|\tilde{\mathcal{Y}}(0)\| + \frac{k_0}{\lambda_0}\left[N\lambda_{\max}(L^{-1})\right]s_{\max,i}(t)
\end{aligned} \tag{8.33}$$

where $\lambda_{\max}(\cdot)$ is the maximum eigenvalue of the matrix.

Noting the above equation and that

$$\int_0^t \mathrm{e}^{-\lambda_0(t-\tau)}\left\|\bar{b}(L^{-1}s)\right\|\mathrm{d}\tau = \int_0^{T_1} \mathrm{e}^{-\lambda_0(t-\tau)}\left\|\bar{b}(L^{-1}s)\right\|\mathrm{d}\tau + \int_{T_1}^t \mathrm{e}^{-\lambda_0(t-\tau)}\left\|\bar{b}(L^{-1}s)\right\|\mathrm{d}\tau$$

We have (8.33) as follows:

$$
\begin{aligned}
\|\tilde{\mathcal{Y}}(t)\| &\le k_0 \mathrm{e}^{-\lambda_0 t}\|\tilde{\mathcal{Y}}(0)\| + k_0 \mathrm{e}^{-\lambda_0 t}\frac{\mathrm{e}^{\lambda_0 T_1}-1}{\lambda_0}\beta_s(T_1) + k_0 \mathrm{e}^{-\lambda_0 t}\frac{\mathrm{e}^{\lambda_0 t_0}-\mathrm{e}^{\lambda_0 T_1}}{\lambda_0}\beta_{s_T} \\
&\le k_0 \mathrm{e}^{-\lambda_0 t}\left(\|\tilde{\mathcal{Y}}(0)\| + \frac{\mathrm{e}^{\lambda_0 T_1}}{\lambda_0}\beta_s(T_1)\right) + \frac{k_0}{\lambda_0}\beta_{s_T} \qquad (8.34)
\end{aligned}
$$

This completes the proof. □

Now we will show that for a proper choice of the control parameters, the trajectories of each vehicle do remain in the compact set. From the fact that $L^{-1}s = ([\Lambda^{\mathrm{T}}\ 1] \otimes I_{N+1})\tilde{\bar{X}}$, where $\tilde{\bar{X}} = [\tilde{X}^{\mathrm{T}}\ \ \tilde{x}_\rho^{\mathrm{T}}]^{\mathrm{T}}$, we can see that $\tilde{x}_\rho = L^{-1}s - (\Lambda^{\mathrm{T}} \otimes I_N)\tilde{X}$. Therefore,

$$
\begin{aligned}
\|\tilde{\bar{X}}\| &\le \|\tilde{X}\| + \|\tilde{x}_\rho\| \\
&\le (1 + \|\Lambda\|)\|\tilde{X}\| + \|L^{-1}\|\|s\| \\
&\le (1 + \|\Lambda\|)\|\tilde{\mathcal{Y}}\| + \lambda_{\max}(L^{-1})\|s\|
\end{aligned}
$$

It follows from (8.34) and the fact that $s_i$ will converge to $\Omega_{si}$, we know that $\|\tilde{\bar{X}}\| \le k_a\|\tilde{\mathcal{Y}}(0)\| + k_b\beta_{s_T} + k_c$, $\forall t \ge T_1$, with $k_a = (1 + \|\Lambda\|)k_0$, $k_b = (k_a/\lambda_0) + 1$ and $k_c = k_a(\mathrm{e}^{\lambda_0 T_1}/\lambda_0)\beta_s(T_1)$. Hence,

$$
\begin{aligned}
\|\bar{X}(t)\| &\le \|\tilde{\bar{X}}(t)\| + \|\bar{X}_d(t)\mathbf{1}\| \\
&\le k_a\|\tilde{\mathcal{Y}}(0)\| + k_b\beta_{s_T} + k_c + c, \quad \forall t \ge T_1 \qquad (8.35)
\end{aligned}
$$

We now provide the conditions which guarantee $\bar{X} \in \Omega_{\bar{X}}$, $\forall t \ge 0$. Define the compact set

$$
\Omega_0 := \left\{\bar{X}(0)\,\middle|\{\bar{X}|\|\bar{X}(t)\| < k_a\|\tilde{\mathcal{Y}}(0)\|\} \subset \Omega_{\bar{X}}, \lambda_{\max}(L^{-1})\|s(0)\| < \beta_{s_T}\right\}
$$

and the positive constant

$$
c^* := \sup_{c\in\mathbb{R}+}\left\{c\,\middle|\left\{\bar{X}|\|\bar{X}\| < k_a\|\tilde{\mathcal{Y}}(0)\| + k_c + c,\ \bar{X}(0) \in \Omega_0\right\} \subset \Omega_{\bar{X}}\right\} \qquad (8.36)
$$

We summarize our results for the full-state feedback case in the following theorem.

**Theorem 8.15.** *Consider a group of helicopters dynamics (8.1) and the communication graph containing a spanning tree which the root helicopter can access for the desired trajectory, with Assumptions 8.1–8.5, under the action of the control (8.19) and parameters update law (8.20) for each helicopter. For initial conditions* $\bar{X}(0), \eta(0), \tilde{\theta}_i(0)$ *and* $\tilde{\varphi}_i(0)$ *starting in any compact set, and the desired trajectory* *with its derivations up to $\rho$-th bounded, all closed signals of the system are Semi-Globally Uniformly Ultimately Bounded (SGUUB), and the total tracking error of the helicopters $\bar{\bar{X}}$ converges to a neighborhood of the origin.*

*Proof.* From (8.35), we know that the overall system state $\bar{X}(t)$ will stay in $\Omega_{\bar{X}}$ for all time. Furthermore, because the NN weights have been proven bounded for any bounded $\hat{\theta}_i(0)$ and $\hat{\varphi}_i(0)$, and due to Lemma 8.11, it can be seen that $\eta_i$ is bound if $x_i$ is bounded. As a result, the states of the internal dynamics of the helicopter will converge to the compact set $\Omega_{\eta_i} = \{\eta_i \in \mathbb{R}^p | \, \|\eta_i\| \leq a_1(\sqrt{2c_2/c_1} + \|X_d\|) + a_2\}$, where $a_1 = \lambda_b a_x/\lambda_a$ and $a_2 = \lambda_b a_q/\lambda_a$ are positive constants. Because the control signal $u_i(t)$ is a function of the weights $\hat{\theta}_i$ and $\hat{\varphi}_i$, the states $\eta_i$, $x_i$, and the filtered tracking error $s_i$, we know that it is also bounded. Therefore, we know that all the closed-loop signals are SGUUB. This completes the proof. □

## 8.4 Control with Partial Information

From the definition (8.11) of reference states of each helicopter, we know that not all of the helicopters can access the desired altitude and its derivation. For each helicopter in the team, its reference output at time $t$ is the weighted average of its neighbors' outputs at the same time, and in the control design, each helicopter needs to use its neighbors' states $y_{ir}^{(k)}(t)$, $k = 1, \ldots, \rho$, which are not easy for them to access. In this section, we assume that each helicopter can only access its neighbors' output information $y_{ir}$, and use high observer to estimate $y_{ir}^{(k)}(t)$, $k = 1, \ldots, \rho$.

In the following lemma, high gain observer used in [7] is presented, which will be used to estimate the neighbors' states.

**Lemma 8.16.** *[35][102] Consider the following linear system:*

$$\begin{aligned}
\epsilon \dot{\pi}_i &= \pi_{i+1} \quad i = 1, 2, \ldots, \rho - 1 \\
\epsilon \dot{\pi}_\rho &= -\bar{\gamma}_1 \pi_\rho - \bar{\gamma}_2 \pi_{\rho-1} - \cdots - \bar{\gamma}_{\rho-1} \pi_2 - \pi_1 + \chi(t)
\end{aligned} \tag{8.37}$$

*where $\epsilon$ is a small positive constant and the parameters $\bar{\gamma}_1$ to $\bar{\gamma}_{\rho-1}$ are chosen such that the polynomial $s^\rho + \bar{\gamma}_1 s^{\rho-1} + \ldots + \bar{\gamma}_{\rho-1} s + 1$ is Hurwitz. Suppose the states $\chi(t)$ and its first n derivatives are bounded, so that $\chi^{(k)} < \varpi_k$ with positive constants $\varpi_k$. Then the following property holds:*

$$\tilde{\chi}^{(k)} := \frac{\pi_k}{\epsilon^{k-1}} - \chi^{(k)} = -\epsilon \zeta^{(k)}, \quad k = 1, 2, \ldots, \rho \tag{8.38}$$

*where $\zeta := \pi_p + \bar{\gamma}_1 \pi_{\rho-1} + \cdots + \bar{\gamma}_{\rho-1} \pi_1$ and $\zeta^{(k)}$ denotes the $k$th derivative of $\zeta$. Furthermore, there exist positive constants $h_k$ and $t^*$ such that for all $t > t^*$ we have $|\zeta^{(k)}| \leq h_k$, $k = 2, 3, \ldots, \rho$.*

Note that $\pi_{k+1}/\epsilon^k$ asymptotically converges to $\zeta^{(k)}$, with a small time constant provided that $\zeta$ and its $k$ derivatives are bounded. Hence, $\pi_{k+1}/\epsilon^k$ for $k = 1, \ldots, \rho$ is a suitable observer to estimate the output derivatives up to the $\rho$-th order.

To prevent peaking [52], saturation functions are employed on the observer signals whenever they are outside the domain of interest $\Omega$ as follows:

$$\pi_{i,j}^s = \bar{\pi}_{i,j}\phi\left(\frac{\pi_{i,j}}{\bar{\pi}_{i,j}}\right), \quad \bar{\pi}_{i,j} \geq \max_{(\tilde{y}_i, s_i, \tilde{\theta}_i, \tilde{\varphi}_i) \in \Omega} (\pi_{i,j})$$

$$\phi(a) = \begin{cases} -1, & \text{for } a < -1 \\ a, & \text{for } |a| < 1 \\ 1, & \text{for } a > 1 \end{cases} \tag{8.39}$$

Now, we revisit the control law (8.19) and adaption laws (8.20) for the full-state feedback case. Via the certainty equivalence approach, we modify them by replacing the partially available quantities with their estimates, which can be written as

$$u_i = -\hat{\theta}_i^{\mathrm{T}} \psi_i(\hat{Z}_i) - k_i \hat{s}_i - \frac{1}{2}\hat{\varphi}_i \hat{s}_i, \quad i = 1, \ldots, N \tag{8.40}$$

And the update law of parameters is designed as

$$\begin{aligned} \dot{\hat{\varphi}}_i &= -\gamma_1 \left[ -\frac{1}{2}(1 - \varpi_{\varphi})\hat{s}_i^2 + \sigma_1 \hat{\varphi}_i \right] \\ \dot{\hat{\theta}}_i &= -\gamma_2 \left( -\psi_i \hat{s}_i + \sigma_2 \hat{\theta}_i \right) \end{aligned} \tag{8.41}$$

where $\gamma_1$, $\gamma_2$, $\sigma_1$ and $\sigma_2$ are positive constants, and

$$\varpi_{\varphi_i} = \begin{cases} 0, & \text{if } |\hat{\varphi}_i| \leq M_{\varphi_i} \\ 1, & \text{otherwise} \end{cases} \tag{8.42}$$

where $M_{\varphi_i}$ is a designed positive constant.

Select Lyapunov function candidate

$$V_{ie} = \frac{1}{2}s_i^2 + \frac{1}{2\gamma_2}\tilde{\theta}_i^{\mathrm{T}}\tilde{\theta}_i + \frac{1}{2\gamma_1}\tilde{\varphi}_i^2 \tag{8.43}$$

And the following lemma is useful for handling the terms containing the estimation errors.

**Lemma 8.17.** *There exist positive constants $F_{ik}$ which are independent of $\epsilon_i$, such that for $t > t^*$, the estimate $\hat{y}_{ir}^{(k)}$, $i = 1, \ldots, N$, $k = 1, \ldots, \rho$, satisfy the following inequalities:*

$$|\tilde{y}_{ir}^{(k)}| = |\hat{y}_{ir}^{(k)} - y_{ir}^{(k)}| \le \epsilon_i F_{ik} \tag{8.44}$$

Since $s_i$ is the linear combination of $Y_i$ and $Y_j$, $j \in \mathcal{N}_i$, we know that there exist positive constants $G_{is}$ which are independent of $\epsilon_i$ such that $|\tilde{s}_i| \le \epsilon_i G_{is}$.

Taking the time derivative of $V_i$ along the closed-loop trajectory and using the property $\psi_i(\hat{Z}_i) - \psi_i(Z_i) = \epsilon_i \psi_{ti}$, where $\psi_{ti}$ is a bounded vector function [30], we have

$$\begin{aligned}
\dot{V}_{ie} &= -\left(\frac{\dot{g}_i}{2g_i^2} + g_0\right) s_i^2 - k_i s_i^2 - k_i s_i \tilde{s}_i - s_i \hat{\theta}_i^{\mathrm{T}} \psi_i(\hat{Z}_i) - \frac{1}{2}\hat{\varphi}_i s_i \hat{s}_i + s_i(d_i + \bar{\varepsilon}_i) \\
&\quad + s_i \theta^{*\mathrm{T}} \psi_i(Z_i) + \frac{1}{\gamma_2}\tilde{\theta}_i \dot{\hat{\theta}}_i + \frac{1}{\gamma_1}\tilde{\varphi}_i \dot{\hat{\varphi}}_i \\
&\le -\frac{k_i}{2}s_i^2 + \frac{k_i}{2}\tilde{s}_i^2 - \frac{1}{2}\hat{\varphi}_i s_i \hat{s}_i + \frac{1}{2}\varphi_i s_i^2 + \frac{1}{2}\tilde{\varphi}_i \hat{s}_i^2 - s_i \hat{\theta}_i^{\mathrm{T}} \psi_i(\hat{Z}_i) + s_i \theta^{*\mathrm{T}} \psi_i(Z_i) \\
&\quad + \hat{s}_i \tilde{\theta}_i^{\mathrm{T}} \psi_i(\hat{Z}_i) - \sigma_2 \tilde{\theta}_i^{\mathrm{T}} \hat{\theta}_i - \sigma_1 \tilde{\varphi}_i \hat{\varphi}_i + \frac{1}{2}
\end{aligned}$$

For the term $-s_i^2 \tilde{\varphi}_i - s_i \tilde{s}_i \hat{\varphi}_i + \hat{s}_i^2 \tilde{\varphi}_i$, we have

$$\begin{aligned}
-s_i^2 \tilde{\varphi}_i - s_i \tilde{s}_i \hat{\varphi}_i + \hat{s}_i^2 \tilde{\varphi}_i &= \tilde{s}_i (s_i \tilde{\varphi}_i + \tilde{s}_i \tilde{\varphi}_i - s_i \varphi_i) \\
&\le \epsilon G_{is} |s_i \tilde{\varphi}_i| + \epsilon^2 G_{is}^2 |\tilde{\varphi}_i| + \epsilon G_{is} |s_i \varphi_i| \\
&\le \frac{1}{2}(s_i^2 + \epsilon^2 G_{is}^2 |\tilde{\varphi}_i|^2) + \frac{1}{2}\epsilon^2 G_{is}^2 \tilde{\varphi}_i^2 + \frac{1}{2}\epsilon^2 G_{is}^2 + \frac{1}{2}s_i^2 \\
&\quad + \frac{1}{2}\epsilon^2 G_{is}^2 \varphi_i^2 \\
&= s_i^2 + \epsilon_i^2 G_{is}^2 \tilde{\varphi}_i^2 + \frac{1}{2}\epsilon_i^2 G_{is}^2 \varphi_i^2 \\
&\le s_i^2 + \epsilon_i^2 G_{is}^2 \left(\tilde{\varphi}_i^2 + \frac{3}{2}\varphi_i^2\right)
\end{aligned} \tag{8.45}$$

For the term $-s_i \hat{\theta}_i^{\mathrm{T}} \psi_i(\hat{Z}_i) + s_i \theta^{*\mathrm{T}} \psi_i(Z_i) + \hat{s}_i \tilde{\theta}_i^{\mathrm{T}} \psi_i(\hat{Z}_i)$, we have

$$\begin{aligned}
&-s_i \hat{\theta}_i^{\mathrm{T}} \psi_i(\hat{Z}_i) + s_i \theta^{*\mathrm{T}} \psi_i(Z_i) + \hat{s}_i \tilde{\theta}_i^{\mathrm{T}} \psi_i(\hat{Z}_i) \\
&= -s_i \tilde{\theta}_i^{\mathrm{T}} \psi_i(\hat{Z}_i) - s_i \theta_i^{*\mathrm{T}} \psi_i(\hat{Z}_i) + s_i \theta^{*\mathrm{T}} \psi_i(Z_i) + \hat{s}_i \tilde{\theta}_i^{\mathrm{T}} \psi_i(\hat{Z}_i) \\
&= \tilde{s}_i \tilde{\theta}_i^{\mathrm{T}} \psi_i(\hat{Z}_i) - \epsilon_i s_i \theta_i^{*\mathrm{T}} \psi_{ti} \\
&\le \frac{1}{2}\tilde{\theta}_i^{\mathrm{T}} \tilde{\theta}_i + \frac{1}{2}\epsilon_i G_{is}^2 \|\psi_{ti}\|^2 + \frac{1}{2}s^2 + \frac{1}{2}\epsilon_i^2 \|\psi_{ti}\|^2 \|\theta_i^*\|^2
\end{aligned} \tag{8.46}$$

Then,

$$\dot{V}_{ie} \leq -\frac{1}{2}(k_i-2)s_i^2 - \frac{\sigma_2-1}{2}\tilde{\theta}_i^{\mathrm{T}}\tilde{\theta}_i - \frac{\sigma_1}{2}\tilde{\varphi}_i^2 + \frac{1}{2}\epsilon_i^2 G_{is}^2\hat{\varphi}_i^2 + \frac{1}{2}\epsilon_i G_{is}^2\|\psi_{ti}\|^2$$
$$+\frac{\epsilon_i^2\|\psi_{ti}\|^2+\sigma_2}{2}\|\theta_i^*\|^2 + \left(\frac{3}{4}\epsilon_i^2 G_{is}^2+\frac{\sigma_1}{2}\right)\varphi_i^{*2} + \frac{k_i}{2}\epsilon_i^2 + \frac{1}{2} \tag{8.47}$$

Then,

$$\dot{V}_{ie} \leq -c_{1ie}V_{ie} + c_{2ie} \tag{8.48}$$

$$c_{1ie} = \min\left\{\frac{1}{2}(k_i-2), (\gamma_2-1)\sigma_2, \gamma_1\sigma_1\right\} \tag{8.49}$$

$$c_{2ie} = \frac{1}{2}\epsilon_i^2 G_{is}^2\hat{\varphi}_i^2 + \frac{1}{2}\epsilon_i G_{is}^2\|\psi_{ti}\|^2 + \frac{\epsilon_i^2\|\psi_{ti}\|^2+\sigma_2}{2}\|\theta_i^*\|^2$$
$$+\left(\frac{3}{4}\epsilon_i^2 G_{is}^2+\frac{\sigma_1}{2}\right)\varphi_i^{*2} + \frac{k_i}{2}\epsilon_i^2 + \frac{1}{2} \tag{8.50}$$

Now define

$$\Omega_{sie} = \left\{ s_i \,\middle|\, |s_i| \leq \sqrt{\frac{2c_{2ie}}{c_{1ie}}} \right\} \tag{8.51}$$

$$\Omega_{\theta_{ie}} = \left\{ (\tilde{\theta}_i, \tilde{\varphi}_i) \,\middle|\, \|\tilde{\theta}_i\| \leq \sqrt{\frac{2c_{2ie}}{\sigma_2}}, |\tilde{\varphi}_i| \leq \sqrt{\frac{2c_{2ie}}{\sigma_1}} \right\} \tag{8.52}$$

$$\Omega_{eie} = \left\{ (s_i, \tilde{\theta}_i, \tilde{\varphi}_i) \,\middle|\, k_i s_i^2 + \frac{\sigma_2}{2}\tilde{\theta}_i^{\mathrm{T}}\tilde{\theta}_i + \frac{\sigma_1}{2}\tilde{\varphi}_i^2 \leq c_{2ie} \right\} \tag{8.53}$$

Since $c_{1ie}$, $\sigma_1$, $\sigma_2$, and $k_i$ are positive constants, we know that $\Omega_{sie}$, $\Omega_{\theta_ie}$ and $\Omega_{eie}$ are compact sets. Equation (8.48) shows that $\dot{V}_{ie} \leq 0$ once the errors are outside the compact set $\Omega_{ei}$. According to the standard Lyapunov theorem, we conclude that $s_i$, $\tilde{\theta}_i$, and $\tilde{\varphi}_i$ are bounded. From (8.48) and (8.51), it can be seen that $V_{ie}$ is strictly negative as long as $s_i$ is outside the compact set $\Omega_{sie}$. Therefore, there exists a constant $T_1$ such that for $t > T_1$, the filtered tracking error $s_i$ converges to $\Omega_{sie}$, that is to say, $s_i \leq \beta_{sie}$, with $\beta_{sie}(k_i, \gamma_1, \gamma_2, \sigma_1, \sigma_2, \theta_i^*, \varphi_i^*, \epsilon_i) = \sqrt{2c_{2ie}/c_{1ie}}$.

We can conclude the following theorem.

**Theorem 8.18.** *Consider a group of helicopters dynamics (8.1) and the communication graph containing a spanning tree with the leader as the root, with Assumptions 8.1–8.5, under the action of the control law (8.40), parameters update law (8.41), and the high gain observer (8.37), which is turned on at time $t^*$ in advance. For initial conditions $\bar{X}(0)$, $\eta(0)$, $\tilde{\theta}_i(0)$ and $\tilde{\varphi}_i(0)$ starting from any compact set, and the desired trajectory with its derivations up to $\rho$-th bounded,* *all closed signals of the system are SGUUB, and the total tracking error of the helicopters $\bar{\bar{X}}$ converges to a neighborhood of origin.*

*Proof.* We have concluded that $s_i$ will converge to a compact set $\Omega_{sie}$, then following Lemma 8.17, it can be concluded that $\|\tilde{\mathcal{Y}}\| \leq k_0 e^{-\lambda_0 t}\left(\|\tilde{\mathcal{Y}}(0)\| + \frac{e^{\lambda_0 T_1}}{\lambda_0}\beta_s(T_1)\right) + \frac{k_0}{\lambda_0}\beta_{sT}$, and from (8.35), we can find that $\|\bar{X}\|$ is also bounded. Following the same procedure in the full-state feedback control, we can complete the proof. □

*Remark 8.19.* It is shown in (8.50) that the smaller $c_{2ie}$ might be obtained by choosing a smaller $\sigma_1$ and $\sigma_2$, which may lead to a smaller tracking error. Nevertheless, from (8.52) it can be seen that the smaller $\sigma_1$ and $\sigma_2$ may cause large NN weight and disturbance compensation errors. If $\sigma_1$ and $\sigma_2$ are chosen to be very large, it will lead to a large tracking error. Hence, the parameters $\sigma_1$ and $\sigma_2$ should be adjusted carefully in practical implementations.

## 8.5 Simulation Study

In this section, we consider the synchronized altitude tracking of 6 X-cell 50 helicopters whose communication graph is shown in Fig. 8.2. The dynamics of the helicopter can be written as follows as in Sect. 4.5.2

$$\begin{aligned}
\dot{\zeta}_1 &= \zeta_2 \\
\dot{\zeta}_2 &= a_0 + a_1\zeta_2 + a_2\zeta_2^2 + \left(a_3 + a_4\zeta_4 - \sqrt{a_5 + a_6\zeta_4}\right)\zeta_3^2 \\
\dot{\zeta}_3 &= a_7 + a_8\zeta_3 + (a_9 \sin\zeta_4 + a_{10})\zeta_3^2 + a_{\text{th}} \\
\dot{\zeta}_4 &= \zeta_5 \\
\dot{\zeta}_5 &= a_{11} + a_{12}\zeta_4 + a_{13}\zeta_3^2 \sin\zeta_4 + a_{14}\zeta_5 - K_1 u
\end{aligned} \tag{8.54}$$

where $\zeta_1$ denotes altitude (m), $\zeta_2$ the height rate of the altitude rate (m/s), $\zeta_3$ the rotational speed of the rotor blades (rad/s), $\zeta_4$ the collective pitch angle (rad), $\zeta_5$ the collective pitch rate (rad/s), $a_{\text{th}} = 111.69\,\text{s}^{-2}$ the constant input to the throttle, and $u$ the input to the collective servomechanisms.

Let $y$ be the altitude $\zeta_1$. By restricting the throttle input to be constant, we obtain a SISO in which $u$ is the only input variable forcing the output $y$ to track a desired trajectory $y_d$, which is generated by

$$y_d = \frac{150.056}{s^4 + 12.6s^3 + 64.19s^2 + 154.35s + 150.056} h_{\text{ref}} \tag{8.55}$$

where

$$h_{\text{ref}}(t) = 5.5 - 0.5 \sin t \tag{8.56}$$

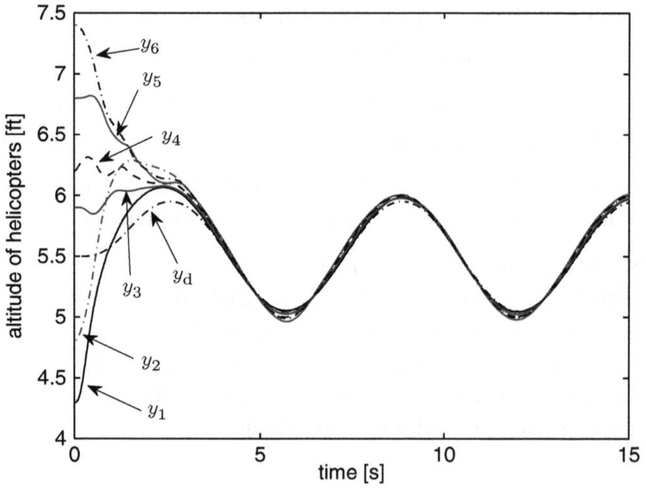


**Fig. 8.3** Altitude of all helicopters with output feedback control

The nominal values for constants $K_1$ and $a_i$ are given to be: $K_1 = 0.25397\text{s}^{-2}$, $a_0 = -17.67\text{m/s}^2$, $a_1 = a_2 = -0.1\,\text{s}^{-2}$, $a_3 = 5.31 \times 10^{-4}$, $a_4 = 1.5364 \times 10^{-2}$, $a_5 = 2.82\times10^{-7}$, $a_6 = 1.632\times10^{-5}$, $a_7 = -13.92\text{s}^{-2}$, $a_8 = -0.7\,\text{s}^{-2}$, $a_9 = a_{10} = -0.0028$, $a_{11} = 434.88\,\text{s}^{-2}$, $a_{12} = -800\,\text{s}^{-2}$, $a_{13} = -0.1$ and $a_{14} = -65\,\text{s}^{-2}$.

It can be shown that the system has strong relative degree 4 as in Sect. 4.5.2. The control parameters are chosen as $\Lambda = [64, 48, 12]^{\text{T}}$, $k_i = 3$, $i = 1, \ldots, 6$, while the NN parameters for each helicopter are chosen as $\sigma_1 = 0.05$, $\gamma_1 = 1$, $\sigma_2 = 0.01$, $\gamma_2 = 100$. For high gain observer, we choose $\epsilon_i = 0.08$, $\bar{\gamma}_1 = 4$, $\bar{\gamma}_2 = 6$, $\bar{\gamma}_3 = 4$, $\bar{\pi}_2 = 0.1$, $\bar{\pi}_3 = 0.15$, $\bar{\pi}_4 = 0.025$. The saturation limits of the control are $\pm400$ mrad. The initial conditions are $\zeta_1(0) = [4.3, 0.0, 95.3567, 0.222, 0.0]^{\text{T}}$, $\zeta_2(0) = [4.8, 0.0, 95.3567, 0.3, 0.0]^{\text{T}}$, $\zeta_3(0) = [5.9, 0.0, 95.4, 0.22, 0.0]^{\text{T}}$, $\zeta_4(0) = [6.2, 0.0, 95.3567, 0.3, 0.0]^{\text{T}}$, $\zeta_5(0) = [6.8, 0.0, 95.3567, 0.22, 0.0]^{\text{T}}$, $\zeta_6(0) = [7.4, 0.0, 95.4, 0.21, 0.0]^{\text{T}}$, $\hat{\theta}_i(0) = 0$, and $\hat{\varphi}_i(0) = 0$ for each helicopter.

Simulation results are shown in Figs. 8.3–8.6. From Fig. 8.3, we can find that good tracking performance is achieved for each helicopter by the proposed control. The tracking performance for full-state and output feedback cases are similar for the choice of $\epsilon_i$ made. The initial errors of all helicopters are sufficiently reduced and the altitude trajectories all lie in close proximity of the desired sinusoidal trajectory. Meanwhile, the internal dynamics and the NN weights are all bounded, as shown in Figs. 8.5 and 8.6. From Fig. 8.4, we can find that the control input of the helicopters are bounded, both in the full-state feedback and the output feedback cases.

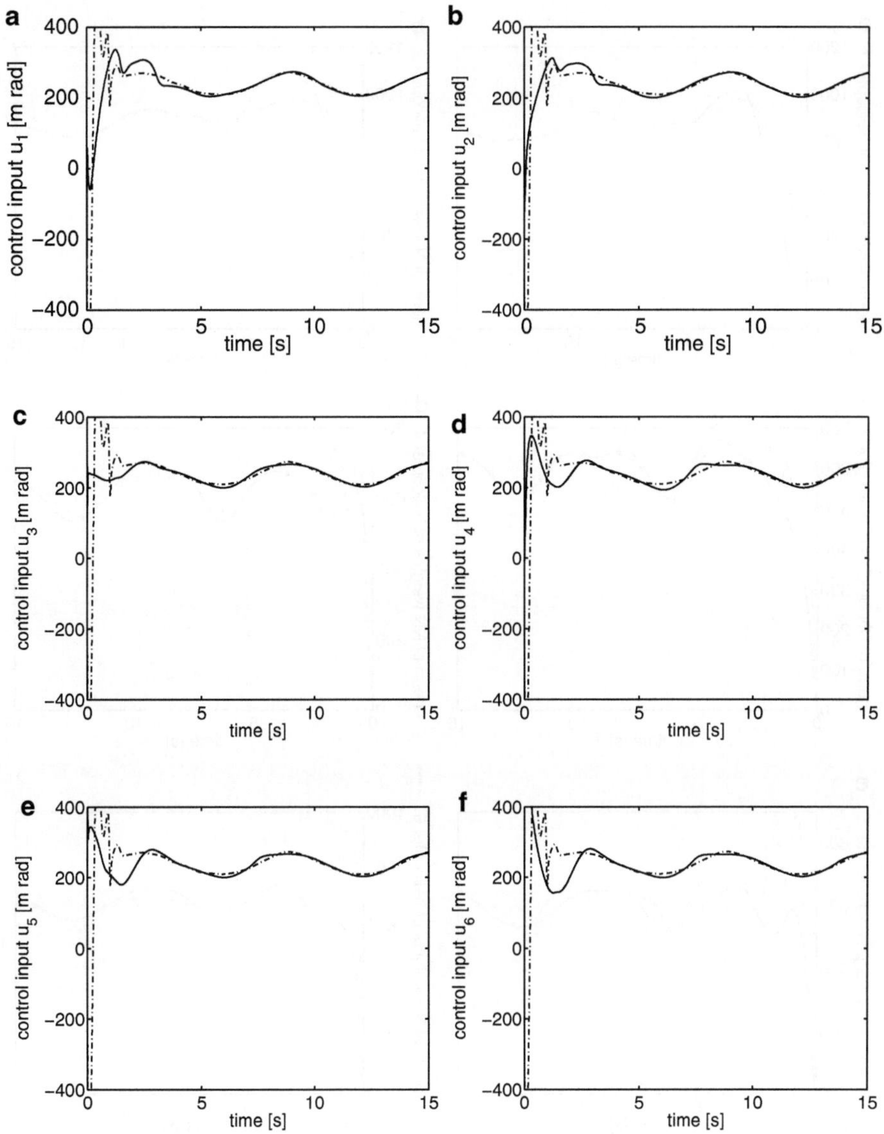


**Fig. 8.4** Control input of helicopters under full-state (*solid*) and output (*dash-dot*) feedback control

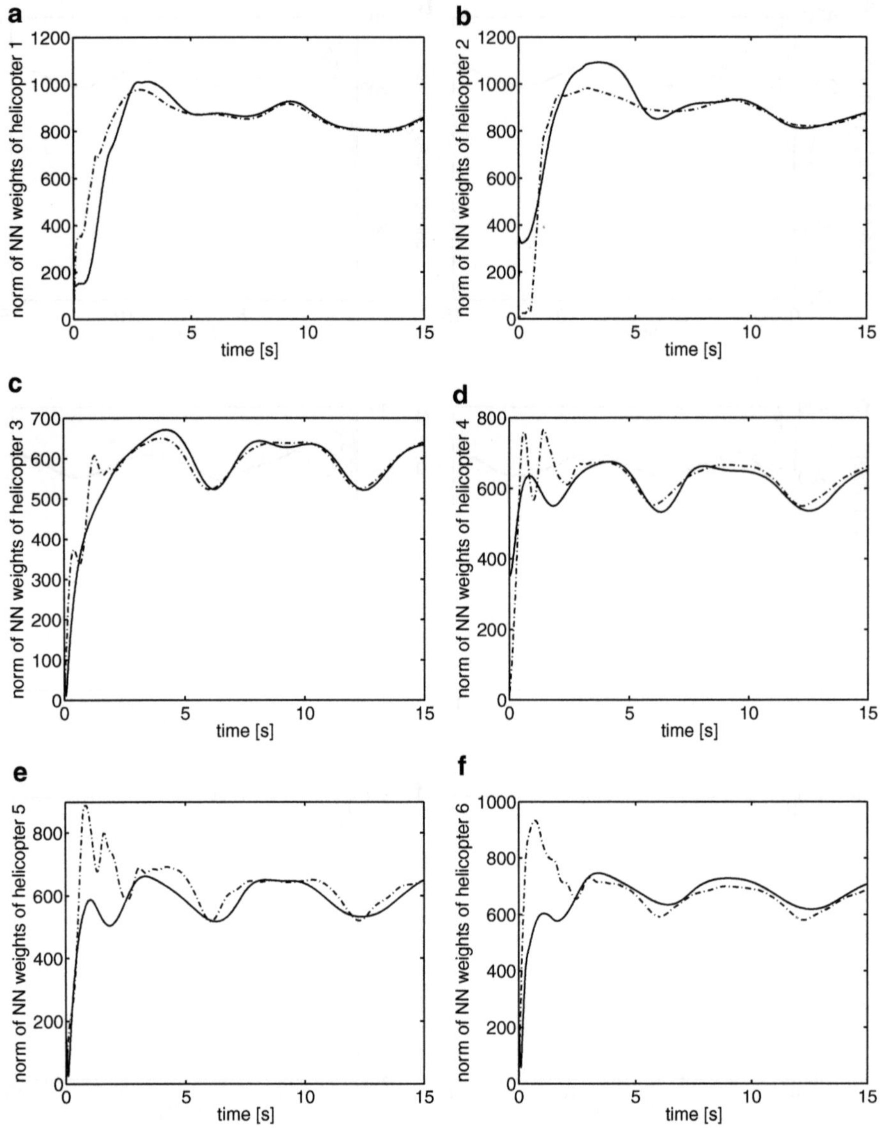


**Fig. 8.5** Norm of neural weights under full-state (*solid*) and output (*dash-dot*) feedback control. Norm of neural weights (**a**) $\|\hat{\theta}_1\|$, (**b**) $\|\hat{\theta}_2\|$, (**c**) $\|\hat{\theta}_3\|$, (**d**) $\|\hat{\theta}_4\|$, (**e**) $\|\hat{\theta}_5\|$, (**f**) $\|\hat{\theta}_6\|$

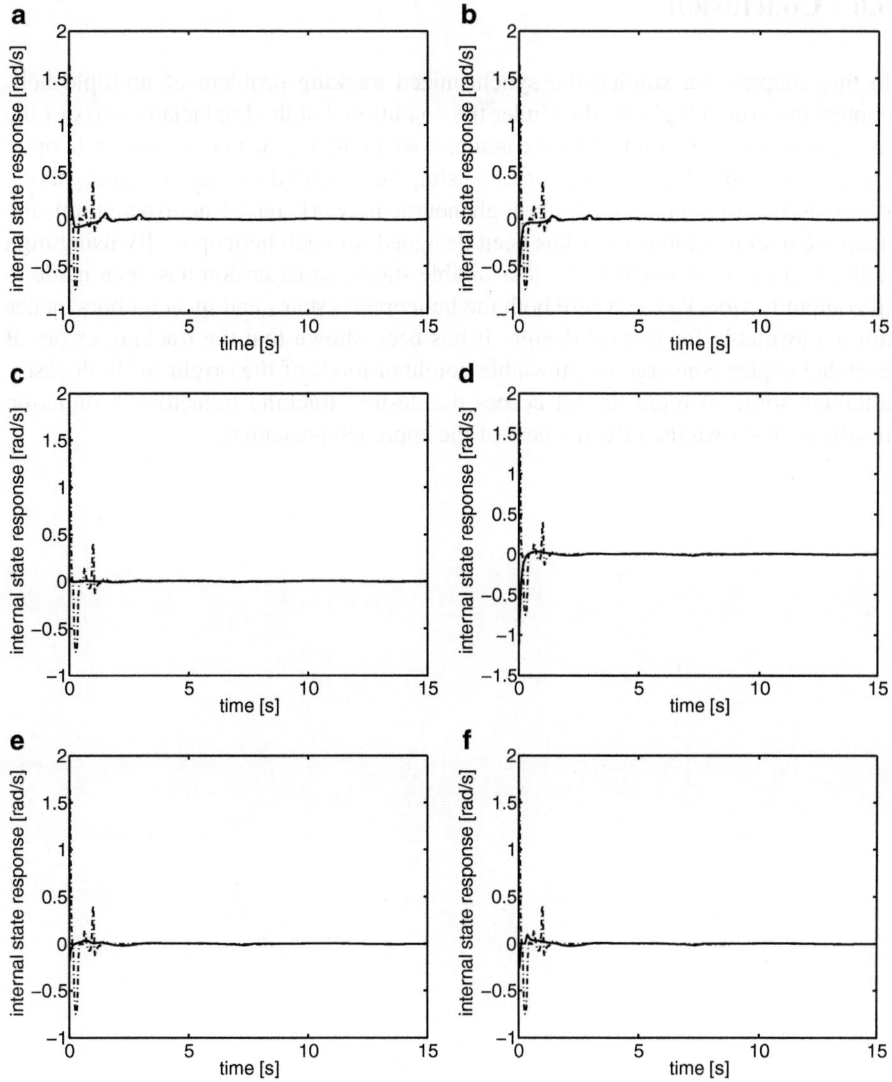


**Fig. 8.6** Internal state response under full-state (*solid*) and output (*dash-dot*) feedback control. (**a**) Internal state of helicopter 1, (**b**) Internal state of helicopter 2, (**c**) Internal state of helicopter 3, (**d**) Internal state of helicopter 4, (**e**) Internal state of helicopter 5, (**f**) Internal state of helicopter 6

## 8.6 Conclusion

In this chapter, we studied the synchronized tracking problem of multiple helicopters in vertical flight mode. Under the condition that the Laplacian matrix of the extended formation graph, which contains a spanning tree which the root helicopter can access for the desired trajectory, by using the weighted average of its neighbors' states as its reference signal, through neural network based approximation, the adaptive tracking control law has been designed for each helicopter. By using high gain observer to reconstruct the unavailable states, an extension has been made to the output feedback case where both the helicopter's states and its neighbors' states are not available for control design. It has been shown that the tracking errors of each helicopter converge to adjustable neighborhoods of the origin for both cases, although some of them do not access the desired tracking trajectory. Simulation results have shown the effectiveness of the approach presented.

# References


1. Apostol TM (1964) Mathematical analysis. Addison-Wesley, Reading
2. Axelsson H, Muhammad A, Egerstedt M (2003) Autonomous formation switching for multiple mobile robots. IFAC Conference on analysis and design of hybrid systems, Sant-Malo, Brittany, France, June 2003
3. Balch T, Arkin RC (1998) Behavior-based formation control for multirobot teams. IEEE Trans Robot Autom 14(6):926–939
4. Balch T, Hybinette M (2000) Social potentials for scalable multi robot formations. Proceedings of the IEEE international conference on robotics and automation, April 2000, pp 73–80
5. Balch T, Parker LE (eds) (2001) Robot teams: from diversity to polymorphism. A K Peters Ltd., Natick, MA
6. Barron AR (2001) Autonomous helicopter control using fuzzy gain scheduling. IEEE International Conference on Robotics and Automation, Seoul, Korea, 21–26 May 2001, pp 2980–2985
7. Behtash S (1990) Robust output tracking for nonlinear system. Int J Contr 51(6):1381–1407
8. Tischler MB, Mettler B, Kanade T (2002) System indentification modeling of a small-scale unmanned rotorcraft for flight control design. J Am Helicop Soc 47(1):50–63
9. Bramwell ARS, Done G, Balmford D (2001) Bramwell's helicopter dyamics, 2nd edn. Botterworth-Heinemann, Boston
10. Cai G, Chen BM, Lee TH (2011) Unmanned Rotorcraft Systems, Springer, New York
11. Carpin S, Parker LE (2002) Cooperative leader following in a distributed multi-robot system. Proceedings of the international conference on robotics and automation, May 2002, pp 2994–3001
12. Castles W, Leeuw JJ (1953) The normal component of the induced velocity in the vicinity of a lifting rotor and some examples of its application. NACA TN 2912
13. Chen C (2006) Development of a simplified inflow model for a helicopter rotor in descent flight. Georgia Institute of Technology, PhD's Thesis
14. Chengjian He (2004) *Flightlab Theory Manual*. Advanced Rotorcraft Technology, Inc.
15. Chen C, Prasad JVR (2007) A simplified rotor inflow model for descent flight. J Aircraft 44(3):936–944
16. La Civita M, Papageorgiou G, Messner W, Kanade T (2002) Design and flight testing of a high-bandwidth H-$\infty$ loop shaping controller for a robotic helicopter. Proceedings of the AIAA guidance, navigation, and control conference and exhibit, August 2002
17. Das AK, Fierro R, Kumar V, Ostrowski JP, Spletzer J, Taylor CJ (2002) A vision based formation control framework. IEEE Trans Robot Autom 18:813–825
18. Desai JP, Ostrowski JP, Kumar V (2001) Modeling and control of formations of nonholonomic mobile robots. IEEE Trans Robot Autom 17:905–908



B. Ren et al., *Modeling, Control and Coordination of Helicopter Systems*,
DOI 10.1007/978-1-4614-1563-3, © Springer Science+Business Media, LLC 2012

19. Devasia S (1997) Output tracking with nonhyperbolic and near nonhyperbolic internal dynamics: helicopter hover control. J Guidance, Control Dyn 20(3):573–580
20. Egerstedt M, Muhammad A, Hu X (2002) Formation control under limited sensory range constraints. 10th Mediterranean conference on control and automation, Lisbon, Portugal, July 2002
21. Enns R, Si J (2003) Helicopter trimming and tracking control using direct neural dynamic proramming. IEEE Trans Neural Networks 14(4):929–939
22. Eren T, Belhumeur PN, Morse AS (2002) Closing ranks in vehicle formations based on rigidity. Proceedings of the IEEE conference on decision and control, December 2002, pp 2959–2964
23. Farrell JA, Polycarpou MM (2006) Adaptive approximation based control. Wiley, Hoboken, NJ
24. Fax JA, Murray RM, Syst NGE, Woodland Hills CA (2004) Information flow and cooperative control of vehicle formations. IEEE Trans Automatic Control 49(9):1465–1476
25. Fua CH (2003) Intelligent helicopter control. Final Year Project Thesis, National University of Singapore
26. Fua CH, Ge SS, Do KD, Lim KW (2007) Multirobot formations based on the Queue-formation scheme with limited communication. IEEE Trans Robot 23(6):1160–1169
27. Funahashi KI (1989) On the approximate realization of continuous mappings by neural networks. Neural Networks 2:183–192
28. Ge SS, Cui YJ (2002) Dynamic motion planning for mobile robots using potential field method. Auto Robots 13:207–222
29. Ge SS, Fua CH (2005) Queues and artificial potential trenches for multi-robot formations. IEEE Trans Robot 21(3):646–656
30. Ge SS, Hang CC, Lee TH, Zhang T (2002) Stable adaptive neural network control. Kluwer Academic Publisher, Boston
31. Ge SS, Hang CC, Zhang T (1999) Adaptive neural network control of nonlinear systems by state and output feedback. IEEE Trans Syst, Man, Cybern B 29(6):818–828
32. Ge SS, Lee TH, Harris CJ (1998) Adaptive neural network control of robotic manipulators. World Scientific, London
33. Ge SS, Wang C (2002) Adaptive nn control of uncertain nonlinear pure-feedback systems. Automatica, 38(4):671–682
34. Ge SS, Wang C (2002) Direct adaptive NN control of a class of nonlinear systems. IEEE Trans Neural Networks 13(1):214–221
35. Ge SS, Zhang J (2003) Neural-network control of nonaffine nonlinear system with zero dynamics by state and output feedback. IEEE Trans Neural Networks 14(4):900–918
36. Gerkey BP, Vaughan RT, Howard A (2003) The player/stage project: tools for multi-robot and distributed sensor systems. Proceedings of the international conference on advanced robotics, June–July 2003, pp 317–323
37. Hartman EP, Biermann D (1982) The aerodynamic characteristics of full-scale propeller having 2, 3, and 4 Blades of Clary Y and R.A¿F. 6 Airfoil Sections. NASA Report 640, November 1982
38. Hovakimyan N, Nardi F, Calise A, Nakwan K (2002) Adaptive output feedback control of uncertain nonlinear systems using single-hidden-layer neural networks. IEEE Trans Neural Networks 13(6):1420–1431
39. Isidori A (1995) Nonlinear control systems, 3rd edn. Springer, Berlin, New York
40. Isidori A, Marconi L, Serrani A (2003) Robust autonomous guidance – an internal model approach. Springer, New York
41. Isidori A, Marconi L, Serrani A (2003) Robust nonlinear motion control of a helicopter. IEEE Trans Automatic Control 48(3):413–426
42. Jewel JW, Heyson HH (1959) Charts of induced velocities near a lifting rotor. NASA MEMO 4-15-59L, May 1959
43. Johnson EN, Kannan SK (2005) Adaptive trajectory control for autonomous helicopters. J Guidance Control Dyn 28(3):524–538

44. Johnson W (1980) Helicopter theory. Dover Publications, Inc., New York
45. Kadmiry B, Driankov D (2001) Autonomous helicopter control using linguistic and model-based fuzzy control. In: Proceedings of IEEE international symposium on intelligent control, México City, México, pp 348–352
46. Kadmiry B, Driankov D (2004) A fuzzy gain-scheduler for the attitude control of an unmanned helicopter. IEEE Trans Fuzzy Syst 12(4):502–515
47. Kadmiry B, Driankov D (2004) A fuzzy flight controller combining linguistic and model-based fuzzy control. Fuzzy Sets Syst 146:313–347
48. Kaloust J, Ham C, Qu Z, Syst LMV, Dallas TX (1997) Nonlinear autopilot control design for a 2-DOF helicopter model. IEE Proc Control Theory Appl 144(6):612–616
49. Kang W, Xi N (1999) Control and adaptation of multi vehicle formations. Proceedings of international conference on intelligent robots and systems, pp 1155–1160
50. Kang W, Xi N, Zhao Y, Tan J, Wang Y (2002) Formation control of multiple autonomous vehicles: theory and experimentation. IFAC 15th Triennial World Congress, pp 1155–1160
51. Kannan SK, Johnson EN (2002) Adaptive trajectory-based flight control for autonomous helicopters. Proceedings of the 21st digital avionics systems conference, October 2002, pp 8.D.1.1–8.D.1.12
52. Khalil HK (2002) Nonlinear systems, 3rd edn. Prentice Hall, NJ
53. Kim BS, Calise AJ (1997) Nonlinear flight control using neural networks. AIAA J Guidance Control Dyn 20:26–33
54. Kim N, Calise AJ, Corban JE, Prasad JVR (2004) Adaptive output feedback for altitude control of an unmanned helicopter using rotor RPM. Proceedings of AIAA guidance, navigation and control conference, August 2004, pp 3635–3640
55. Koo TJ, Sastry S (1998) Output tracking control design of a helicopter model based on approximate linearization. In: Proceedings of IEEE conference on decision and control, Tampa, FL
56. Kostelnik P, Šamulka M, Jánošík M (2002) Scalable multi-robot formations using local sensing and communication. Third international workshop on robot motion and control, November 2002, pp 319–324
57. Kutay AT, Calise AJ, Idan M, Hovakimyan N (2005) Experimental results on adaptive output feedback control using a laboratory model helicopter. IEEE Trans Fuzzy Syst 13(2):196–202
58. Lee S, Lee KC, Lee MH, Harashima F (2002) Integration of mobile vehicles for automated material handling using profibus and IEEE 802.11 networks. IEEE Trans Industr Electron 49(3):693–701
59. Leonard NE, Fiorelli E (2001) Virtual leaders, artificial potentials and coordinated control of groups. Proc IEEE Conf Decision Control 3:2968–2973
60. Levin AU, Narendra KS (1996) Control of nonlinear dynamical systems using neural networks – Part II: observability, identification, and control. IEEE Trans Neural Networks 7(1):30–42
61. Lewis FL, Jagannathan S, Yesilidrek A (1999) Nerual network control of robot manipulators and nonlinear systems. Taylor and Francis, Philadelphia
62. Lewis FL, Yesildirek A, Liu K (1996) Multilayer neural network robot controller with guaranteed tracking performance. IEEE Trans Neural Networks 7(2):388–399
63. Lewis MA, Tan KH (1997) High precision formation control of mobile robots using virtual structures. Autonomous Robots 4(4):387–403
64. Luo CC, Liu RF, Yang CD, Chang YH (2003) Helicopter $H_\infty$ control design with robust flying quality. Aerospace Sci Technol 7(2):159–169
65. MacLennan B (1991) Synthetic ethology: an approach to the study of communication. Artificial life II: $2^{nd}$ workshop on synthesis and simulation of living systems, vol. 11, pp 631–658
66. Marconi L, Naldi R (2007) Robust full degree-of-freedom tracking control of a helicopter. Automatica 43(11):1909–1920
67. Marconi L, Naldi R (2008) Aggressive control of helicopter in presence of parametric and dynamical uncertainties. Mechatronics 18(7):381–389

68. McCormick BW (1995) Aerodynamics, aeronautics, and flight mechanics, 2nd edn. Wiley, New York
69. Muhammad A, Egerstedt M (2005) Connectivity graphs as models of local interactions. J Appl Math Comput 168(1):243–269
70. Muhammad A, Ji M, Egerstedt M (2005) Applications of connectivity graph processes in networked sensing and control. In: Workshop on networked embedded sensing and control, Notre Dame, IN, USA, October 2005
71. Narendra KS, Annaswamy AM (1987) A new adaptive law for robust adaptation without persistent excitation. IEEE Trans Automatic Control 32:134–145
72. Narendra KS, Annaswamy AM (1989) Stable adaptive systems. Prentice Hall, Englewood Cliffs, NJ
73. Narendra KS, Parthasarathy K (1990) Identification and control of dynamic systems using neural networks. IEEE Trans Neural Networks 1(1):4–27
74. Ogata K (1997) Modern control engineering. Prentice-Hall, Englewood Cliffs
75. Ögren P, Leonard NE (2003) Obstacle avoidance in formation. IEEE international conference on robotics and automation, September 2003, pp 2492–2497
76. Olfati-Saber R, Murray RM (2002) Distributed structural stabilization and tracking for formations of dynamic multi agents. Proceedings of IEEE conference on decision and control, December 2002, pp 209–215
77. Padfield GD (2007) Helicopter flight dynamics. Oxford OX4 2QD, 2nd edn. Blackwell, UK
78. Padfield GD (1995) Helicopter flight dynamics: the theory and application of flying qualities and simulation modeling. AIAA Education Series
79. Parker LE (1993) Designing control laws for cooperative agent teams. Proc. Int Conf Robotics Automation 3:582–587
80. Parker LE (1995) The effect of action recognition and robot awareness in cooperative robotic teams. Proc. Int Conf Intelligent Robots Syst 1:212–219
81. Prasad JVR (2003) Helicopter stability and control – class notes at georgia institute of technology
82. Prempain E, Postlethwaite I (2005) Static $H_\infty$ loop shaping control of a fly-by-wire helicopter. Automatica 41(9):1517–1528
83. Prouty R (1997) Helicopter performance, stability, and control. Robert E. Krieger Publishing Company, Malabar, Florida
84. Prouty RW (1990) Helicopter performance, stability, and control. Robert E. Krieger, Malabar, FL
85. Ren W, Beard RW (2004) A decentralized scheme for spacecraft formation flying via the virtual structure approach. AIAA J Guidance Control Dyn 27(1):73–82
86. Rimon E, Koditschek DE (1992) Exact robot navigation using artificial potential functions. IEEE Trans. Robotics Automation 8(5):501–518
87. Rysdyk RT, Calise AJ (1998) Nonlinear adaptive flight control using neural networks. IEEE Control Syst Magzine 18:14–25
88. Saber RO, Murray RM (2003) Flocking with obstacle avoidance: cooperation with limited communication in mobile networks. Proc. IEEE Conf Decision Control 2:2022–2028
89. Saffarian M, Fahimi F (2009) Non-iterative nonlinear model predictive approach applied to the control of helicopters' group formation. Robotics Autonomous Syst 57(6–7):749–757
90. Sanner RM, Slotine JE (1992) Gaussian networks for direct adaptive control. IEEE Trans. Neural Networks 3(6):837–863
91. Schrage DP, Yillikci YK, Liu S, Prasad JVR, Hanagud SV (1999) Instrumentation of the yamaha R-50/rmax helicopter testbeds for airloads identification and follow-on research. In: 25th European rotorcraft forum, Rome, Italy, pp P4-1–P4-13
92. Shivakumar PN, Chew KH (1974) A sufficient condition for nonvanishing of determinants. Proceedings of the American mathematical society, pp 63–66
93. Sira-Ramirez H, Zribi M, Ahmad S (1994) Dynamical sliding mode control approach for vertical flight regulation in helicopters. IEE Proc Control Theory Appl 141:19–24
94. Slotine J-JE, Li W (1991) Applied nonlinear control. Prentice-Hall, Englewood Cliffs, NJ

95. Smerlas AJ, Walker DJ, Postlethwaite I, Strange ME, Howitt J, Gubbels AW (2001) Evaluating $H_{\infty}$ on the NRC Bell 205 fly-by-wire helicopter. Control Eng Pract (9):1–10
96. Song P, Kumar V (2002) A potential field based approach to multi robot manipulation. Proceedings of IEEE international conference on robotics and automation, May 2002, pp 1217–1222
97. Talbot PD, Tinling BE, Decker WA, Chen RTN (1982) A mathematical model of a single main rotor helicopter for piloted simulation. NASA TM 84281
98. Tanner H, Kumar A (2005) Towards decentralization of multi-robot navigation functions. Proceedings of the IEEE international conference on robotics and automation, April 2005, pp 4143–4148
99. Tanner HG, Jadbabaie A, Pappas GJ (2007) Flocking in fixed and switching networks. IEEE Trans Automatic Control. Submitted 2004/2005, pp 863–868
100. Tanner HG, Pappas GJ, Kumar V (2004) Leader-to-formation stability. IEEE Trans Robotics Automation 20(3):443–455
101. Tee KP, Ge SS (2006) Control of fully actuated ocean surface vessels using a class of feedforward approximators. IEEE Trans Control Syst Technol 14(4):750–7560
102. Tee KP, Ge SS, Tay EH (2008) Adaptive neural network control for helicopters in vertical flight. IEEE Trans Control Syst Technol 16(4):753–762
103. The Player and Stage Project. Available from *http://playerstage.sourceforge.net/*. Accessed on 2007
104. Vilchis JCA, Brogliato B, Dzul A, Lozano R (2003) Nonlinear modelling and control of helicopters. Automatica 39:1583–1596
105. Walker DJ (2003) Multivariable control of the longitudinal and lateral dynamics of a fly-by-wire helicopter. Control Eng Pract (11):781–795
106. Walker DJ, Voskuijl M, Manimala BJ (2008) Nonlinear attitude control laws for the Bell 412 helicopter. J Guidance Control Dyn 31(1):44–52
107. Wang PKC (1991) Navigation strategies for multiple autonomous mobile robots moving in formation. J Robotic Syst 8(2):177–195
108. Yaggy PF, Mort KW (1962) Wind-tunnel tests of two VTOL propellers in descent. NACA TN 1766

# Index

**A**
Actuator dynamics. *See* Attitude control, uncertain helicopters
Aerodynamic interference
  power penalties, 31
  superposition method, 31
AF25B helicopter, forward flight
  coupled linearized model, 55–56
  eigenvalue distribution, 57–58
  sanity check, 56–57
Altitude and yaw control
  control design
    MNN, 106–114
    RBFNN, 97–105
  formulation and preliminaries
    dynamics structure, 94–95
    Lagrangian formulation, 94
    simulation section, 95
    stability analysis, 96
    VARIO scale model, 94
    vertical flight mode, 95
  MIMO model, 93
  MNN and RBFNN, 93–94
  simulation
    adaptive neural control, 114
    aerodynamical disturbances, 116
    altitude tracking performance, 117
    control inputs, altitude and yaw angle, 117, 119
    internal dynamics stability analysis, 115–116
    norm, neural weights, 117, 118
    trajectories, 114
    yaw angle tracking performance, 117, 118
Altitude tracking
  proposed controller, 84
  rotor RPM and neural weights, 84, 85
  Yamaha R50 helicopter, 83
APID MK-III helicopter
  approximation-based attitude control
    angular velocity signals, 145
    command control signals, 145
    nodes, 146
    parametric uncertainties, 142
  model-based attitude control
    angular velocity signals, 141
    command control signals, 141
    tracking performance, 140
  parameters, 139
  robust attitude control
    function uncertainties and external disturbance, 142
    uncertainties and disturbances, 143
  saturation values, 139
Approximation-based attitude control
  adaptive laws, 135, 137
  APID MK-III helicopter
    angular velocity signals, 145
    command control signals, 145
    nodes, 146
    parametric uncertainties, 142
  backstepping technique, 133
  closed-loop stability, 138
  Lyapunov function candidate, 133, 136
  NNs, 136–137
  RBFNN, 134
  stability, error signals, 135, 137
  time derivative, 137
  unknown moment coefficients, 136, 138


DOI 10.1007/978-1-4614-1563-3, 

Approximation-based attitude control (*cont.*)
variables, 136
virtual control law, 133, 134
Attitude control, uncertain helicopters
approximation-based (*see* Approximation-based attitude control)
model-based, nominal plant
bounded initial condition, 128–129
chattering phenomenon, 129
Lyapunov analysis, 127–128
moment coefficients and mass, 126
problem formulation
coefficients, moment equations, 122–123
continuous functions, 126
control gain parameters, 125
forces and torques, 123
model-based control, 126
model uncertainty and external disturbance, 124–125
nonlinear form, 122
propulsions, 123
system modeling uncertainties, 124
thrusts, main and tail rotors, 123
robust, uncertainties and disturbances
closed-loop system, 132
error variables, 129
Lyapunov function candidate, 129–131
RBFNN (*see* Radial basis function neural network)
simulation
APID MK-III model, 139
approximation-based, 142, 145–146
control gain and design matrix, 139
model-based, 140–141
nominal parameters, 139
robustness, 142–144

**B**

Blade flapping dynamics
body angular rates, 14
low frequency, 15
mass distribution, 13
parameters, 13
second-order differential equation, 12–13
teetering configuration, 16

**C**

Control design
adaptive neural control, 96
computable signals, 96
MNN-based (*see* Multilayer neural network)
RBFNN-based (*see* Radial basis function neural network)
Controllability matrix, 46, 49

**D**

Diagonally dominant matrix, 198
Directed graph, 197–198
Division, information flow
fast-time scale, 150
(re)allocation, 150
slow-time scale, 150

**E**

*e*-modification law, 98
Equations of motion
force equation, 7–8
kinematic equation, 9
moment equation, 8–9
navigation equation, 9–10

**F**

Filtered tracking error, 96, 201, 203, 207, 210
Flight control system
cyclic controls, 33
first order dynamic system, 34
fly bar, teetering rotor configuration, 33
rotor blades, 34
Force equation, 7–8
Formations, 3, 151–153, 164–166. *See also* Q-structures
Formulation control
graph edge definitions, 197–198
Laplacian matrix, 199
positive constants, 200
synchronized tracking problem, 198–199
trajectory model, 200
Full information control
closed-loop trajectory, 203
compact set and positive constant, 206
dynamics theorem, 206–207
error dynamics, 205
extended formation graph, 204
filtered error, 204
Hurwitz equation, 205
invertible matrix, 204
maximum eigenvalue, 205–206
NN control scheme, 200
nonlinear function, 202
standard Lyapunov theorem, 203

synchronized tracking problem, 201
trajectories, 206
Young's inequality, 202
Full state feedback control
adaptation laws, 69
compact set, SGUUB and tracking error, 71–72
Lyapunov function, 68
Mean Value Theorem, 67–68
trajectories, 70–71
virtual control function, 66–67
Function approximation
MNN (*see* Multilayer neural network)
RBFNN, 64–65
Fuselage
aerodynamic database, 30
velocity components, local body axes system, 30

**G**
Graph theory
altitude, output feedback control, 212
formulation, problem
bounded vector function, 209
communication graph, 210–211
control, helicopters, 197–200
filtered tracking error, 208
helicopter dynamics, 196–197
information control, 200–207
linear system, 207
Lyapunov function, 208
positive constants, 209
saturation functions, 208
internal state response control, 212, 215
kinematic control scheme, 195
neural weights, feedback control, 212, 214
solid and dash-dot feedback control, 212, 213
strong relative degree, 212
synchronized tracking, 195–196
throttle restriction, 211
6 X-cell tracking, 211

**H**
Helicopter dynamics
SISO systems description, 197
smooth functions, 198
zero dynamics, 197
Helicopter model in hover
aerodynamic parameters, 47–48
MATLAB, 49
symbols and subscripts, 46, 47
Taylor series expansion, 48–49
High gain observer, 85, 196, 207, 212
Horizontal tail
lift and drag, local wind axes system, 28
pitch moment, 28
velocity components, local body axes system, 27–28
Hub-wind axes system
horizontal and side force, 19–20
rolling and pitching moment, 21
rotor torque, 22
thrust equation, 18, 19

**I**
Imperfect communication, helicopter team
changing formations
separation and control forces, 189, 193
spikes, graphs, 189
switching, 192
formation convergence and scaling
deactivation/removal, 186, 187
distance, queue's encapsulating area, 189
minimum inter-helicopter separation, 186, 188
potential field forces, 188
link breakdowns, 192
moving formations
minimal inter-helicopter distance and control signals, 191
queues and virtual formation vertex, 190
navigation (*see* Navigation)
target determination, queue
algorithm, 177
constant targets, 177–178
position, queue, 175
Implicit function theorem, 60, 63, 92
Information flow, 150
Internal dynamics stability analysis
main rotor angular velocity, 115, 116
RBFNN and MNN based control, 115

**J**
Jacobian matrix, 42, 43, 62

**K**
Kinematic equation, 9
Kinematic formation control
approaches, 147
multi-helicopters cooperation, 147

Kinematic formation control (*cont.*)
Q-structure
assumptions, 148–150
elements, 150–153
imperfect communication (*see* Imperfect communication, helicopter team)
information flow division, 150
perfect communication (*see* Perfect communication, helicopter team)
properties, 153–156

**L**

Lagrangian formulation, 94

**M**

Main rotor helicopters
blade flapping, 11
hub-wind axes system, 18–21
quasi-steady flapping dynamics, 11
rigid blade, 11
small angle approximation, 10
Mean Value Theorem, 60, 63, 67, 92
MNN. *See* Multilayer neural network
$\sigma$-Modification law, 98
Moment equation, 8–9
Momentum theory
flow interaction, 17
vortex rings model, 17–18
VRS, 17
Multilayer neural network (MNN)
approximation error, 110
constant vectors, 111
continuous function, 65
estimation errors, 106
ideal weights, 65–66
inequalities, 112
Lyapunov function, 107
nonlinear parameterization, 106
positive definite functions, 110
residual term, 110
robust controller, 68
SGUUB system, 113
sigmoidal functions, 66
Taylor series expansions, 106
Young's inequality, 109

**N**

Navigation
approximation error, 185
attractive potentials, 178
closed loop dynamics, 180
collision avoidance behavior, 178–179
equilibrium points, 181, 182
linear-time-invariant (LTI), 182
Lyapunov candidate, 182–183
non-collision, agents, 181
subspace, 184
time derivative, 179
unstable equilibrium points, 183, 184
Navigation equation, 9–10
Neural network (NN) control
altitude tracking, 83–85
approximate dynamic inversion, 60–61
full state feedback control, 68–72
MNN, 65–66
modeling errors, 60
nonlinear model
PD control and norm of neural weights, 90, 91
rotor speed and collective pitch angle, 90, 91
SISO system and relative degree, 88
tracking performance, 89–90
X-cell 50 helicopter, 88
output feedback control (*see* Output feedback control)
pitch tracking, 85–87
RBFNN, 64–65
SISO helicopter systems, 61–63
unrestrained helicopter motion, 59
Nonaffine, 4, 60, 63, 82, 83
Nonlinear rotary-wing aircraft model
aerodynamic interference, 31
blade flapping dynamics (*see* Blade flapping dynamics)
equations of motion
force equation, 7–8
kinematic equation, 9
moment equation, 8–9
navigation equation, 9–10
features, 38
flight control system, 33–34
fuselage, 30
horizontal tail, 27–28
main rotor
blade flapping, 11
forces and moments, hub-wind axes system, 18–22
quasi-steady flapping dynamics, 11
rigid blade, 11
small angle approximation, 10
Momentum Theory, 17–18
performance prediction
endurance and range, 37

main rotor power, 34
profile power, 37
tail rotor and ancillary power, 35
Yamaha RMAX helicopter, power prediction, 36
propeller, 24–26
rotorcraft model, 10, 11
rotor rotational degree of freedom, 31–32
tail rotor, 23–24
transformation, body to hub, 12, 22
vertical tail, 29
wing, 29
Nonsingular, 128, 129
Normal form, 62

**O**

Observability matrix, 46, 49
Output feedback control
bounding control, 76
closed loop trajectories, 75
control law and adaptation laws, 74–75
Hurwitz, 73–74
inequalities, 76–79
saturation functions, 74
SGUUB, 80–82

**P**

Perfect communication, helicopter team
changing queues
evaluation process, 157
status, algorithm, 157
switching, 157
convergence
circle formation, 164, 165
column formation, 166
scaling, 166
wedge formation, 164, 165
disruption, wireless communications, 171, 172, 174, 175
double column formation, 171
maneuvers, confined spaces
corridor and waypoints, 166, 167
deformation, turn, 169
deformed wedge, 167, 171
distance measure, 166
movement, 5m and 3m wide corridor, 170
wedge and column formation movement, 168
potential trench functions
forces acting, 160
formation adaptation and deformation, 160–162
top view, plane, 160
reaction, obstacles
$\delta$ *vs.* time, 173
size and shape, 169
type I and II field, 168, 172
regular formations, 164
representative formations, 163
wireless communication disruption
degrees, communication breakdown, 174, 176
different frequencies, error, 171, 174
$P_{txloss}$ and $I_{loss}$, 174, 175
queue status, 172, 175
Performance prediction
endurance and range, 37
main rotor power, 34
profile power, 37
tail rotor and ancillary power, 35
Yamaha RMAX helicopter, power prediction, 36
Pitch tracking
control input and norm of neural weights, 87
tracking errors and observer errors, 86
Yamaha R50 helicopter, 85
Potential trench functions
forces acting, 160
formation adaptation and deformation, 160–162
top view, plane, 160
Propeller
closed-form equations, 24
Momentum-Blade Element Theory and Vortex Theory, 26, 27
thrust and torque calculation procedure, 24–26

**Q**

Q-structures
assumptions
graphical notation, 149
stability and convergence, 149–150
vectors, 148, 149
consistency, formation representation
addition/removal, nodes, 154
connectivity graphs, 154, 155
division, information flow
allocation, 150
fast-time and slow-time scale, 150
efficiency and optimality, 155–156

Q-structures (*cont.*)
elements
closed queues, 152
formation, 151
open queues, 152–153
queues, 151–152
formation decomposition and computation, 155
graphical representation
triangular and three column formation, 154
virtual Q-vertices, 153
imperfect communication (*see* Imperfect communication, helicopter team)
perfect communication (*see* Perfect communication, helicopter team)
robustness, 156
Queues
changing, 156–158
closed, 152
elements, 151
open, 152

**R**
Radial basis function neural network (RBFNN)
bounded system, 105
closed loop stability, 100
continuous function, 64
$e$-modification law, 98
Gaussian function, 64
hyperbolically minimum-phase, 102
ideal weight vector, 65
inequalities, 102
internal dynamics, 103
Lyapunov synthesis, 97
$\sigma$-modification law, 98
positive constant, 99–100
positive definite functions, 100
stability analysis, internal dynamics, 104
subsystem, closed-loop, 103
time derivative, 97
time-invariant functions, 100
weight adaption law, 101
Young's inequality, 99
zero dynamics, 102
RBFNN. *See* Radial basis function neural network
Relative degree, 61, 62, 88, 197, 212
Rotary-wing aircraft, stability analysis
Copterworks AF25B helicopter, forward flight, 55–58
linearization process
control matrix, 46
drag damping, 44
gravitational and inertial terms, 45
helicopter model, hover, 46–49
pitch attitude, 43
stability and control derivatives, 44
sub-matrices, 44–45
stability and control derivatives
angle of attack stability, 50
drag damping, 49–50
pitch damping and dihedral effect, 50
roll damping and weathercock stability, 51
speed stability, 50
trim, 41–43
Yamaha R50 helicopter at hover, 51–54
Rotor rotational degree of freedom, 31–32

**S**
Saturation function, 74
Semi-globally uniformly ultimately bounded (SGUUB), 63, 71, 72, 80, 82
SGUUB. *See* Semi-globally uniformly ultimately bounded
Single-input-single-output (SISO) helicopter systems
implicit function theorem, 63
Jacobian matrix, 62
Mean Value Theorem, 63
positive function, 62–63
reference trajectory, 61
SGUUB, 63
zero dynamics of system, 62
Singularity, 9, 42, 89
SISO helicopter systems. *See* Single-input-single-output helicopter systems
Spanning tree, 198–200
Substochastic matrix, 198

**T**
Tail rotor
features, 23
teetering configuration, 23–24
Taylor series expansion, 48–49, 63
Trim calculation
Jacobian matrix, 42
$Matlab^{TM}$, 41
Newton's method, 42–43
variables and targets, 43

**U**
Unmanned helicopters
  control techniques, 2
  linearized models, 1–2
  MIMO nonlinear dynamic systems, 5
  nonlinear flight dynamic modeling, 3–4
  Q-structure formation control, 5–6
  robust adaptive NN control, 4–5
  stability analysis, 4
  UAV formation control, 2–3
  underactuated configuration, 1

**V**
VARIO scale model helicopter, 94
Vertical tail, 29
Vortex rings model
  convection speed and velocity, 17
  downward velocity, 18
  effect, 18

**W**
Weighted adjacency matrix, 197
Wing, 29
Wireless communication disruption
  degrees, communication breakdown, 174, 176
  different frequencies, error, 171, 174
  $P_{txloss}$ and $I_{loss}$, 174, 175
  queue status, 172, 175

**X**
X-cell 50 helicopter, 88, 211

**Y**
Yamaha R50 helicopter
  eigenvalue plot, 52
  longitudinal linearized model, 51
  main rotor, 52
  pitch damping, 53, 54
  speed stability, 53, 54

**Z**
Zero dynamics, 59, 62, 66

MIX
Papier aus verantwortungsvollen Quellen
Paper from responsible sources
FSC® C105338

If you have any concerns about our products,
you can contact us on
**ProductSafety@springernature.com**

In case Publisher is established outside the EU,
the EU authorized representative is:
**Springer Nature Customer Service Center GmbH**
**Europaplatz 3, 69115 Heidelberg, Germany**

Printed by Libri Plureos GmbH
in Hamburg, Germany